To my good Colleague Prof Bhavik Bakshi
with my best wishes

May 20, 2011

CHEMICAL LOOPING SYSTEMS FOR FOSSIL ENERGY CONVERSIONS

CHEMICAL LOOPING SYSTEMS FOR FOSSIL ENERGY CONVERSIONS

LIANG-SHIH FAN
University Distinguished Professor
C. John Easton Professor in Engineering, and
Professor of Chemical and Biomolecular Engineering
The Ohio State University
Columbus, Ohio

AIChE®

A JOHN WILEY & SONS, INC., PUBLICATION

A Joint Publication of the American Institute of Chemical Engineers and John Wiley & Sons, Inc.

Published by John Wiley & Sons, Inc., Hoboken, New Jersey
Published simultaneously in Canada

For general information on our other products and services or for technical support, please contact our Customer Care Department within the United States at (800) 762-2974, outside the United States at (317) 572-3993 or fax (317) 572-4002.

Wiley also publishes its books in a variety of electronic formats. Some content that appears in print may not be available in electronic formats. For more information about Wiley products, visit our web site at www.wiley.com.

Library of Congress Cataloging-in-Publication Data:

Fan, Liang-Shih.
    Chemical looping systems for fossil energy conversions/Liang-Shih Fan.
        p. cm.
    Includes index.
        ISBN 978–0–470–87252–9 (hardback)
    1. Fluidized-bed combustion. 2. Fossil fuels–Combustion. 3. Energy conversion. I. Title.
    TP156.F65F348 2010
    621.402′3—dc22

                                                                                    2010010530

Printed in the United States of America

10 9 8 7 6 5 4 3 2 1

Dedicated to

Fan Club Members
for Their Indefatigable Spirit and Devotion
to the Exploration of Engineering Science and Technology
and for Their Friendship

In Cooperation with
Professor Fan's Research Group Members

Fanxing Li
Ah-Hyung Alissa Park
Liang Zeng
Shwetha Ramkumar
Puneet Gupta
Deepak Sridhar
Luis G. Velazquez-Vargas
Mahesh Iyer
Hyung Ray Kim
Fei Wang

CONTENTS

PREFACE

A given chemical reaction often can be decomposed into multiple subreactions using chemical intermediates. When this reaction is carried out in a process system, the subreaction schemes can be designed to minimize the exergy (available energy) loss in this reactive process system, whereas the desired products or undesirable by-products generated from the reaction can be separated with ease, yielding an economically viable process system. A reaction scheme of this nature is called chemical looping. The processes associated with chemical looping are called the chemical looping processes.

The concept of chemical looping has been widely applied in chemical industries, for example, in the production of hydrogen peroxide (H_2O_2) from hydrogen and oxygen using 9,10-anthraquinone as the looping intermediate. Fundamental research on chemical looping reactions also has been applied to energy systems, for example, the splitting of water (H_2O) to produce oxygen and hydrogen using ZnO as the looping intermediate. The chemical looping applications to fossil energy systems, specifically coal-based systems, were practiced commercially with the Steam Iron Process in the 1900s–1940s and were demonstrated at a pilot scale with the Carbon Dioxide Acceptor Process in the 1960s and 1970s. Currently, no chemical looping processes use fossil fuels in commercial operation. The key factors hampering the continued use of these earlier processes were that the looping particles were not sufficiently reactive or recyclable and that the energy conversion efficiencies were low. These factors led to higher product costs for using the chemical looping processes, compared with the other direct fossil fuel conversion processes.

With CO_2 emission control now being a requirement, interest in chemical looping technology has resurfaced. In particular, chemical looping processes are appealing because of their unique ability to generate a sequestration-ready CO_2 stream. Renewed fundamental and applied research since the early 1980s has emphasized improvement over the earlier shortcomings. New techniques have been developed for direct processing of coal or other solid carbonaceous feedstock in chemical looping reactors. With a significant progress under way in particle design and chemical looping reactor development, as demonstrated in the operation of several pilot- or subpilot-scale units worldwide, the chemical looping technology can be commercially viable in the future for processing fossil fuels or carbonaceous fuels in general.

This book contains six chapters and a CD. Chapter 1 presents an overview of the global energy demand and supply from fossil, nuclear and renewable sources, and of the current or conventional fossil energy conversion processes along with the chemical looping processes. It also discusses both coal combustion and coal gasification with an emphasis on the conversion efficiencies of their process applications and gaseous pollutant control methods, including CO_2 capture and sequestration. The exergy and its conversion efficiency concepts are described, as well as their applications to the selection of coal conversion process alternatives. Examples of chemical looping reactions and chemical looping processes using fossil fuels and their conversion to hydrogen, chemicals, liquid fuels, and electricity are elaborated.

Chapter 2 is devoted to the subjects of solid particle design, synthesis, properties, and reactive characteristics as the looping media employed in the processes are in solid form and the success of the chemical looping technology applications strongly depends on the performance of the particles. The looping processes can be applied for combustion and gasification using, directly or indirectly, gaseous carbonaceous fuels such as natural gas and syngas or solid carbonaceous fuels such as coal, petroleum coke, and biomass as feedstock.

Chapters 3–5 describe the design, analysis, optimization, energy conversion efficiency and economics of the looping processes, as well as current or conventional processes for combustion and gasification applications. Chapter 3 illustrates the chemical looping processes for combustion with gaseous or solid carbonaceous fuels as feedstock. Chapter 4 describes chemical looping processes for gasification using gaseous fuels as feedstock. Discussion includes applications of the reducer in these processes that serve as a reformer for treating C_1–C_4+ hydrocarbons. Chapter 5 elucidates the chemical looping processes for gasification using solid carbonaceous fuels as feedstock, in which the reducer serves the function of a gasifier used in the traditional processes.

Chapter 6 presents potential chemical looping applications, including hydrogen storage and onboard hydrogen production, the Carbonation-Calcination Reaction (CCR) Process for carbon dioxide capture, carbon dioxide and hydrogen reaction processes, chemical looping integrated with solid oxide fuel cells, enhanced steam methane reforming, tar sand digestion, and oxygen uncoupling. The CD appended to the book provides the detailed procedure and results of the simulation of various chemical looping reactors and processes based on the ASPEN PLUS (Aspen Technology, Inc., Houston, TX) software. Note that, unless otherwise stated, the unit of tonnage used in the book is the metric ton.

The book can be used as a research reference and/or used for teaching courses on energy and environmental reactions and process engineering with a focus on carbonaceous energy processing. Part of the book has been used for the senior process design course at The Ohio State University, in which student groups are asked to choose between a traditional process and a chemical looping process for coal conversion to hydrogen. The key part of the problem statement for this course is given as follows:

> ABC Energy Ltd., a major hydrogen producer in Ohio, is evaluating the feasibility of a potential process to produce hydrogen and electricity from coal using a traditional coal gasification reactor (gasifier). The coal processing rate of the gasifier is set to be 1,000 MW$_{th}$. Your group in the R&D division of the company is going to compare two processes that produce hydrogen from coal. You will need to submit a comprehensive evaluation report to the division manager addressing the technological and economic advantages and disadvantages of the traditional process versus one of the two new chemical looping processes, that is, the Syngas Chemical Looping Process or the Calcium Looping Process. You will also need to recommend to the manager which process, that is, the traditional process or chemical looping process, is a better choice for the new coal-to-hydrogen process.

This book will offer students an in-depth understanding of the technical aspects of the process options along with scale-up challenges in their deliberation about the process choice. This book also can be used to supplement undergraduate or graduate courses in chemical reaction engineering, thermodynamics, and process analysis/simulation. Material of particular relevance to these courses includes the contact mode of multiphase reactors, multiphase reactor design, reaction kinetics, thermodynamic analysis, process integration, and ASPEN Plus® process simulation.

This book was written in collaboration with ten of my graduate students who had worked or are working on chemical looping technology or related subjects as their dissertation research topics at The Ohio State University. The students who participated in writing this book and are still in the program are Dr. Fanxing Li (Post-Doctoral Research Associate), Shwetha Ramkumar, Liang Zeng, Deepak Sridhar, Hyong Ray Kim, and Fei Wang. The students who had participated, but left the program are Dr. Ah-Hyung Alissa Park (currently Lenfest Junior Professor in Applied Climate Science at Columbia University), Dr. Luis Velazquez-Vargas (currently Research Engineer at Babcock and Wilcox Company), Dr. Puneet Gupta (currently Research Engineer at CRI/Criterion), and Dr. Mahesh Iyer (currently Group Leader of Hydrogen Technology at Shell Global Solutions). Completion of this book would not have been possible without their extensive knowledge and insights into the intricate fundamental and applied nature of the chemical looping problems, as well as without their enthusiasm and commitment to this book, as it is apparent from the citations in the text. Dr. Alissa Park and Dr. Fanxing Li devoted an enormous amount of time to coordinating its writing. Their gracious and helpful efforts are deeply appreciated. I am, however, solely responsible for the presentation of this book including its scope, topic selection, structure, logic sequence, format and style.

My team and I are very grateful to Dr. Chau-Chyun Chen, Eric Sacia, Siwei Luo, Zhenchao Sun, Prof. Sankaran Sundaresan, Prof. Arvind Varma, Prof. Da-Ming Wang, William Wang, Wan-Yi Wu, Cindy Yizhuan Yan, and Fu-Chen Yu who reviewed one or more chapters of the book or the CD and provided valuable suggestions and comments. We would also like to express our gratitude to Debbie Bauer, Francis Coppa, Prof. Luis F. de Diego, Mark Hornick, Svein Harald Ledaal, Dr. Shiying Lin, Dr. Ke Liu, Dr. Adam Luckos, Prof. Anders Lyngfelt, Sylfest Myklatun, Dr. Tobias Pröll, George Rizeq, Dr. Ho-Jung Ryu, Dag Schanke, Prof. Laihong Shen, Dr. Tore Torp, and Greg Wolf who kindly granted permission to use their photos and valuable information in this book. Our special thanks are extended to those industrial and governmental colleagues, with whom we have had the privilege of working for many years, for generously sharing their knowledge and insights into the industrial applications: Jackie Bird, John Bloomer, Lou Benson, Robert Brown, Daniel Cicero, Dan Connell, James Derby, Jeff Gerken, Mark Golightley, Steve Grall, Robert Morris, Bartev Sakadjian, Hamid Sarv, Mark Shanagan, Kevin Smith, Dr. Robert Statnick, Gary Stiegel, Dr. Ted Thomas, and Richard Winschel. Over the years, several agencies including the Ohio Coal Development Office, U.S. Department of Energy, and U.S. Air Force, along with an industrial consortium consisting of Babcock and Wilcox, Consol Energy, Air Products, Shell, CRI/Criterion, AEP, Duke Energy, First Energy, and Specialty Minerals provided significant funding in support of my team's research, development, and demonstration efforts. Much of the results reported in this book are based on these efforts. Their support is most gratefully acknowledged. Our thanks also go to Samuel Bayham for his excellent figure drawings, Carol Mohr for her outstanding editorial assistance in the preparation of the manuscript, and Haeja Han for her help in the preproduction of the book.

LIANG-SHIH FAN

Columbus, Ohio

CHAPTER 1

INTRODUCTION

L.-S. FAN

1.1 Background

Energy is the backbone of modern society. A clean, relatively cheap, and abundant energy supply is a prerequisite for the sustainable economic and environmental prosperity of society. With the significant economic growth in the Asia Pacific region and the expected development in Africa, the total world energy demands are projected to increase from 462.4 quadrillion BTU in 2005 to much more than 690 quadrillion BTU by 2030,[1] as shown in Figure 1.1. The projected energy supply through 2030 will be drawn from oil, coal, natural gas; renewable forms of energy; and nuclear energy, in that order. Figure 1.1 reveals that fossil fuels account for more than 86% of the world's energy supply.[1]

The impact of the global warming induced by the CO_2 emission from fossil energy conversion processes has been an issue of international concern. An energy solution prompted by the combination of ever-increasing energy consumption and rising environmental concerns thus requires a consideration of coupling fossil energy conversion systems with economical capture, transportation, and safe sequestration schemes for CO_2. A long-term energy strategy for low or zero carbon-emission technologies also would include nuclear energy and renewable energy. Nuclear power is capable of generating electricity at a cost comparable with the electricity generated from fossil fuels.[2] Other than electricity, the heat generated from a nuclear plant can be used for hydrogen generation in hydrogen-producing thermochemical or high-temperature electrolysis plants.[3] A variety of social and political issues, as well as

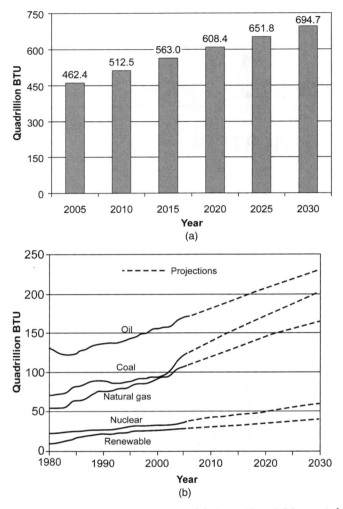

Figure 1.1. Projected global energy (a) demand and (b) supply.[1]

operational safety and permanent waste disposal concerns, however, would limit nuclear energy's widespread utilization in overall energy production.[2,4]

1.1.1 Renewable Energy

Renewable energy (i.e., hydro, wind, solar, biomass, and geothermal) is attractive because of its regenerative and environment-friendly nature. Among the renewable energy sources, hydraulic power provides the largest share of the total renewable energy supply.[5] For example, in 2005, the world's total hydroelectricity capacity was 750 GW accounting for 63% of the total renewable energy supply or 19% of the world's electricity supply for that year.[5] In the United States, hydropower electricity generation is equivalent to 95,000 MW and accounts for 7% of total U.S. electricity generation, supplying 28 million households.[6] Hydraulic power is less costly compared with other renewable energy sources. Once the hydroelectric plant is built, it does not require a significant amount of operational costs or raw materials. Also, when the overall energy life cycle is considered, a hydroelectric plant emits less greenhouse gases compared with a fossil fuel power plant. Despite these advantages, hydraulic power is subject to geological constraints. Moreover, the construction of hydroelectric plants often can be disruptive to the surrounding ecosystems and the life of inhabitants.[7]

Biomass, defined as "any organic matter, which is available on a renewable basis, including agricultural crops and agricultural wastes and residues, wood and wood wastes and residues, animal wastes, municipal wastes, and aquatic plants"[8], is a versatile renewable resource that can be converted into electricity, H_2, and biofuels. Biomass can be directly combusted to generate heat or electricity; it also can be gasified to produce syngas for power generation or liquid fuel synthesis. Other than direct combustion and gasification, alcohol-based biofuels such as bioethanol and biobutanol can be produced from sugar or starch crops through fermentation; biodiesel is produced from vegetable oils and animal fats through transesterification; and biogas can be produced from anaerobic digestion of waste or other organic material.[9] Biomass is often considered to be a carbon neutral energy source, because plant matter is produced from atmospheric CO_2 through photosynthesis. Although biomass is advantageous in several ways, the carbon savings, which are affected by the agricultural practices, for starch-crop-based bioethanol fermentation are doubtful. The life-cycle analysis indicates that the energy consumed in cultivation, transportation and conversion of biomass into bioethanol may be greater than the energy contained in the resulting fuel.[10,11] According to a more recent study, bioethanol has a positive net energy when the conversion by-products are accounted for properly.[12] The study, however, estimated that ~42 joules (J) of energy from fossil fuels will be consumed to produce 100 J of ethanol from corn through fermentation.

The more advanced cellulosic ethanol production technique, which is under development, converts cellulose such as agricultural residues, forestry wastes, municipal wastes, paper pulp, and fast-growing prairie grasses into ethanol. As a result of the significantly reduced feedstock cost, the cellulosic ethanol production process can be more economical than traditional biofuel production techniques. Moreover, the energy input for generating the biofuel potentially can be reduced, leading to reduction in net CO_2 emissions.[13] There are, however, technological challenges in the cellulosic biomass pretreatment prior to fermentation. The thermochemical conversion, such as gasification (as in coal gasification), of cellulosic biomass has been considered an attractive approach to generating not only biofuels but also biochemicals and electricity.

Wind, as another source of renewable energy, is used mainly for electricity generation through turbines on wind farms. Wind power can generate electricity at ~5.6 cents per kilowatt-hour.[4] Under the current tax credit subsidy of 1.9 cents per kilowatt-hour (in 2006 dollars) in the United States, wind power is economically competitive with coal-fired power plants. Within the last decade, the global generation of wind power has increased at an average rate of more than 25% per year.[14] The global capacity of wind turbines reached a record high of 73.9 GW at the end of 2006.[15] The major drawbacks of wind power are its intermittency and highly variable nature, which imposes difficult challenges on grid management. Wind turbines also can be noisy and visually intrusive, but the offshore installation option (turbines installed more than 10 km away from shore) and the recently proposed airborne turbine option[16] would eliminate these aesthetic concerns.

The sum of the solar power that reaches the earth's surface is estimated to be more than 5,000 times the current world energy consumption, and solar electric generation has a higher average power density than any other renewable energy source.[17] Solar energy often is used to generate electricity via photovoltaic cells (solar cells) or heat engines (steam engines or Stirling engines). Apart from electricity generation, solar energy also can be converted into chemical energy such as hydrogen and syngas through solar chemical processes.[18] The major disadvantage of solar power, however, is the high cost of the photovoltaics, which leads to the electricity generated from solar power costing approximately four times that of electricity generated from fossil fuels.[14] The solar chemical processes are still under demonstration in the prototype phase.

Although renewable energy sources are attractive from the environmental viewpoint, they face complex constraints for large-scale application. Even when both the decrease in renewable energy costs and the increase in fossil fuel prices are taken into account, only approximately 8.5% (see Figure 1.1) of the total energy demands in 2030 are projected to come from renewable sources. For primarily economic reasons, fossil fuels, including crude oil, natural gas, and coal, will continue to play a dominant role in the world's energy supply for the foreseeable future.

1.1.2 Fossil Energy Outlook

Among various fossil energy sources, oil is expected to maintain its leading status and its consumption will increase for the foreseeable future. Compared with natural gas and coal, crude oil is relatively easy to be pumped, transported, and processed into high-energy density liquid fuels and chemicals. The 2008 British Petroleum (BP) energy review[19] reported the estimated total oil reserves to be 1.24 trillion barrels, with more than 60% of these reserves located in the Middle East, but the high crude oil prices[1] and limited oil reserves are expected to decrease the share of crude oil in the overall energy supply from 37% to 33%.[1]

Dramatic technological advancements have improved the economics of synthetic crude oil production from nonconventional fossil fuels, such as tar sands, extra heavy oil, and oil shale. These new fuels are abundant compared with conventional crude oil reserves. Among them, tar sands and extra heavy oil are relatively cheap to process. The conversion of oil shale into syncrude is economically viable only when crude oil prices are higher than $70/barrel (in 2004 dollars) using current technologies.[4] The reserves of tar sands are estimated to be equivalent to about 2.7 trillion barrels of oil, and 81% of the reserves are in Canada; however, only 315 billion barrels are economically recoverable by current technologies.[4] The estimated tar sand reserves in the United States are between 60 and 80 billion barrels, and approximately 11 billion barrels are recoverable.[20] In addition, the Orinoco belt in Venezuela contains large quantities of extra heavy oil. Approximately 270 billion barrels of oil can be recovered from extra heavy oil in this region.[21] According to the estimates by the U.S. Energy Information Administration, 2.9 trillion barrels of recoverable oil are available from oil shale worldwide. The United States, which has the largest oil shale reserves, has approximately 750 billion barrels of recoverable oil.[4]

The proven natural gas reserves are 180 trillion m^3 worldwide, which is equivalent to 1.12 trillion oil barrels. Among these reserves, 0.34 trillion barrels lie in the United States.[19] Natural gas has been a common energy source for heating, electricity generation, and hydrogen production. Major research, development, and commercial efforts in recent years have included an interest in liquefied natural gas and natural gas conversion to chemicals and liquid fuels such as methanol, dimethyl ether, ethylene, propylene, gasoline, and diesel. The efforts devoted to liquid fuel production from natural gas are manifested by the major ongoing gas-to-liquid (GTL) commercial activities with an overall liquid production capacity of more than 70,000 bbl/day (barrels per day) in 2006.[22] Figure 1.2 shows the commercial activities in GTL technology using mostly stranded natural gas. The GTL technology is less carbon intensive and is less sensitive to crude oil prices, but the commercial GTL technology is very expensive. For example, the Quarter Petroleum sources in 2009 indicated that the estimated final project cost for the Pearl GTL plant in Qatar that was designed to produce 140,000 bbl/d of the petroleum liquids in 2011 was projected to be more than three times of the original estimate.

The price of natural gas varies significantly with locations and sources. The average natural gas price has been rising during the last several years and is projected to increase continuously for the foreseeable future. Despite the projected price increase, the share of natural gas in world energy consumption is expected to remain at 24% from 2005 to 2030.[1]

The demand for coal is expected to increase faster than the demand for both crude oil and natural gas because of coal's relatively low price and abundant supply in some of the largest energy-consuming and developing countries, including China and India. Figure 1.3 and Table 1.1 provide a comparison of world fossil fuel reserves by geological regions. It is expected that the share

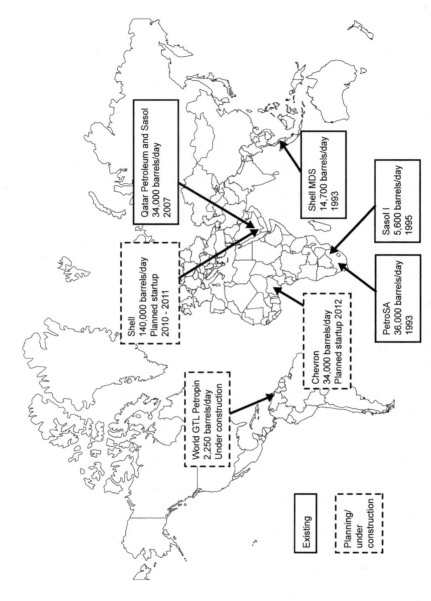

Figure 1.2. Locations and capacities of ongoing commercial technology developments directly associated with natural gas conversion to liquid phase products such as paraffin and olefin.[22]

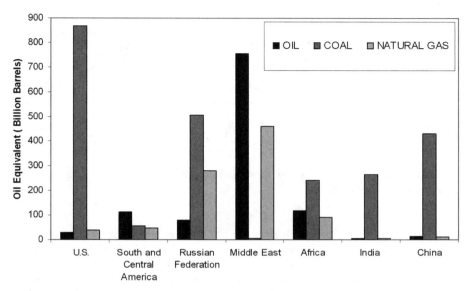

Figure 1.3. World fossil fuel reserve distributions (1 million ton of oil equivalent equals 1.111 billion m^3 of natural gas, 1.5 million ton of hard coal, or 3 million ton of lignite).[19]

TABLE 1.1. World Fossil Fuel Reserves[19]

	Oil (10^9 Barrels)	Natural Gas (10^{12} m^3)	Coal (10^6 t)
United States	29.4	5.98	242,721
South and Central America	111.2	7.73	16,276
Russian Federation	79.4	44.65	157,010
Middle East	755.3	73.21	1,386
Africa	117.5	14.58	49,605
China	15.5	1.88	114,500
India	5.5	1.06	56,498
Rest of the World	124.1	28.27	209,492
Total	1,237.9	177.36	847,488

of the energy supplied by coal will increase from 26% to 29% in the next three decades (2005–2030).[1]

The United States possesses more than 26% of the total coal reserves in the world, and, based on the current consumption rate, these reserves will last for approximately 250 more years.[23] Currently, more than a third (40%) of the electricity generated worldwide comes from coal, whereas in the United States about half of the electricity is generated from coal. Most of the current coal-based power plants are combustion power plants. With improvements over traditional coal combustion technology and the development of new technologies, such as the integrated gasification combined cycle (IGCC), the dominance

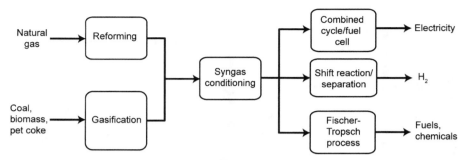

Figure 1.4. Cogeneration of electric power, chemicals, and fuels from coal gasification or natural gas reforming.

of coal in electricity generation is expected to continue well into the 21st century.[4] In the United States, the Energy Vision 2020 of the U.S. Department of Energy (DOE) has embraced clean coal technology as part of the response to the urgent need for independence from nondomestic energy resources.

Similar to natural gas, coal also can be used for liquid fuel and chemical synthesis. In the well-known commercial indirect coal-to-liquid (CTL) process, coal is first gasified into syngas, a gas mixture that contains H_2 and CO. The coal-derived syngas is then conditioned and liquefied through Fischer–Tropsch synthesis. An alternative method for coal liquefaction involves direct liquefaction of coal using hydrogen under very high pressure (up to 210 atm) in the presence of catalysts. The first direct coal liquefaction commercial plant was commissioned for operation at 6,000 ton of coal per day in 2009 in China. The cost of such a direct coal liquefaction process, however, is higher than that of the indirect CTL process, but the direct coal liquefaction process produces more liquid fuels.[24] Figure 1.4 presents the process flow diagrams to coproduce various products from different carbonaceous feedstock from coal gasification or natural gas reforming. Although liquid fuel synthesis processes using coal, biomass, or natural gas are much more capital intensive than oil refining, they can become economical if the crude oil price is high. As shown in Figure 1.4, syngas derived from coal gasification or natural gas reforming shares a similar downstream processing scheme in the cogeneration of electricity, hydrogen, chemicals, and/or liquid fuels. There are processing similarities of biomass, petroleum coke, and other carbonaceous feedstock to coal. In the next section, the energy conversion processes concerned with coal as feedstock are discussed.

1.2 Coal Combustion

In a power plant, coal can be burned in a variety of combustors or boilers such as a pulverized combustor, a stoker combustor, or a fluidized bed combustor.[25]

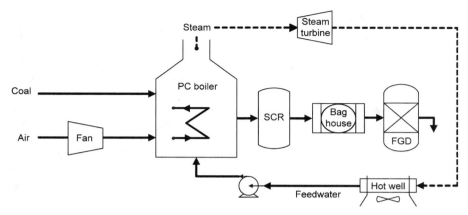

Figure 1.5. Simplified schematic diagram of a pulverized coal combustion process for power generation (PC: pulverized coal; SCR: selective catalytic reduction; FGD: flue gas desulfurization; - - - steam).

The heat of combustion is used to generate high-pressure, high-temperature steam that drives the steam turbine system to generate electricity. Currently, pulverized coal (PC) fired power plants account for more than 90% of the electricity generated from coal.[26] The schematic flow diagram of a PC power plant is illustrated in Figure 1.5.

In a PC power plant, coal is first pulverized into fine powder with more than 70% of the particles smaller than 74 μm (200 mesh). The pulverized coal powder is then combusted in the boiler in the presence of ~20% excess air.[25] The heat of combustion is used to generate high-pressure, high-temperature steam that drives the steam turbine system based on a regenerative Rankine cycle for electricity generation. Although the underlying concept seems to be simple, the following challenges need to be addressed for modern PC power plants: enhancement of energy conversion efficiency; effective control of hazardous pollutants emission; and CO_2 capture (and sequestration).

1.2.1 Energy Conversion Efficiency Improvement

An increase in combustion process efficiency leads to reduced coal consumption, reduced pollutant emissions, and potential cost reduction for electricity generation. The first-generation coal-fired power plants constructed in the early 1900s converted only 8% of the chemical energy in coal into electricity (based on the higher heating value, HHV).[27] Since then, a significant improvement in plant efficiency has been made. Thermodynamic principles require higher steam pressures and temperatures for a higher plant efficiency. The corrosion resistance of the materials for boiler tubes, however, constrains the maximum pressure and temperature of the steam. Most of the PC power plants currently under operation use subcritical PC (Sub-CPC) boilers that produce steam with pressures up to 22 MPa and temperatures around 550°C. The energy conversion efficiencies of traditional sub-CPC power plants typically range from 33% to 37% (HHV).[28]

With an increase in the steam pressure, supercritical PC (SCPC) power plants were first introduced in the early 1960s in the United States.[27] Supercritical power plants use steam with a typical pressure of ~24.3 MPa and temperatures up to 565°C, leading to a plant efficiency of 37%–40%.[28] Many supercritical power plants were constructed in the 1960s and 1970s in the United States. However, because of the low reliability of the boiler materials, the further application of the SCPC technology essentially was halted in the United States in the early 1980s. The development of high-performance super alloys coupled with increasing environmental concerns and the rising cost of coal during the last two decades has stimulated the revival of supercritical technology, especially in Europe and Japan, leading to the reduction of sub-critical boilers in newly installed plants.

Recent advances in coal combustion technologies are highlighted by the generation of ultra-supercritical (USCPC) steam conditions that can provide even higher process efficiencies. The ultra-supercritical condition refers to operating steam-cycle conditions above 565°C (>1,050°F).[28] The pressure and temperature of the steam generated from existing ultra-supercritical power plants can reach 32 MPa and 610°C, which corresponds to an energy conversion efficiency of more than 43%.[28,29] The global ongoing research and development (R&D) activities on PC boilers focus on the development of superalloys that can sustain steam pressures up to 38.5 MPa and temperatures as high as 720°C. It is expected that a plant efficiency of more than 46% can be achieved under such conditions.[28–30] Other efforts in ultra-supercritical technology include minimizing the usage of superalloys, improving the welding technique, and optimizing the boiler structure design to minimize the length of the steam line to the steam turbine.[29]

Apart from PC boilers, fluidized bed combustors (FBCs) using either turbulent fluidized beds or circulating fluidized beds (CFBs) also are being used for steam and power generation worldwide. In these processes, limestone is injected to capture SO_x formed during coal combustion. Compared with PC boilers, the FBC has lower SO_x and NO_x emissions.[31,32] Furthermore, it has superior fuel flexibility.[33] Most commercial FBC plants operate under atmospheric pressure, with energy conversion efficiencies similar to sub-CPC power plants. Higher efficiencies can be achieved by operating the FBC at increased pressures.[33–35] The pressurized fluidized-bed combustor (PFBC) generates a high-temperature, high-pressure exhaust gas stream, that drives a gas-turbine/steam-turbine combined cycle system for power generation. In an advanced PFBC configuration, fuel gas is generated from coal via particle oxidation and pyrolysis. The fuel gas is combusted to drive a gas turbine (topping cycle). Such a process has the potential to achieve an energy conversion efficiency of more than 46%.[34] The capital investment for PFBC is higher than PC power plants with a similar efficiency.[36] Other potential challenges to the PFBC technology

TABLE 1.2. Energy Conversion Efficiencies (HHV) of Various Coal Combustion Technologies and Energy Penalty for CO$_2$ Capture Using MEA[28,38–43]

Technology	Sub-CPC	SCPC	USCPC	AFBC	PFBC
Base case efficiency (%) HHV	33–37	37–40	40–45	34–38	38–45
MEA retrofit derating (%)[a]	30–42	24–34	21–30	~35[b]	~30[b]

[a]Decrease in energy conversion efficiency when a retrofit MEA system is used to capture 90% of the CO$_2$ in the flue gas.
[b]Estimate based on ASPEN simulation.

include scale-up, high-temperature particulates/alkali/sulfur removal for gas turbine operation and mercury removal from the flue gas.[33,37] Currently, the usage of the PFBC is low. Table 1.2 compares the performance of different coal combustion technologies.

1.2.2   Flue Gas Pollutant Control Methods

Modern coal-combustion power plants need to be able to capture environmentally hazardous pollutants released from coal combustion. Such pollutants include sulfur oxides, nitrogen oxides, fine particulates, and trace heavy metals such as mercury, selenium, and arsenic. Methods for capturing these contaminants from the flue gas streams abound. The challenges, however, lie in the efficient and cost-effective removal of these contaminants.

The traditional method for SO$_x$ removal uses wet scrubbers with alkaline slurries. The wet scrubber is effective, but it is costly and yields wet scrubbing wastes that must be disposed. Alternative methods include more cost-effective lime spray drying and dry sorbent duct injection. The lime spray drying method employs a slurry alkaline spray yielding scrubbing wastes in solid form, which eases waste handling. Dry sorbent duct injection employs a dry alkaline sorbent for direct in-duct injection, circumventing the use of the scrubber. Recent pilot tests have used reengineered limestone sorbents (i.e., PCC precipitated calcium carbonate) of high reactivity, yielding a sorbent sulfation efficiency of more than 90%, compared with less than 70% with ordinary limestone sorbent. The result indicates a viability of the dry sorbent duct injection method with very active sorbents.[44–46] The NO$_x$ is commonly removed by selective catalytic reduction (SCR). Other methods that can be employed include a low NO$_x$ burner and O$_3$ oxidation. The recent pilot testing of the CARBONOX process using coal char impregnated with alkaline metal revealed a high NO$_x$ removal efficiency at low flue gas temperatures.[47] The trace heavy metals such as mercury, selenium, and arsenic can be removed by calcium-based sorbents and/or activated carbon.[44,48]

The techniques to control the flue gas pollutants indicated above are well developed. An effective capture (and sequestration) of CO$_2$, an important greenhouse gas (GHG) that accounts for 64% of the enhanced greenhouse effect[49] is, however, a challenging task.

1.3   CO$_2$ Capture

As noted, excessive accumulation of CO_2 in the atmosphere can cause serious
climate changes. Anthropogenic activities, primarily resulting from fossil fuel
usage, have contributed to the increase of the atmospheric CO_2 concentration
from the preindustrial level of 280 ppm to the current value of 380 ppm.[50]
Moreover, under the current carbon emission growth rate, the atmospheric
CO_2 concentration could reach 580 ppm—a threshold value to trigger severe
climatic changes—within just 50 years.[51]

After years of controversy, the Intergovernmental Panel on Climate Change
(IPCC) conclusively determined in 2007 that the increase in anthropogenic
greenhouse gases was the cause for the worldwide temperature rise. Because
of its relative abundance compared with other greenhouse gases, CO_2 is by far
the most important of these gases and accounts for up to 64% of the enhanced
greenhouse effect.[49] The CO_2 represents ~15% of the flue gas stream from
coal-combustion power plants and the total amount of CO_2 produced from
coal-based power plants accounts for more than 40% of all anthropogenic
CO_2 emissions.[52] Proper CO_2 management is, therefore, important. In a fossil-
fuel-based power plant, the CO_2 management involves three steps: capture
including separation and compression; transportation; and sequestration.
Among the three steps, the CO_2 capture is the most energy-consuming.[28,53]

The existing techniques for CO_2 capture from PC power plants include the
well-established monoethanolamine (MEA) scrubbing technology. Figure
1.6(a) shows the schematic diagram of the MEA scrubbing process with key
stream conditions.[54–56] In this process, the flue gas is first cooled to ~40°C
before entering the absorber where fresh amine solvent is used to absorb CO_2
in the flue gas stream. The spent amine solvent with a high CO_2 concentration
is then regenerated in the stripper under a higher temperature (100–150°C),
and CO_2 is then recovered at low pressures (1–2 atm). A large amount of high-
temperature steam is required to strip the CO_2 in the regeneration step.[56] Thus,
although the theoretical minimum work for MEA scrubbing is as low as
0.11 MWh/ton CO_2 or equivalently, a parasitic energy consumption of ~12%[57]
used for steam generation and CO_2 compression for transportation, it is estimated that the actual energy consumption using amine scrubbing can reduce
the power generated from the entire plant by as much as 42%.[40] This energy
consumption amounts to ~70%–80% of the total cost in the overall three steps
of carbon management, that is, carbon capture, transportation, and sequestration.[53,58] As a result, a process that can reduce the energy consumption in the
CO_2 capture step will be vital for CO_2 management in coal-fired power plants.

The chilled ammonia process, illustrated in Figure 1.6(b), is another solvent-
based CO_2 capture technology, in which ammonium carbonates and bicarbonate slurries are used to capture the CO_2 in the flue gas stream at 0–10°C and
atmospheric pressure. The CO_2-rich solvent is then regenerated at 110–125°C
and 20–40 atm. The capability to regenerate CO_2 at increased pressures reduces
the energy consumption for CO_2 compression. Based on the studies by the

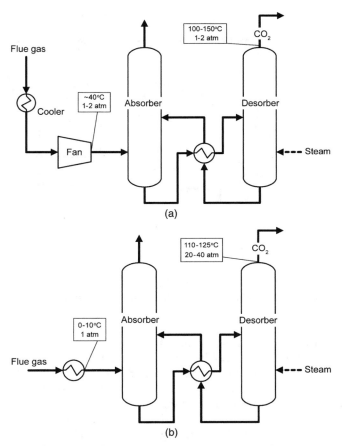

Figure 1.6. Conceptual schematic of (a) MEA scrubbing technology for CO_2 separation and (b) chilled ammonia technology for CO_2 separation.

Electric Power Research Institute (EPRI) and ALSTOM, the overall energy penalty for CO_2 capture is estimated to be lower than 16% when the chilled ammonia process is used.[59,60] A 5-MW$_{th}$ (megawatts thermal) equivalent chilled ammonia process demonstration plant, jointly supported by ALSTOM and EPRI, is currently under construction at We Energies' Pleasant Prairie Power Plant in Wisconsin.[61] American Electric Power (AEP) is also planning to demonstrate the chilled ammonia process at the 20-MW$_e$ (megawatts electricity) scale, starting in 2009, before building a 200-MW$_e$ commercial-level chilled ammonia retrofit system in 2012.[62]

Similar to solvent-based CO_2 scrubbing techniques, high-temperature sorbents such as limestone, potassium carbonates, lithium silicates, and sodium carbonates can be used to capture CO_2 in the flue gas at increased temperatures.[63,64] With good heat integration, these solid sorbents can be effective in

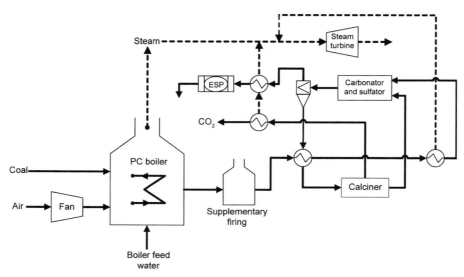

Figure 1.7. Conceptual scheme of carbonation–calcination reactions integrated with a 300-MWe coal-fired power plant depicting heat integration strategies (- - - steam).

the CO_2 capture. One heat integration scheme is of particular relevance, which is based on the carbonation–calcination reactions using hydrated lime, natural limestone, or reengineered limestone sorbents at 600–700°C for CO_2 capture.[65] Figure 1.7 delineates the heat integration scheme for retrofitting the carbonation–calcination reactions to an existing PC power plant. In the process, both CO_2 and SO_2 in the flue gas are captured by the calcium-based sorbents in the carbonator operated at ~650°C, forming $CaCO_3$ and $CaSO_3/CaSO_4$. The carbonated sorbent, $CaCO_3$, is then regenerated to calcium oxide (CaO) sorbent in the calciner at 850–900°C, yielding a pure CO_2 stream. In the Carbonation–Calcination Reaction (CCR) Process developed at the Ohio State University (OSU), the regenerated CaO may or may not be followed by the hydration reaction before it is injected to the flue gas stream. The sulfated sorbent and fly ashes are removed from the system by means of a purge stream. With an optimized energy management scheme, the CCR Process consumes 15%–22% of the energy generated in the plant.[66,67] The Process is being demonstrated in a 120-KW$_{th}$ (kilowatts thermal) plant at OSU (see Section 6.3). Another calcium-based process using $CaCO_3$ is being demonstrated at CANMET Energy Technology Center in Canada.[68] The CANMET process captures CO_2 at a substantially higher Ca:C molar ratio compared with that in the CCR Process. At low temperatures (~590°C), separation of CO_2 also can be carried out by an adsorption approach using a K_2CO_3 promoted hydrotalcite sorbent through a pressure/temperature swing process.[69]

To generalize, several retrofit systems under different stages of development can be used to capture CO_2 from existing power plants. As PC power plants will continue to provide a significant portion of the electricity needs well into the 21st century,[70] these CO_2 capture systems are essential to mitigate the environmental impact from coal burning. As noted, CO_2 capture and compression from coal combustion flue gas are costly and energy intensive. Promising approaches to reduce the overall carbon footprint of a coal-based plant are to devise coal conversion processes that are intrinsically advantageous from a carbon management and energy conversion standpoint. Such approaches are described in Sections 1.5–1.7.

In addition to the previous CO_2 capture methods, oxy-fuel combustion provides another means for carbon management in coal-fired power plants. In this method, pure oxygen, instead of air, is used for coal combustion. As a result, a concentrated CO_2 stream is generated, avoiding the need for CO_2 separation, but the energy-consuming cryogenic air separation step will reduce the overall plant efficiency by 20%–35%.[28,41,53] This process has been demonstrated by the Babcock & Wilcox Company on a 1.5-MW$_{th}$ pilot scale PC unit. Demonstration on a 30-MW$_{th}$ unit is under way. The ongoing pilot-scale studies on oxy-fuel combustion include those carried out by ALSTOM, Foster-Wheeler, CANMET Energy Technology Center, Vattenfall, and Ishikawajima-Harima Heavy Industries (IHI).[71]

1.4 CO₂ Sequestration

Although the cost of CO_2 sequestration represents a relatively small portion of the total carbon management cost, it is necessary to achieve permanent and efficient sequestration of the stream containing CO_2. For storage capacities, geological,[72] mineral[73] and ocean[74] based sequestration—each capable of storing much more than ~10^3 Gt (gigatons)—could reduce significantly current anthropogenic carbon emission levels, which are estimated to be 6 to 8 Gt of carbon per year, for several hundred years.[75]

The oceans, being natural absorbers of CO_2, are already processing a large fraction of the emission. Various injection methods have been investigated to enhance oceanic CO_2 uptake, including dissolution of CO_2 at a depth below the mixed layer of oceans and bottom injection of CO_2, forming lakes of dense CO_2. This CO_2 reacts with seawater to form a clathrate, a cage structure with approximately six water molecules per CO_2. Although ocean sequestration is promising in terms of carbon sequestration capacity, the environmental impact caused by changed ocean chemistry may be the most significant factor determining the acceptability of oceanic storage.

Currently, considerable interest in storage means is focused on geological sequestration. The ideal geological formations that can be used for geological CO_2 sequestration include depleted oil and gas reservoirs (200–500 Gt of carbon, GtC), deep unmineable coal seams (100–300 GtC), and deep saline aquifers (10^2–10^3 GtC). Mined salt domes and rock caverns also are considered to be viable options.[75] Injecting CO_2 into such depleted reservoirs is similar to the well-established and effective practice of using supercritical CO_2 for enhanced oil and gas recovery (EOR). The uncertainty of this practice for CO_2

storage, however, lies in the long-term stability of the reservoirs.[76] When injected into the coal bed, CO_2 can replace adsorbed CH_4 in a molar ratio of ~2:1. By doing so, coal beds can serve as both a CO_2 reservoir and a source for methane production.[74] The first large project that applies geological seques-tration to contain carbon emissions uses a saline formation as the CO_2 reposi-tory. In this project, CO_2 from Statoil's Sleipner West gas reservoir in the North Sea is injected into an aquifer under the sea floor.[77] The graphic representa-tion, given in Figure 1.8, illustrates the operation of the aquifer sequestration.

The Sleipner field – CO_2 Treatment and Injection

StatoilHydro

(a)

Sleipner CO_2 Injection

StatoilHydro

(b)

Figure 1.8. (a) The Sleipner gas field center in the North Sea. (b) the Sleipner CO_2 injection into the Utsira Saline formation at 1,000 m below sea bottom.[77] Photos: cour-tesy of StatoilHydro.

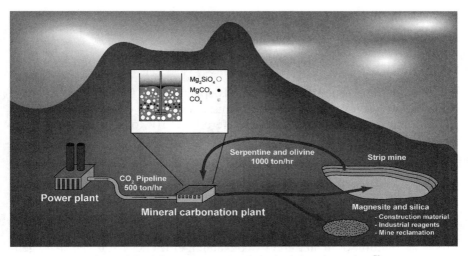

Figure 1.9. CO$_2$ sequestration by mineral carbonation.[79]

Geological sequestration is attractive because of the technological readiness and low injection costs. There are, however, some concerns such as the consideration that most of the actual costs will be associated with future monitoring of the injected CO$_2$ because the accidental release of CO$_2$ could result in the devastating suffocation of humans and animals similar to the incidents that occurred in Nyos, Cameroon, in 1984 and 1986. Thus, monitoring technology will play a key role in the success of the geological sequestration method.[78]

Mineral sequestration of CO$_2$ can be conducted in a chemical reactor, as shown in Figure 1.9, in order to achieve a high reaction rate of immobilization of carbon in a mineral matrix. The reaction between CO$_2$ and most minerals is slow in nature. Serpentine (Mg$_3$Si$_2$O$_5$(OH)$_4$) and olivine (Mg$_2$SiO$_4$), however, were found to be two of the most promising minerals that could be used in the mineral sequestration because of their ability to form carbonates and their abundant availability.[79] In the mineral sequestration process, Mg in serpentine is dissolved in solution and reacts with dissolved CO$_2$ to form MgCO$_3$, which is environmentally benign. A pH swing process that dissolves Mg at a low pH and precipitates MgCO$_3$ under a high pH offers a viable approach for mineral sequestration of CO$_2$, while producing value-added solid products.[80] The cost of the carbon mineral sequestration process is still higher than that of the geological sequestration process. Thus, optimization is ongoing to improve its economic feasibility.

Another important sequestration method is bio-based. This method also mimics the natural capture of carbon. One of the largest natural carbon flows through the environment is driven by photosynthesis. Plants take up CO$_2$ and water, and turn them into reduced carbon compounds such as starch and cellulose. Thus, forestation and agricultural fixation of carbon, either in biomass or soil, can play a role in carbon sequestration. Currently, researchers are investigating various biosystems such as ocean fertilization and algae production to capture CO_2 at a rapid rate.

The preceding discussion suggests that there are several carbon sequestration methods currently being investigated and developed. It is desirable to consider multiple sequestration methods concurrently for carbon sequestration resulting from its massive flux of emission. A life-cycle analysis is needed to ascertain the amount of CO_2 sequestered relative to the cost and energy requirements for each sequestration method employed.

1.5 Coal Gasification

For many years, commercial efforts on clean coal processes have focused on coal combustion for power generation. Nonetheless, the development of new processes mindful of higher energy conversion efficiencies for electricity generation, as well as of variability in product formation, has generated considerable interest. Because of the unpredictable crude oil price and concern over carbon management, the development of clean coal technology (CCT) has led to serious consideration of gasification as a commercial coal processing route. Coal gasification schemes can provide a variety of products in addition to electric power. It is noted that from the carbon management viewpoint, gasification also is a preferred scheme over combustion because of a higher CO_2 concentration and a lower volumetric gas flow stream from which CO_2 is to be separated.

The commercial operation for gasification dates back to the late 18th century when coal was converted into town gas for lighting and cooking. Since the 1920s, there have been numerous different gasification processes developed for the production of syngas, hydrogen, chemicals, liquid fuels, and/or electricity.[82,83] Figure 1.10 shows a general scheme of the coal gasification process.[81] The heart of a coal gasification process is the coal gasifier. Figure 1.11 shows several types of gasifiers that can be classified generally based on the state of fluidization of the particles, including a moving bed gasifier, a bubbling/turbulent bed gasifier, and an entrained flow gasifier. The moving bed gasifier is characterized by dense downward movement of particles with a significant axial temperature gradient. The bubbling/turbulent bed gasifier is operated under the bubbling or the turbulent fluidization regime in dense-phase fluidization with a uniform temperature distribution in the bed.[84] The entrained flow gasifier, however, is operated under the pneumatic conveying regime in dilute-phase fluidization with a uniform temperature in the gasifier. The bubbling fluidization regime is characterized by the flow of gas bubbles with an increased tendency of bubble coalescence when the gas velocity increases, whereas the turbulent fluidization regime is characterized by the flow of gas bubbles with an increased tendency of bubble breakup when the gas velocity increases. The pneumatic conveying regime in dilute-phase fluidization is represented by

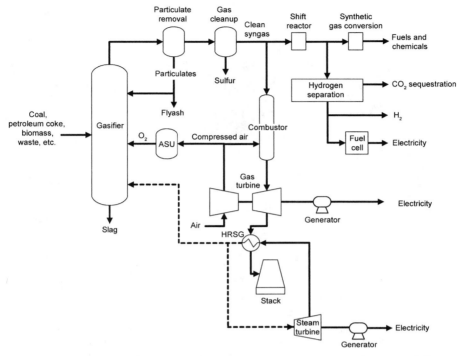

Figure 1.10. Schematic diagram of coal gasification processes (- - - steam).[81]

dilute transport of solid particles in clustering or nonclustering form.[84] Most modern gasifiers adopt an entrained-flow design thanks to better fuel flexibility, carbon conversion, and syngas quality.[82]

Figure 1.12 shows examples of three different designs for the entrained bed gasifier. The design varies with the manner in which coal/coal slurry and gaseous reactants are introduced and heat is removed. Furthermore, different gasifiers have different feeding rates of water/oxygen, and as a result, the composition of the syngas at the outlet of the gasifier varies. The Shell gasifier, shown in Figure 1.13, uses water to cool the membrane-based gasifier walls, and the dry coal powder feedstock is introduced along with oxygen at the side of the gasifier. In addition, the Shell gasifier typically adopts a gas quench configuration, in which a recycled cold syngas stream is introduced to the gasifier at the raw syngas exit to quench the molten ash entrained in the raw syngas. The quenching would prevent the molten ash from agglomeration.

The common features of entrained flow gasifiers include high operating temperature, low water feed rate, and a high CO/H_2 ratio in the raw syngas. In a typical gasification process using an entrained-flow gasifier, the high-temperature, high-pressure raw syngas (~1,300°C, ~25 atm) is produced from the gasifier and contains such pollutants as particulates, H_2S, COS, and mercury.

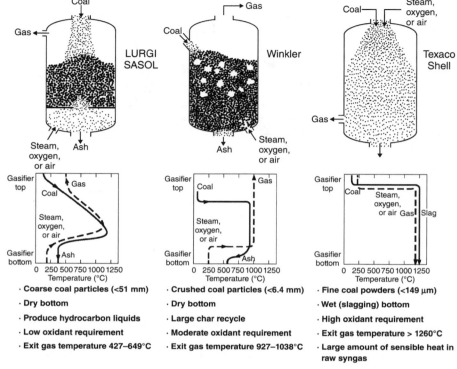

Figure 1.11. Schematic diagrams and temperature distributions of various gasifiers.[25]

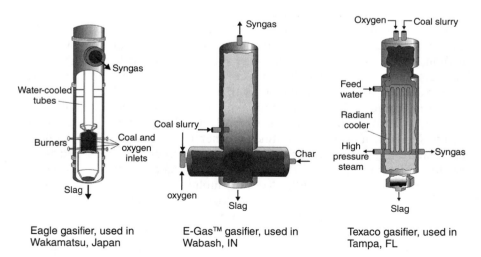

Figure 1.12. Schematic configurations of various entrained flow gasifiers.[85–87]

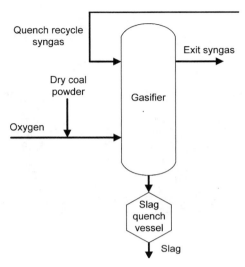

Figure 1.13. Schematic diagram of the key flow streams around the Shell gasifier.[25]

The raw syngas is first cooled by water quenching or by passing it through heat exchangers. It then flows through a dry ceramic filter system or a wet scrubber system where fly ash is removed. A hydrolysis unit then is used to convert COS in the syngas into H_2S and CO_2. This syngas stream is further cooled to near ambient or even lower temperature and passes through a solvent-based sulfur removal unit.[25] Mercury and other trace metals in the H_2S cleaned syngas are removed subsequently by activated carbon using a fixed bed operated at ~35°C and 60 atm. The activated carbon usually is impregnated with 10–15 wt.% sulfur, which would react with mercury to form HgS, a stable compound that is suitable for long-term storage.[88] Different processing routes can be followed after the syngas is cleaned up. For electricity generation, the syngas can be introduced directly into a combustion turbine to generate electricity. The high-temperature exhaust gas from the combustion turbine is then sent to a heat recovery steam generator (HRSG) to produce high-temperature and high-pressure steam to drive a steam turbine to generate more electricity. The gasification process that involves both gas turbines and steam turbines for electricity generation is known as the IGCC process. The overall energy conversion efficiency for the IGCC can exceed ~45% when carbon capture is not required.[89] For hydrogen generation, the syngas is introduced to the water–gas shift (WGS) reactor that converts CO into hydrogen and CO_2. The gas then passes through the acid gas removal units and pressure swing adsorption (PSA) units to separate out H_2 from CO_2. The PSA tail gas with low H_2 concentration can be combusted in a gas turbine to generate electricity for parasitic energy use. The hydrogen can be a fuel source for either fuel cell or gas turbine operations. The hydrogen also can be used as a makeup gas to tune

Figure 1.14. Tampa Electric's Polk IGCC Power Station Unit 1. Photo courtesy of Mr. Mark J. Hornick of Tampa Electric.

Figure 1.15. Air Products's LPMEOH process demonstration unit for methanol synthesis using syngas from coal gasification with slurry bubble column reactor, located at the Eastman Chemical cite in Kingsport, Tennessee. U.S. DOE Topical Report No. 11, April 1999, under DOE contract #DE-FC22-92PC90543. Photo: coursey of Debbie Bauer, Corporate Communications of Air Products.

up the CO/H_2 ratio to 1/2 so that it can be used for Fischer–Tropsch synthesis, which converts CO and H_2 to liquid hydrocarbons or chemicals. It is noted that the raw syngas from the gasifier usually has a CO/H_2 ratio ranging from 1.4/1 to 2/1.[25] The key to innovative gasification technology development lies in well-conceived process integration and intensification.

The current development of coal gasification processes varies from conceptual to pilot demonstration to commercial scale. Figure 1.14 shows the coal/petcoke IGCC plant at Polk County, FL, which has been producing 260-MW electricity on grid. The plant uses the Texaco gasifier shown in Figure 1.12. The coal-to-liquid technology can be exemplified by the production of methanol from syngas derived from coal gasification. The commercial plant of Eastman Chemicals, which produces methanol at a capacity of 5,000 ton/d, is shown in Figure 1.15. Figure 1.16(a) shows a commercial slurry bubble column reactor with 5-m ID developed by Sasol for the production of gasoline, diesel, and wax at a capacity of 2,500 bbl/d using syngas from coal or natural gas as feedstock.

(a)   (b)

Figure 1.16. (a) Sasol's 5-m ID F-T reactor for coal-to-liquid synthesis. (b) Sasol's Oryx gas-to-liquid plant in Qatar, which consists of two low-temperature Fischer–Tropsch SPD (slurry-phase distillate) reactors with a capacity of 17,000 bbl/d each, producing a primary product of low-sulfur diesel oil using the feedstock of reformed natural gas. It shows one reactor of 10-m ID in front, with another of the same size behind it. Photo: courtesy of Sasol Limited.

A slurry bubble column reactor of 10-m ID for Fischer–Tropsch synthesis, which is shown in Figure 1.16(b), is currently operated at a capacity of 17,000 bbl/d for GTL applications.

Challenges to existing coal gasification processes include significant energy consumption in the alteration of temperatures and pressures of the syngas in the pollutant removal process. Technologies that can enhance pollutant and product separation and chemical reactions at increased temperatures are being developed. These technologies include high-temperature sorbents for H_2S removal; sulfur-tolerant catalysts for water-gas shift reactions; metallic, ceramic, polymeric, or hybrid membranes for CO_2, H_2 and/or H_2S separation; and catalysts with high selectivity for the Fischer–Tropsch synthesis. For the membrane technology, when applied to the water-gas shift reaction with *in situ* removal of either CO_2 or H_2 from the reacting product gases, it can enhance the water-gas shift reaction conversion significantly. The variables that dictate the membrane performance include the types of the materials used (organic, inorganic, or hybrid), the permeability and selectivity properties, the tolerance level toward pollutants, and the physical strength in enduring the high-pressure drop across the membrane. Factors that affect the economical viability of the membrane-based gasification technologies include the cost and reliability of the membranes. Membranes can be used in product separation when coupled with reactor operation. Process simulations indicate that by using H_2- or CO_2-selective membranes, the energy penalty for CO_2 capture in a coal-to-hydrogen process can be reduced.[90–92]

In addition to vigorous commercial activities on the IGCC process, novel clean coal processes, characterized by high-energy conversion efficiencies and ease in implementing product and/or pollutant separation strategies, are now at various stages of development. The novel coal reforming processes being developed worldwide include the Zero Emission Coal Alliance (ZECA) Process, the HyPr-Ring Process, the GE Fuel-Flexible Process, the ALSTOM Process, the chemical looping combustion processes, and the OSU's Syngas Chemical Looping Process, CDCL Process, and Calcium Looping Process. These new processes employ the chemical looping concept, and some are at the demonstration level. In the next section, the fundamental concept and characteristics of chemical looping technology and its process applications are discussed.

1.6 Chemical Looping Concepts

A given reaction can be decomposed into multiple subreactions in a reaction scheme using chemical intermediates that are reacted and regenerated through the progress of the subreactions. A reaction scheme of this nature is referred to as chemical looping. An ideal chemical looping scheme is to design the subreactions in such a manner that the exergy loss of the process resulting from this reaction scheme can be minimized while allowing the separation of the products and/or pollutants generated from the reactions to be accomplished with ease, thereby yielding an overall efficient and economical process system.

Achieving a high process efficiency for the energy conversion to products such as electricity, H_2, and liquid fuels while maintaining low pollutant emissions represents a major challenge for any fossil fuel conversion system. This is particularly the case for coal, which is the most carbon intensive among all the available fossil energy resources. To achieve a high process efficiency, a proper energy management strategy is necessary. Furthermore, to assess the effect of the strategy, it is desirable to resort to the analysis of the energy conversion process based on the thermodynamic concept of exergy, or availability. Exergy and its application to such an analysis are briefly explained in the following discussion.

The exergy of a system is defined as the maximum amount of usable work extractable from the system during a transformation that brings the system into equilibrium with a reference state.[93] The ambient environment commonly is used as the reference state. The exergy rate is the ratio between the available work and the total amount of energy. The second law of thermodynamics indicates that energy degrades in all irreversible processes, or in equivalent terms, exergy loss will occur in any nonideal process. Although it is not possible to eliminate the energy degradation, the exergy loss in a process can be minimized through strategic energy management. A useful approach to improving an existing system is to identify the steps where the largest irreversibility occurs and to minimize these losses.

Energy with a low exergy rate can be integrated into energy with higher exergy rates to improve the overall energy conversion efficiency of the process.[94] The following two examples illustrate how energy integration can be used to minimize exergy loss whose results are calculated with the assumptions follows:

1. The environmental temperature is $T_0 = 273.15\,K$, and the environmental pressure is $P_0 = 101{,}325\,Pa$.
2. The reference substance for carbon and CO is 400 ppm CO_2, the reference substance for H_2 is pure water, and the reference substance for Fe, FeO, and Fe_3O_4 is pure Fe_2O_3.
3. The upper limit for the gas turbine operating temperature is $1{,}800\,K$.
4. Coal is considered as pure carbon.
5. Heat can be integrated with a 100% efficiency whenever feasible.
6. The enthalpy of devaluation is defined as the enthalpy change for a substance from the current state to its reference state.
7. The feedstock in the outside of the reaction loop is at the environmental state.

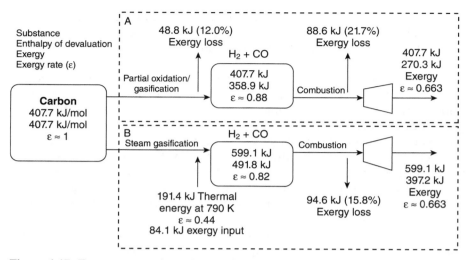

Figure 1.17. Exergy recovering schemes for carbon gasification/gas turbine system, showing scheme A (top) and scheme B (bottom).

Scheme A in Figure 1.17 shows two major steps of the traditional IGCC process. In the first step, coal reacts with oxygen and steam to form syngas; in the second step, syngas generated in the first step is combusted to drive the gas turbine/steam turbine combined cycle. As shown Figure 1.17, both steps in scheme A are highly irreversible and involve inevitable exergy loss. Specifically, in the first step, carbon is oxidized partially to provide the heat needed in the steam carbon reaction. Although no heat loss is assumed, 12% exergy is lost from the formation of syngas, which has a lower exergy rate ($\varepsilon = 0.88$) than that of carbon ($\varepsilon = 1$). In the second step, 21.7% exergy is lost from the limitation on the gas turbine operating temperature. An operating temperature of 1,800 K would correspond to an exergy rate of 0.663 from combustion, which is much lower than the exergy rate of the syngas ($\varepsilon = 0.88$). Thus, with a total of 33.7% exergy loss, the energy conversion efficiency of scheme A, which is equivalent to the exergy conversion efficiency in this specific case, cannot exceed 66.3% for electricity generation. To improve the theoretical energy conversion efficiency, the exergy loss in scheme A should be minimized. One option for reducing the exergy loss is to incorporate the low exergy rate energy into carbon. This option leads to scheme B.

In scheme B, the thermal energy with a low exergy rate ($\varepsilon = 0.44$ at 790 K), which is abundant in a typical IGCC process, is integrated into the gasification step. The thermal energy would provide the heat needed for the steam carbon reaction to form H_2 and CO_2. This approach serves two purposes: (1) Exergy loss can be avoided by the heat integration through incorporation of the low exergy rate thermal energy into the endothermic steam carbon reaction; and (2) less exergy loss would occur in the combustion step because of a smaller

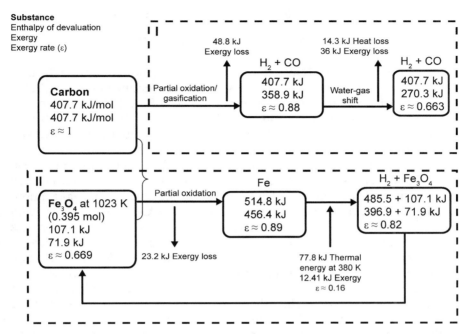

Figure 1.18. Exergy recovering scheme for carbon gasification-water–gas shift process, scheme I (top) and scheme II (bottom: the reaction temperature is assumed to be 1,123 K for scheme II).

exergy rate difference between H_2 and the thermal energy at the combustion temperature. By recovering the thermal energy, more than half of the exergy loss can be avoided in scheme B.

Other than heat recovery, the choice of suitable chemical intermediates and optimization of the chemical reaction scheme often can decrease drastically the exergy loss for an energy conversion process. Figure 1.18 illustrates two schemes in the second example for H_2 generation from coal.

Scheme I represents a traditional route where carbon is converted into syngas followed by the water–gas shift reaction to produce hydrogen. Similar to the previous example, the gasification step would lead to 12% exergy loss from partial oxidation. The second step in scheme I, the water–gas shift reaction, will lead to another 8.8% exergy loss from the conversion of CO into H_2, which has both lower enthalpy devaluation and a lower exergy rate. In the previous calculation, the exergy in the low-temperature waste heat generated in the WGS reaction is not taken into account. The energy required for separating hydrogen from CO_2 also is neglected.

Scheme II offers an alternative approach using the chemical looping concept. Here, Fe is used as a chemical intermediate to convert carbon into hydrogen. Also, two steps are involved in scheme II. The reactions are as follows:

$$\text{Step 1: } C + 0.395Fe_3O_4 + 0.21O_2 \rightarrow 1.185Fe + CO_2 \qquad (1.6.1)$$

$$\text{Step 2: } 3Fe + 4H_2O \rightarrow Fe_3O_4 + 4H_2 \qquad (1.6.2)$$

A small amount of O_2 is introduced in the first reaction to provide the heat needed to reduce Fe_3O_4 to Fe. Much less exergy loss occurs in step 1 (23.1 kJ) compared with the traditional gasification step (48.8 kJ). This is from the recovery of a low-exergy-rate chemical, Fe_3O_4 ($\varepsilon = 0.669$), which reacts with carbon ($\varepsilon = 1$) to form a chemical, Fe ($\varepsilon = 0.89$), of a medium exergy rate. With H_2O introduced into the system under ambient conditions, the steam–iron reaction in step 2 of scheme II requires only a small amount of low grade heat, rendering the ease in balancing the heat required of this step. A zero exergy loss is achievable in step 2 of scheme II. As a result, a traditional gasification-water–gas shift process would lead to at least 68.7-kJ exergy loss for every mole of H_2 generated. On the other hand, when the chemical looping process is used with $Fe-Fe_3O_4$ as the looping media, the exergy loss can be reduced to 14.7 kJ for each mole of H_2 generated. Thus, the exergy loss for H_2 production is reduced by more than four times using the chemical looping process.

Note that the results presented in the examples are based on a set of assumptions given earlier. Thus, they represent only the upper bound of the energy conversion efficiencies in the conversion process. Nevertheless, because of the common assumptions used in assessing the process options, the results obtained serve as a good guide for relative comparisons of these options. To generalize, an efficient energy conversion process should comprise steps that are less irreversible. The following are general guidelines in devising reaction schemes that are thermodynamically advantageous:

1. Highly exothermic reactions that take place at low temperatures should be avoided.
2. Whenever products with lower exergy rates are produced, energy with an even lower exergy rate than the products should be incorporated into the reactants to minimize the exergy loss. The carrier of the low grade energy can be either heat or chemical substances.
3. The excessive heat generated from exothermic reactions should be integrated into chemical energy by integration with endothermic reactions whenever possible; this integration requires that the exothermic reactions operate under a similar or higher temperature than do the endothermic reactions.
4. One single reaction with high exergy loss can be dissociated into a set of reactions employing appropriate chemical intermediates. This new reaction scheme may be subject to less exergy loss.

As illustrated in Figure 1.18, the choice of the reaction scheme is based on one that can minimize the exergy loss. In the rest of the section, examples of chemical reaction schemes that encompass reactions in a looping manner are described. Rather than directly converting feedstock (reactants) into products, the chemical looping strategy uses a series of reaction schemes to generate end products through transformations of certain chemical intermediates. By doing so, as noted, the exergy loss in the conversion process potentially can be minimized. The ideal approach to conduct the chemical looping reaction scheme is to employ a self-sustaining medium that can assist the chemical transformation during the conversion process. The medium also is intended to alleviate the reaction barrier and minimize energy loss in the formation of the product.

An ideal medium, in either solid or fluid form, to be selected for use should possess such properties as high reactivity at proper temperature and pressure ranges, physical integrity, chemical stability, favorable equilibrium toward intermediate product formation, spontaneity of intermediate reaction scheme, ease in intermediate product separation, and moderate exothermic to endothermic heat of reactions. For the chemical looping system to be applicable to process operation, other factors to be considered include simplicity of the chemical looping scheme, ease in heat integration, and process economics. The chemical looping principles discussed earlier can be applied to reaction systems for chemical synthesis or to separation systems for chemical separation. Salient reaction looping examples that may not be related to energy conversion are described as follows:

A. Water Splitting for Hydrogen and Oxygen Production

A-1. I_2 and SO_2 Loopings

$$SO_2(g) + I_2(l) + 2H_2O(l) \rightarrow 2HI + H_2SO_4 \qquad (1.6.3)$$

$$2HI(aq) \rightarrow H_2(g) + I_2(g) \qquad (1.6.4)$$

$$H_2SO_4(aq) \rightarrow H_2O(g) + SO_2(g) + 0.5O_2(g) \qquad (1.6.5)$$

$$\text{Net reaction}: H_2O \rightarrow H_2 + 0.5O_2 \qquad (1.6.6)$$

$$\text{Looping medium}: I_2 \leftrightarrow HI, SO_2 \leftrightarrow H_2SO_4 \qquad (1.6.7)$$

The water-splitting iodine–sulfur (IS) process produces hydrogen and oxygen. The gaseous SO_2 and I_2 liquid react with water to produce two acids, HI and H_2SO_4. This reaction is called the Bunsen reaction. Aqueous HI and H_2SO_4 are separated using a liquid–liquid separator, and then each acid is decomposed at 500°C and 850°C, respectively. I_2 and SO_2 obtained after the decomposition of the acids are recycled to the Bunsen reaction.[95]

A-2. CaO, $FeBr_2$, and Br_2 Loopings

$$CaBr_2 + H_2O \rightarrow CaO + 2HBr \qquad (1.6.8)$$

$$CaO + Br_2 \rightarrow CaBr_2 + 0.5O_2 \qquad (1.6.9)$$

$$3FeBr_2 + 4H_2O \rightarrow Fe_3O_4 + 6HBr + H_2 \qquad (1.6.10)$$

$$Fe_3O_4 + 8HBr \rightarrow 3FeBr_2 + 4H_2O + Br_2 \qquad (1.6.11)$$

$$\text{Net reaction:} \ H_2O \rightarrow H_2 + 0.5O_2 \qquad (1.6.12)$$

$$\text{Looping medium:} \ CaO \leftrightarrow CaBr_2, FeBr_2 \leftrightarrow Fe_3O_4, Br_2 \leftrightarrow HBr \qquad (1.6.13)$$

A bromine system produces hydrogen using metal bromides and bromine as looping media. Calcium oxide and hydrogen bromide are obtained from a high-temperature steam reaction with calcium bromide. Hydrogen can be recovered from a mixture of steam and hydrogen bromide generated from the steam and iron bromide reaction.[96]

A-3. Zn Looping

$$ZnO \rightarrow Zn + \frac{1}{2}O_2 \qquad (1.6.14)$$

$$Zn + H_2O \rightarrow ZnO + H_2 \qquad (1.6.15)$$

$$\text{Net reaction:} \ H_2O \rightarrow H_2 + 0.5O_2 \qquad (1.6.16)$$

$$\text{Looping medium:} \ ZnO \leftrightarrow Zn \qquad (1.6.17)$$

For this process, parabolic mirrors concentrate sunlight over a small area to achieve high-temperature heat. This heat can be transferred to chemical energy by dissociating ZnO into Zn and O at ~1,700°C in a solar reactor. Zinc then reacts with steam at ~400°C to regenerate ZnO, yielding hydrogen. Also, different metal oxides other than zinc oxide also can be used. Compared with other processes of hydrogen production, such as chemical looping, given in cases A-1 and A-2, this process requires the reaction to occur at a higher temperature; however, renewable solar energy can be used as the energy source for ZnO dissociation.[97]

B. Production of Hydrogen Peroxide

$$(1.6.18)$$

9,10-Anthraquinone   9,10-Anthrahydroquinone

$$\text{(1.6.19)}$$

$$\text{Net reaction: } H_2 + O_2 \rightarrow H_2O_2 \qquad (1.6.20)$$

Looping medium: $(1.6.21)$

9,10-Anthraquinone   9,10-Anthrahydroquinone

Hydrogen peroxide is formed by the reaction of oxygen with the looping intermediate, 9,10-anthrahydroquinone. Another looping intermediate, 9,10-anthraquinone, is formed which is then separated and reacts with H_2 to form 9,10-anthrahydroquinone. The looping process is then repeated.[98]

C. Production of Maleic Anhydride from Butane

Oxidized VPO + Butane → Maleic anhydride + Reduced VPO   (1.6.22)

Reduced VPO + Air → Oxidized VPO + N_2 and remaining O_2 (1.6.23)

Net reaction: Butane + Air → Maleic anhydride + N_2 and remaining O_2
                                                                                                        (1.6.24)

Looping medium: Oxidized VPO ↔ Reduced VPO   (1.6.25)

The maleic anhydride production from butane catalyzed by oxidized vanadium phosphorous oxide (VPO), which also serves as an oxygen carrier, was developed by DuPont in the 1990s. Maleic anhydride is an essential component for spandex fibers and copolyester elastomers. The butane oxidation and conversion to maleic anhydride occur in a circulating fluidized bed system, as shown in Figure 1.19. In this system, butane is oxidized in a riser with oxidized VPO to form maleic anhydride. Reduced VPO is then oxidized using air in a turbulent fluidized bed regenerator. Thus, the VPO particles loop around the reactors in an oxidized or reduced state. The particles, that usually contain inert support, require sustaining mechanical and thermal stresses in the looping operation.[99] This looping process was demonstrated well at the bench and pilot scales. Its commercial operation, however, was hampered, that was noted to be caused by the scale-up effects of the CFB reactor on the reactant contact time and the solid particle holdup in the CFB system.[100]

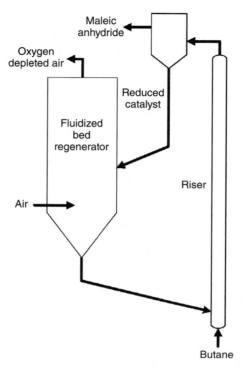

Figure 1.19. Circulating fluidized chemical looping system for production of maleic anhydride.

The inadequacy for sustainable commercial operation principally may be from the chemical and mechanical viabilities of the VPO particles and their associated effects on the transport properties and overall reaction kinetics of the particles in performing their catalytic and looping functions in the circulating fluidized-bed system.

Several examples of chemical looping reactions of relevance to carbonaceous fuel conversions and processes are discussed as follows:

A. CO_2 Separation Using Calcium-Based Sorbent

$$CaO + CO_2 \rightarrow CaCO_3 \qquad (1.6.26)$$

$$CaCO_3 \rightarrow CaO + CO_2 \qquad (1.6.27)$$

Net reaction: CO_2 (mixed with other gases) $\rightarrow CO_2$ (pure) + other gases

$$(1.6.28)$$

Looping medium: $CaO \leftrightarrow CaCO_3 \qquad (1.6.29)$

This reaction scheme can be applied to the separation of CO_2 from a gas mixture where CaO is used as the looping medium. These looping reactions can be conducted by varying the temperature and/or pressure in gasification or combustion systems. As these reactions are carried out at high temperatures, proper heat integration would yield a high thermal efficiency of the process, leading to a low-energy penalty for CO_2 capture.[101]

B. Combustion of Carbonaceous Fuel to CO_2

$$(2x + y/2 - z)MO + C_xH_yO_z \rightarrow (2x + y/2 - z)M + xCO_2 + y/2\,H_2O$$
$$(1.6.30)$$

$$M + Air \rightarrow MO + N_2 \text{ and remaining } O_2 \qquad (1.6.31)$$

$$\text{Net reaction: } C_xH_yO_z + Air \rightarrow xCO_2 + y/2\,H_2O + N_2 \text{ and remaining } O_2$$
$$(1.6.32)$$

$$\text{Looping medium: } M \leftrightarrow MO \qquad \text{(where M is a suitable metal)}$$
$$(1.6.33)$$

Instead of direct combustion of carbonaceous fuel using air to produce heat and hence electricity, this reaction scheme uses metal oxides (MOs) as the looping medium. The first reaction in the scheme uses MO to convert carbonaceous fuel into a steam and CO_2 mixture from which high-purity CO_2 can be separated. The second reaction, which is the combustion of reduced metal (M) with air, liberates heat. This process, by decoupling combustion into two reactions that occur in two different reactors, allow expeditious CO_2 separation from the combustion products.[102,103]

C. Gasification of Carbon to H_2

$$2MO + C \rightarrow 2M + CO_2 \qquad (1.6.34)$$

$$M + H_2O \rightarrow MO + H_2 \qquad (1.6.35)$$

$$\text{Net reaction: } C + 2H_2O \rightarrow CO_2 + H_2 \qquad (1.6.36)$$

$$\text{Looping medium: } M \leftrightarrow MO \qquad \text{(where M is a suitable metal)}$$
$$(1.6.37)$$

The only difference between this looping scheme and the combustion of carbonaceous fuel to CO_2 above is the oxidation agent used for the second reaction. In this example, steam is used instead of air, yielding a reaction product of hydrogen, instead of heat. Furthermore, the advantages of this looping reaction route over the traditional gasification and water–gas shift route for H_2 production include the ease of CO_2 separation. With a suitable metal employed, the reactions can proceed at an optimized temperature with a minimal exergy loss.[104]

D. Gasification of Carbon to Syngas

$$2MO + C \rightarrow 2M + CO_2 \qquad (1.6.38)$$

$$2M + H_2O + CO_2 \rightarrow 2MO + H_2 + CO \qquad (1.6.39)$$

$$\text{Net reaction}: C + H_2O \rightarrow CO + H_2 \qquad (1.6.40)$$

$$\text{Looping medium}: M \leftrightarrow MO \qquad \text{(where M is a suitable metal)} \qquad (1.6.41)$$

Similar to scheme C, when CO_2 is used as the reacting gas along with H_2O in the second reaction, syngas with any combination of concentrations of CO and H_2 can be produced. Thus, the subsequent reactions that synthesize chemicals and fuels based on CO and H_2 can proceed under optimum CO and H_2 concentration conditions.

E. H_2 Production from the Water–Gas Shift Reaction

$$CO(g) + H_2O(g) \rightarrow CO_2(g) + H_2(g) \Big\} \text{Reactor 1} \qquad (1.6.42)$$
$$CO_2(g) + CaO(s) \rightarrow CaCO_3(s) \qquad (1.6.43)$$

$$CaCO_3 \rightarrow CaO + CO_2(g)\} \text{Reactor 2} \qquad (1.6.44)$$

$$\text{Net reaction}: CO + H_2O \rightarrow CO_2 + H_2$$
$$(CO(g) + H_2O(g) + CaO(s) \rightarrow H_2(g) + CaCO_3(s) \text{ for Reactor 1)} \qquad (1.6.45)$$

$$\text{Looping medium}: CaO \leftrightarrow CaCO_3 \qquad (1.6.46)$$

The introduction of CaO looping particles in a water–gas shift reaction would chemically react with CO_2 generated *in situ*, thereby driving the equilibrium toward the formation of H_2. With highly efficient looping particles, nearly 100% conversion of CO can be achieved. Thus, H_2 of high purity can be produced and a sequestration-ready CO_2 stream can be generated from the calciner[105] where the calcination reaction occurs.

F. Separation of H_2S

$$H_2S + MO \rightarrow MS + H_2O \qquad (1.6.47)$$

$$2MS + 3O_2 \rightarrow 2MO + 2SO_2 \qquad (1.6.48)$$

$$\text{Net reaction}: 2H_2S + 3O_2 \rightarrow 2H_2O + 2SO_2 \qquad (1.6.49)$$

$$\text{Looping medium}: MO \leftrightarrow MS \qquad \text{(where M is a suitable metal)} \qquad (1.6.50)$$

In this scheme, a sorbent such as ZnO first is used to react with H_2S in a gas mixture. The spent sorbent, MS, is then regenerated using O_2. This reaction is commonly used in a high-temperature gas cleaning of sulfur in the gasification process.[106]

G. Production of S from FeS or FeS_2

$$FeS_2 + 16Fe_2O_3 \rightarrow 11Fe_3O_4 + 2SO_2 \qquad (1.6.51)$$

$$4Fe_3O_4 + Air \rightarrow 6Fe_2O_3 + N_2 \text{ and remaining } O_2 \qquad (1.6.52)$$

$$\text{Net reaction}: 2FeS_2 + Air \rightarrow 4SO_2 + Fe_2O_3 + N_2 \text{ and remaining } O_2 \qquad (1.6.53)$$

$$\text{Looping medium}: Fe_2O_3 \leftrightarrow Fe_3O_4 \qquad (1.6.54)$$

These reactions are of relevance to mineral processing. In this reaction scheme, sulfur in pyrite (FeS_2) or in pyrrhotite ($Fe_{(1-x)}S, x = 0$ to 0.2) is extracted using hematite (Fe_2O_3) as a looping medium. The reaction scheme generates SO_2, which is then reduced to sulfur in a separate reaction.[107]

H. Combustion with Dodecane Liquid Fuels

$$C_{12}H_{26} + 37NiO \rightarrow 12CO_2 + 13H_2O + 37Ni \qquad (1.6.55)$$

$$Ni + Air \rightarrow NiO + N_2 \text{ and remaining } O_2 \qquad (1.6.56)$$

$$\text{Net reaction}: C_{12}H_{26} + Air \rightarrow 12CO_2 + 13H_2O + N_2 \text{ and remaining } O_2 \qquad (1.6.57)$$

$$\text{Looping medium}: NiO \leftrightarrow Ni \qquad (1.6.58)$$

In these reactions, the carbonaceous fuel used is dodecane that is introduced into the reducer in liquid form. The reduction reactions yield a concentration of CO_2 near 100% without formation of carbon on the surface of the oxygen carrier NiO.[108] The oxygen carrier $NiAl_{0.44}O_{1.67}$ also can be used to yield a similar liquid fuel conversion behavior in the reducer.[109]

1.7 Chemical Looping Processes

With the pressing needs for efficient and cost/energy effective separation scheme for CO_2 emission control in the coal conversion processes, the chemical looping technique has evolved as an important alternative to the traditional techniques. In this section, a historical perspective on the development of the chemical looping processes for carbonaceous fuel conversions is given. Several representative chemical looping processes that currently are being developed are discussed briefly here with details of some of them presented in later chapters. Although gaseous and liquid chemical looping media such as molybdenum oxide vapor and molten iron oxide-tin could be used to convert carbonaceous fuels, commercial operation of such looping systems may not be practical and can be much more challenging than that of using the solid looping medium. Therefore, the chemical looping processes considered in this book involve only the solid chemical looping medium.

The principles of chemical looping for carbonaceous fuel conversion were first applied for industrial practice between the late 19th century and the early 20th century. Howard Lane from England was among the first researchers/engineers who conceived and successfully commercialized the steam–iron process for hydrogen production using the chemical looping principle. With the aid of the iron oxide chemical intermediate, the steam–iron process generates H_2 from reducing gas obtained from coal and steam through an indirect reaction scheme. The reaction scheme for the steam–iron process can be represented by

$$Fe_3O_4 + 4CO \, (or \, H_2) \rightarrow 3Fe + 4CO_2 \, (or \, H_2O) \tag{1.7.1}$$

$$3Fe + 4H_2O \rightarrow 4H_2 + Fe_3O_4 \tag{1.7.2}$$

$$Net \ reaction: CO + H_2O \rightarrow CO_2 + H_2 \tag{1.7.3}$$

$$Looping \ medium: Fe_3O_4 \leftrightarrow Fe \tag{1.7.4}$$

The first commercial steam–iron process based on the Howard Lane design was constructed in 1904. Hydrogen plants based on the same process were then constructed throughout Europe and the United States, producing 850 million ft^3 of hydrogen annually by 1913.[110] After the development of Lane's invention, a German scientist named Anton Messerschmitt simplified the design of the steam–iron process. The improved design reduced the cost of the equipment and maintenance of the steam–iron plant.[110] The Lane Process and Messerschmitt Process diagrams are shown in Figures 1.20(a) and (b), respectively. Although further improvements were made subsequently, the steam–iron process only partially converts the reducing gas. Moreover, the iron-based looping medium has poor recyclability, especially in the presence of sulfur.[110–113] With the introduction of less costly hydrogen production techniques using oil and natural gas as feedstock in the 1940s, the steam–iron process became less competitive and was then phased out. This process is discussed in detail in Chapter 4.

In the 1950s, the chemical looping scheme was proposed for CO_2 generation used for the beverage industry. Oxides of copper or iron were used as the looping particles, and carbonaceous material was used as the feedstock.[114] The Lewis and Gilliland Process for CO_2 production made use of two fluidized bed reactors, the CO_2 generator and metal oxide regenerator, for continuous operation. Both reactors initially were heated using hot gases. After achieving a temperature of ~850°C in the generator, metal oxide flow started. It was then

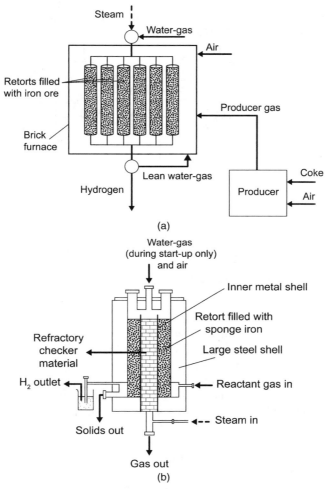

Figure 1.20. Initial steam–iron processes using fixed bed: (a) Lane Process;[113] (b) Messerschmitt Process.[113]

followed by the initiation of the flow of fuels such as methane, syngas, and/or solid carbonaceous fuels. Steam and/or recycled CO_2 are used as fluidizing gases in the generator. The regenerator used fluidizing air, which oxidized the reduced metal oxide. The process flow diagram is shown in Figure 1.21. Alternative designs using moving beds and alternating fixed beds also were considered.[114]

In the early years, the adoption of a chemical looping strategy was mainly prompted by the lack of effective chemical conversion/separation techniques in the generation of the product. In contrast, modern applications of chemical

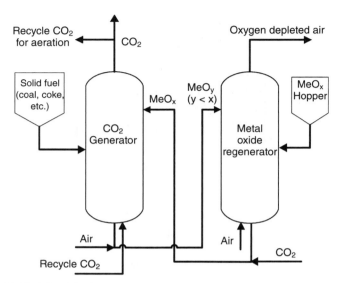

Figure 1.21. Fluidized bed chemical looping for CO_2 generation using solid fuel.[114]

looping processes are prompted by the need for developing an optimized reaction scheme that minimizes the exergy loss involved in the chemical/energy conversion system.[115–118] Also driven by the envisaged CO_2 emission control, the recent developments in chemical looping systems have focused on the efficient conversion of gaseous carbonaceous fuels such as natural gas and coal-derived syngas,[66,104,115–117,119,120] and solid fuels such as petroleum coke and coal,[121–123] whereas CO_2 separation is achieved through the looping reaction scheme.

Several examples of modern chemical looping processes using coal or coal-derived syngas as feedstock are described in this section. These processes include the ZECA Process, chemical looping combustion processes, the Syngas Chemical Looping Process, the Coal-Direct Chemical Looping Process, the GE Fuel-Flexible Process, the ALSTOM Hybrid Combustion-Gasification Process, the HyPr-Ring Process, and the Calcium Looping Process.

In the ZECA Gasification Process, conceived by Los Alamos National Laboratory in the United States, coal reacts with steam and recycled H_2 to produce methane. The methane is subsequently reformed to produce H_2 and finally electricity using a solid oxide fuel cell with a reported overall electricity generation efficiency of ~57%.[124,125]

In the chemical looping combustion (CLC) processes, carbonaceous fuel such as coal-derived syngas or natural gas first reacts with metal oxide (e.g., nickel oxide) in the reducer, where metal oxide is reduced to metal (e.g., nickel). The reaction products are CO_2 and steam, from which CO_2 is readily separable. Metal that exits from the reducer enters the combustor where it reacts with air to regenerate the metal oxide. The metal oxides is then recycled back to the reducer. The heat of oxidation is carried by the high-temperature, high-pressure spent air from the combustor. The spent air is used to drive a steam turbine/gas turbine combined cycle system for electricity generation. Figure 1.22 shows a simplified chemical looping combustion process for electricity generation using coal-derived syngas as feedstock. Figure 1.23 shows a simplified block diagram of the Syngas Chemical Looping (SCL) Process.

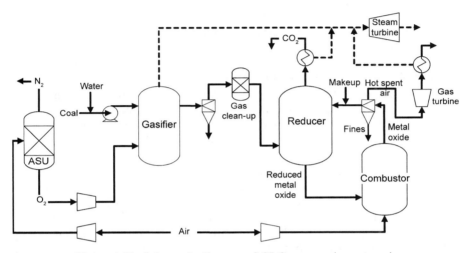

Figure 1.22. Schematic diagram of CLC process (- - - steam).

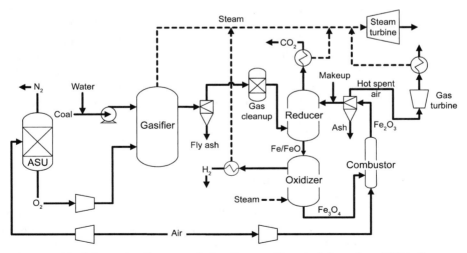

Figure 1.23. Schematic diagram of the Syngas Chemical Looping (SCL) Process (- - - steam).

Compared with the CLC process, the SCL process has the flexibility to copro-duce hydrogen and electricity.[104,126–128]

There are three main reactors in the SCL process: the reducer, the oxidizer, and the combustor or combustion train. In the reducer, coal-derived syngas with moderate levels of pollutants (i.e., HCl, NH_3, sulfur, and mercury) is used to reduce specially tailored iron oxide composite particles that can undergo multiple reduction–oxidation cycles. The syngas is converted completely into carbon dioxide and water, whereas the iron oxide composite particles are reduced to a mixture of Fe and FeO at 750–900°C. The Fe/FeO particles leaving the reducer are then introduced into the oxidizer, which is operated at 500–750°C. In the oxidizer, the reduced particles react with steam to produce a gas stream that contains only H_2 and unconverted steam. The steam easily can be condensed out to obtain a high-purity H_2 stream. Meanwhile, the Fe and FeO are regenerated to Fe_3O_4. The Fe_3O_4 formed in the oxidizer is regen-erated further to Fe_2O_3 in an entrained flow combustor that also transports solid particles discharged from the oxidizer to the reducer inlet, completing the chemical loop. A portion of the heat produced from the oxidation of Fe_3O_4 to Fe_2O_3 can be transferred to the reducer through the particles, whereas at the high pressure and high temperature, spent air produced from the combus-tor can be used to drive a gas turbine/steam turbine combined cycle system to generate electricity for parasitic energy consumption. In yet another configu-ration, a fraction or all of the reduced particles from the reducer can bypass the oxidizer and be introduced directly to the combustor if more heat or elec-tricity is desired. Hence, both chemical looping reforming and chemical looping combustion concepts are applied in the SCL system, rendering it a versatile process for H_2 and electricity coproduction.

Another type of chemical looping processes that drastically simplify the coal conversion scheme is represented by The Ohio State University's (OSU's) Coal-Direct Chemical Looping Process as shown in Figure 1.24. Here, a spe-cially tailored, highly reactive Fe_2O_3 particle similar to that used in the SCL Process is used for converting coal to hydrogen. In this process, Fe_2O_3 particles are introduced into the reducer, together with fine coal powder. By using suit-able gas–solid contacting patterns, coal will be gasified into CO and H_2. The reductive gas will react with Fe_2O_3 particles to form Fe and FeO, while produc-ing a highly concentrated CO_2 and H_2O flue gas stream. H_2O in the flue gas can be condensed readily, leaving a sequestration-ready CO_2 stream. The reduced Fe/FeO particles from the reducer enter the oxidizer to react with steam to generate hydrogen while being oxidized to Fe_3O_4. The resulting Fe_3O_4 exiting from the hydrogen production reactor will be conveyed back to the reducer pneumatically. During the particle conveying, the Fe_3O_4 particle will be oxidized to its original state, that is, Fe_2O_3. Compared with other looping processes, the Coal-Direct Chemical Looping Process characterizes a much simpler coal conversion scheme with a hydrogen production efficiency (on an HHV basis) of close to 80%.[129]

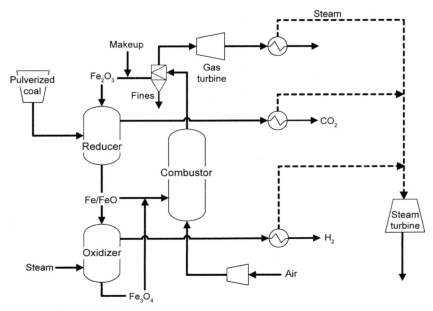

Figure 1.24. Schematic diagram of the Coal-Direct Chemical Looping Process (- - - steam).

Apart from coal or syngas, the chemical looping technique also can process other carbonaceous fuels, making the utilization of by-products from the fossil fuel processing industry possible. Moreover, the generation of H_2 via steam makes the chemical looping process compatible with any existing system that is a net producer of low grade heat such as low-pressure, low-temperature steam. This is possible because water can be used to extract the heat of any exhaust stream with temperatures higher than 100°C to produce steam, which can, in turn, be used for hydrogen production in the chemical looping oxidizer. By doing so, the latent heat of the steam can be stored in the H_2 produced in the form of chemical energy, which has a higher exergy ratio. The latent heat also can be stored in liquid fuels. The key to the high efficiency of this process is the staged reactions in the looping reactors and the design of highly reactive and highly recyclable particles.

The HyPr-Ring Process developed in Japan involves coal gasification using pure oxygen and steam. This process closely follows the concept of the Carbon Dioxide Acceptor Process developed by the Consolidation Coal Company and later the Conoco Coal Development Company in the 1960s and 1970s.[130] Figure 1.25 illustrates the HyPr-Ring Process. In gasification, coal is fed along with calcium oxide, steam, and oxygen to the gasifier. The presence of excess steam in the gasifier drives the reaction toward the formation of H_2. Calcium oxide captures CO_2 generated in the water–gas shift reaction, resulting in a

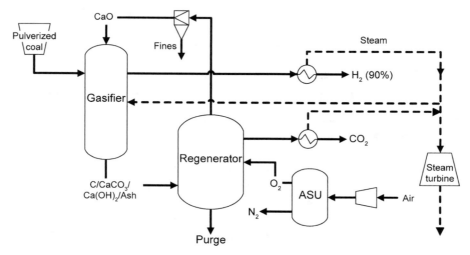

Figure 1.25. Schemcatic diagram of the HyPr-Ring Process.

product gas stream of ~90% H_2 mixed with methane. The solids from the gasifier consist of mostly used CaO sorbents ($CaCO_3$) and some unconverted carbon, which is to be introduced to a regenerator along with oxygen. The heat generated by combusting the unreacted carbon allows the calcination reaction to be carried out for CaO regeneration while producing a high purity of CO_2 for sequestration. It is estimated that a 77% (HHV) H_2 production efficiency can be achieved using this process without taking into account the energy consumption for CO_2 compression.[131]

The Calcium Looping Process (CLP) developed by OSU is closely related to the reactive CO_2 separation strategy from the coal combustion flue gas stream, given in Figure 1.7, using high-temperature sorbents. The process requires the calcium-based sorbent to be of high reaction capacity and to have a high ability for regeneration, while maintaining viability of the sorbent for more than 100 or more cycles of usage. The design of sorbent particles that possess the requisite characteristics is thus crucial to the viability of the process. Figure 1.26 describes a gasification process represented by the CLP.

As shown in Figure 1.26, the CLP is integrated into a coal gasification system capable of electric power generation, H_2 production, and Fischer–Tropsch synthesis for fuels and chemicals. The coal gasifier is operated at high pressures and high temperatures. In the gasifier, reactions of coal, steam, and air/pure oxygen take place in fluidized beds in a variety of contact modes. As the syngas mixture exits from the gasifier, steam is introduced for a water–gas shift reaction downstream for converting CO to H_2, whereas calcium oxide also is introduced to remove CO_2 in the WGS product gas stream, forming calcium carbonate ($CaCO_3$). The removal of CO_2 allows the WGS reaction to proceed with full conversion of CO to H_2 without thermodynamic constraints.

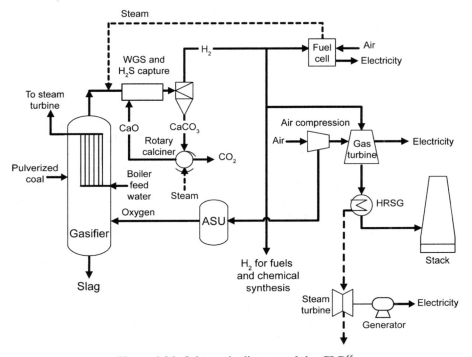

Figure 1.26. Schematic diagram of the CLP.[66]

Thus, this process can be tailored to enhance the H_2 concentration to the maximum extent possible. Calcium oxide also removes acid gases (i.e., H_2S and HCl). At the exit of the WGS reactor, the reacted $CaCO_3$ particles are captured by a high-temperature cyclone, and the spent solids are then sent to a calciner for regeneration that converts $CaCO_3$ back to CaO. The regeneration may also involve a hydration step. In that case, the sorbent introduced to the process will be in hydrate form. The calcined gas mixture will be of high-purity CO_2 with some H_2S that can be economically compressed for sequestration. If necessary, H_2S can be separated from the CO_2 stream before sequestration. The hydrogen-enriched fuel gas can be further purified for fuel cell applications or used for electric power generation without any low-temperature cleanup requirements. The calcium looping scheme achieves a process intensification goal in that it reduces the excess steam requirement, while removing CO_2 and acid gases, including H_2S, providing a sequestrable CO_2 stream, and producing a high-purity H_2 stream.

The GE Fuel-Flexible Process takes different types of feedstock such as coal and biomass to coproduce H_2 and electricity. The fundamental process concept for the fuel-flexible process is similar to the HyPr-Ring Process except that, for conducting the calcination reaction, instead of pure oxygen, metal oxide is used. As a result, the reaction scheme for this process involves two chemical loops and, hence, two different looping media. The two loops are operated using three interconnected fluidized bed reactors. In the first reactor, coal is partially gasified with steam to form a mixture of H_2, CO and CO_2. The CO_2 is captured by calcium-based sorbents to form $CaCO_3$. The depletion of CO_2 results in an enhanced water–gas shift reaction toward the formation of H_2. Moreover, sulfur in the coal also can be captured by the sorbent-forming $CaSO_3$. As a result, a high-purity H_2 stream is obtained from the first reactor. The solids in the first reactor, which mainly consist of reacted sorbents ($CaCO_3$, $CaSO_3$) and unconverted carbon, are introduced to the second reactor where high-temperature steam is injected. In this reactor, the unconverted carbon reacts with a high-temperature oxygen carrier (mainly Fe_2O_3) from the third reactor to form reduced metal; furthermore, the heat carried by the oxygen carrier and the high-temperature steam provide heat to regenerate the spent sorbents coming from the first reactor. Therefore, a high-concentration CO_2/SO_2 gas stream is generated from the second reactor. The third reactor regenerates the reduced oxygen carrier obtained from the second reactor by reacting it with air. Heat from all the hot exhaust gas streams is used for steam generation to drive the turbine system. Thus, the products from this process are pure hydrogen from the first reactor and electricity from the turbines. Meanwhile, the CO_2 stream from the second reactor is ready for sequestration. In this process, part of the solids needs to be discharged during the operation to avoid ash accumulation and to maintain solid reactivity.[132] The overall energy conversion efficiency for the fuel-flexible process is estimated to be 60% (HHV) with 50–50 hydrogen and electricity coproduction. Figure 1.27 illustrates the overall scheme of the process.

The ALSTOM Hybrid Combustion-Gasification Process (Figure 1.28) contains three different operational configurations for the purpose of effective operations: (1) indirect coal combustion for heat generation, (2) coal gasification for producing syngas, and (3) coal gasification for producing hydrogen.[134] For the first and second configurations, one chemical loop is used, whereas for the third configuration, two chemical loops are used. In the first configuration, two main reactors are used with calcium sulfate as the looping medium. The calcium sulfate is reduced to calcium sulfide by coal in the first reactor, forming a high-purity CO_2 stream. The calcium sulfide formed is then combusted in the second reactor with air. Part of the heat generated from the combustor is used to compensate for the heat required for coal gasification in the first reactor, whereas the rest is used to produce high-temperature, high-pressure steam for electricity generation. The second configuration, although similar, uses a much higher coal-to-$CaSO_4$ ratio and a higher steam feed rate for the first reactor. Thus, the reduction of $CaSO_4$ is accompanied by the formation of CO and H_2 resulting from the presence of an excessive amount of carbon and steam. In this configuration, most of the heat generated in the combustor is used to offset the heat required for coal gasification in the first reactor. The product for this configuration is syngas, and most of the carbon in coal is

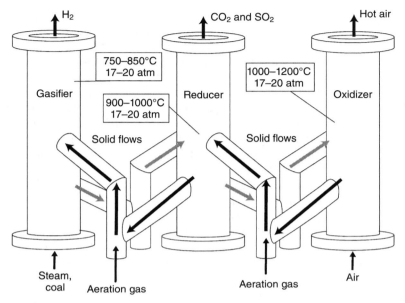

Figure 1.27. Schematic diagram of the fuel-flexible gasification-combustion process.[133]

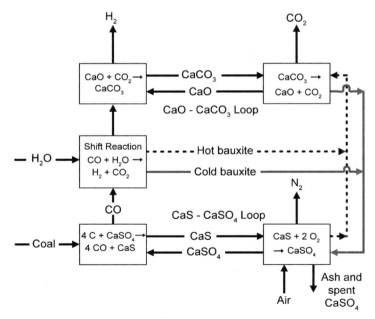

Figure 1.28. Schematic diagram of the ALSTOM Process.[135]

converted to gaseous CO and H_2. Thus, there is no carbon capture necessary. In the third configuration, as shown in Figure 1.28, however, pure hydrogen is produced with the introduction of the third reactor (calciner) and an additional chemical loop—a calcium oxide/calcium carbonate loop. The idea is to introduce even more steam than the second configuration to conduct the WGS reaction in addition to the reduction reaction of $CaSO_4$. Calcium oxide is used in the first reactor to capture the CO_2 generated by the WGS reaction and thus drives the reaction toward the formation of pure H_2 as the product. The heat integration of this configuration includes the utilization of part of the heat generated from calcium sulfide combustion to calcine calcium carbonate in the calciner, forming CO_2.[135] In all the configurations, bauxite is used as the heat carrier, transferring the heat from the exothermic reaction ($CaSO_4$ formation) to the endothermic reaction (calcination reaction).

Note that the processes discussed earlier have high-energy efficiencies for H_2/electricity production. Moreover, they have integrated the CO_2 capture into the process. As an example, the energy conversion efficiencies from the ASPEN simulation for the SCL Process, the Coal-Direct Chemical Looping (CDCL) Process, and the CLP are compared in an efficiency chart with those of the state-of-the-art processes that generate H_2 or electricity from coal in a carbon-constrained scenario as given in Figure 1.29.[136] The figure shows that significantly improved energy conversion efficiencies are realized with near full carbon capture using the novel chemical looping concept.

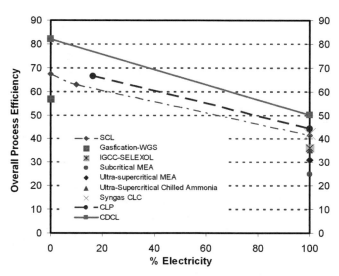

Figure 1.29. Efficiency comparisons among various technologies for H_2/electricity production from coal.

It is, thus, evident from the preceding discussion that coal gasification schemes can be efficient and versatile in generating a variety of products, including hydrogen, liquid fuels, and chemicals in addition to electric power. From a carbon management viewpoint, gasification is a preferred scheme over conventional combustion. The coal conversion processes in the future will inherently be of high energy and economic efficiencies with a reaction pathway that generates readily separable streams of products and pollutants, including carbon dioxide. The chemical looping processes that are being actively pursued worldwide possess such desirable characteristics. Their applications to coal conversion, therefore, represent a promising direction for the optimum coal conversion process development.

1.8 Overview of This Book

In this book, the current carbonaceous fuel conversion technologies based on chemical looping concepts are detailed in the context of traditional or conventional technologies. The key features of the chemical looping processes as well as their ability to generate a sequestration-ready CO_2 stream are discussed thoroughly. The looping media employed in the processes are mainly in solid form, whereas the carbonaceous fuels can be in solid, liquid, or gas form. As the success of the chemical looping technology depends strongly on the performance of the particles, Chapter 2 is devoted entirely to the subjects of solid particle design, synthesis, properties, and reactive characteristics. The looping processes can be applied for combustion and/or gasification of carbon-based material such as coal, natural gas, petroleum coke, and biomass directly or indirectly for steam, syngas, hydrogen, chemical, electricity and liquid fuel production. The details of the energy conversion efficiency and the economics of these looping processes for combustion and gasification applications in contrast to those of the conventional processes are given in Chapters 3–5. Chapter 3 describes the chemical looping processes for combustion using solid carbonaceous fuels, syngas, or natural gas as feedstock. Chapter 4 illustrates chemical looping processes for gasification using gaseous fuels as feedstock such as syngas, natural gas, and light hydrocarbons. Chapter 5 describes the chemical looping processes for gasification using solid carbonaceous fuels as feedstock such as coal. Finally, Chapter 6 presents additional chemical looping applications that are potentially beneficial, including those for H_2 storage and onboard H_2 production, CO_2 capture in combustion flue gas, power generation using fuel cells, steam-methane reforming, tar sand digestion, and chemicals and liquid fuel production. A CD is appended to this book that contains the chemical looping simulation files, and the simulation results based on the ASPEN PLUS software for such reactors as gasifier, reducer, oxidizer, and combustor, and for such processes as conventional gasification processes, SCL Process, CLP, and CCR Process.

References

1. Energy Information Administration, "International Energy Outlook 2008," U.S. Department of Energy, Washington, D.C. (2008).
2. Massachusetts Institute of Technology, "The Future of Nuclear Power," MIT, Cambridge, MA (2003).
3. Yidiz, B. and M. S. Kazimi, "Efficiency of Hydrogen Production Systems Using Alternative Nuclear Energy Technologies," *International Journal of Hydrogen Energy*, 31(1), 77–92 (2006).
4. Energy Information Administration, "International Energy Outlook 2006," U.S. Department of Energy, Washington, D.C. (2006).
5. Martinot, E., "Renewables Global Status Report," REN21, Paris, France (2006).
6. U.S.Department of Energy, "Hydropower," http://www.doe.gov/energysources/hydropower.htm (2008).
7. World Commission on Dams, "Dams and Development: A New Framework for Decision-Making," Earthscan Publications, London, UK (2000).
8. Gera, D., "Biofuels and Bioenergy," Encylopedia of Chemical Processing, edited by S. Lee, CRC Press, Boca Raton, FL (2005).
9. Environment, Food and Rural Affairs Committee, "Climate Change: The Role of Bioenergy," U. K. House of Commons, London (2006).
10. Pimentel, D., "Ethanol Fuels: Energy, Economics and Environmental Impacts are Negative," *Natural Resources Research*, 12(2), 127–134 (2004).
11. Pimentel, D., "Ethanol Fuels: Energy Security, Economics, and the Environment," *Journal of Agricultural and Environmental Ethics*, 4, 1–13 (1991).
12. Farrell, A. E., R. J. Plevin, B. T. Turner, A. D. Jones, M. O' Hare, and D. M. Kammen, "Ethanol Can Contribute to Energy and Environmental Goals," *Science*, 311(5760), 506–508 (2006).
13. Greene, N., and R. Roth, "Ethanol: Energy Well Spent. A Review of Corn and Cellulosic Ethanol Energy Balances in the Scientific Literature to Date," *Industrial Biotechnology*, 2(1), 36–39 (2006).
14. Kammen, D. M., "The Rise of Renewable Energy," *Scientific American*, 295(3), 84–93 (2006).
15. "New World Record in Wind Power Capacity," Press Release, World Wind Energy Association, Bonn, Germany (2007).
16. Gibbs, W., "Plan B for Energy," *Scientific American*, 295(3), 102–108, 110, 112 (2006).
17. Smil, V., "Energy at the Crossroads," Global Science Forum Conference on Scientific Challenges for Energy Research, Paris, France (2006).
18. Kraupl, S., and A. Steinfeld, "Operational Performance of a 5-KW Solar Chemical Reactor for the Co-Production of Zinc and Syngas," *Journal of Solar Energy Engineering*, 125(1), 124–126 (2003).
19. British Petroleum, "Statistical Review of World Energy," http://www.bp.com/productlanding.do?categoryId=6929&contentId=7044622 (2008).

20. "Fact Sheet: U.S. Tar Sands Potential," U.S. Dept. of Energy, Office of Petroleum Reserves, Washington, D.C., http://www.fossil.energy.gov/programs/reserves/npr/Tar_Sands_Fact_Sheet.pdf(2006).

21. Chalot, J. P., "The New Heavy-Oil Economics," *Oil and Gas Investor* (Nov., 2006).

22. Rahmim, I. I., "GTL, CTL Finding Roles in Global Energy Supply," *Oil & Gas Journal*, 106(12), (2008).

23. National Research Council, Coal: Research and Development to Support National Energy Policy, The National Academies Press, Washington, D.C. (2007).

24. Sun, Q. Y., J. J. Fletcher, Y. Z. Zhang, and X. K. Ren, "Comparative Analysis of Costs of Alternative Coal Liquefaction Processes," *Energy & Fuels*, 19(3), 1160–1164 (2005).

25. Stultz, S. C., and J. B. Kitto, Steam, Its Generation and Use, 40th edition, Babcock & Wilcox, Lynchburg, VA (1992).

26. Miller, B. G., Coal Energy Systems, Academic Press, Washington, D.C. (2004)

27. Yeh, S., and E. S. Rubin, "A Centurial History of Technological Change and Learning Curves for Pulverized Coal-Fired Utility Boilers," *Energy*, 32(10), 1996–2005 (2007).

28. Massachusetts Institute of Technology, "The Future of Coal: Options for a Carbon-Constrained World," MIT, Cambridge, MA (2007).

29. Bugge, J., S. Kjaer, and R. Blum, "High-Efficiency Coal-Fired Power Plants Development and Perspectives," *Energy*, 31(10–11), 1437–1445 (2006).

30. Stringer, J., and L. A. Ruth, "Future Perspectives including Fuel Cells, Gas Turbines, USC and HGCU—the US Perspective," *Materials at High Temperatures*, 20(2), 233–239 (2003).

31. Bradley, M. J., and B. M. Jones, "Reducing Global NO$_x$ Emissions: Developing Advanced Energy and Transportation Technologies," *Ambio*, 31(2), 141–149 (2002).

32. Banales-Lopez, S., and V. Norberg-Bohm, "Public Policy for Energy Technology Innovation—A Historical Analysis of Fluidized Bed Combustion Development in the USA," *Energy Policy*, 30(13), 1173–1180 (2002).

33. U.S. Department of Energy, "Fluidized Bed Technology—Overview," http://www.fossil.energy.gov/programs/powersystems/combustion/fluidizedbed_overview.html (2008).

34. Rockey, J. M., and R. E. Weinstein, "Gas Turbines for Advanced Pressurized Fluidized Bed Combustion Combined Cycles (APFBC)," Paper presented at the 15th International Conference on Fluidized Bed Combustion, Savannah, GA (1999).

35. U.S. Clean Coal Technology Demonstration Program, "The JEA Large-Scale CFB Combustion Demonstration Project," Topical Report (22), Clean Coal Technology (2003).

36. Watson, J., "Advanced Cleaner Coal Technologies for Power Generation: Can They Deliver?" British Insitute of Energy Economics(BIEE) Academic Conference, Oxford, U.K. (2005).

37. Minchener, A. J., "Fluidized Bed Combustion Systems for Power Generation and Other Industrial Applications," Proceedings of the Institution of Mechanical Engineers: Part A, *Journal of Power and Energy*, 217(A1), 9–18 (2003).

38. Bohm, M. C., "Capture-Ready Power Plants—Options, Technologies, and Economics," Massachusetts Institute of Technology, Cambridge, MA (2006).

39. Woods, M. C., P. J. Capicotto, J. L. Haslbeck, N. J. Kuehn, M. Matuszewski, L. L. Pinkerton, M. D. Rutkowski, R. L. Schoff, and V. Vaysman, "Cost and Performance Baseline for Fossil Energy Plants," NETL, U.S. Department of Energy, Washington, D.C. (2007).

40. Ramezan, M., T. J. Slone, N. Y. Nasakala, and G. N. Liljedahl, "Carbon Dioxide Capture from Existing Coal-Fired Power Plants," NETL, U.S. Department of Energy, DOE/NETL-401/110907, Washington, D.C. (2007).

41. Lu, Y., S. Chen, M. Rostam-Abadi, R. Varagani, F. Chatel-Pelage, and A. C. Bose, "Techno-Economic Study of the Oxy-Combustion Process for CO_2 Capture from Coal-Fired Power Plants," Paper presented at the International Pittsburgh Coal Conference, Pittsburgh, PA (2005).

42. Rubin, E. S., C. Chen, and A. B. Rao, "Cost and Performance of Fossil Fuel Power Plants with CO_2 Capture and Storage," *Energy Policy*, 35(9), 4444–4454 (2007).

43. Hutchinson, H., "Old King Coal," *Mechanical Engineering*, 124 (8), 41–45 (2002).

44. Gupta, H., T. Thomas, A.-H. A. Park, M. Iyer, P. Gupta, R. Agnihotri, R. A. Jadhav, H. W. Walker, L. K. Weavers, T. Butalia, and L.-S. Fan, "Pilot-Scale Demonstration of the Oscar Process for High-Temperature Multipollutant Control of Coal Combustion Flue Gas, Using Carbonated Fly Ash and Mesoporous Calcium Carbonate," *Industrial & Engineering Chemistry Research*, 46(14), 5051–5060 (2007).

45. Fan, L.-S., A. Ghosh-Dastidar, and S. Mahuli, "Calcium Carbonate Sorbent and Methods of Making and Using Same," U.S. Patent 5,779,464, (1998).

46. Fan, L.-S., and R. A. Jadhav, "Clean Coal Technologies: OSCAR and CARBONOX Commercial Demonstrations," *AICHE Journal*, 48(10), 2115–2123 (2002).

47. Gupta, H., S. A. Benson, L.-S. Fan, J. D. Laumb, E. S. Olson, C. R. Crocker, R. K. Sharma, R. Z. Knutson, A. S. M. Rokanuzzaman, and J. E. Tibbets, "Pilot-Scale Studies of NO_x Reduction by Activated High-Sodium Lignite Chars: A Demonstration of the CARBONOX Process," *Industrial & Engineering Chemistry Research*, 43(18), 5820–5827 (2004).

48. Taerakul, P., P. Sun, D. W. Golightly, H. W. Walker, L. K. Weavers, B. Zand, T. Butalia, T. Thomas, H. Gupta, and L.-S. Fan, "Characterization and Re-Use Potential of By-Products Generated from The Ohio State Carbonation and Ash Reactivation (OSCAR) Process," *Fuel*, 86(4), 541–553 (2007).

49. Bryant, E. A., Climate Process and Change, Cambridge University Press, Cambridge, UK (1997).

50. Caldeira, K., D. Archer, J. P. Barry, R. G. J. Bellerby, P. G. Brewer, L. Cao, A. G. Dickson, S. C. Doney, H. Elderfield, V. J. Fabry, et al., "Comment on Modern-Age Buildup of CO_2 and Its Effects on Seawater Acidity and Salinity by Hugo A. Loaiciga," *Geophysical Research Letters*, 34(18), L18608 1–3 (2007).

51. Stix, G., "A Climate Repair Manual.," *Scientific American*, 295(3), 46–49 (2006).

52. National Council for Science and the Environment, "Energy for a Sustainable and Secure Future: A Report of the Sixth National Conference on Science, Policy and the Environment," edited by D. E. Blockstein and M. A. Shockley, Washington DC (2006).

53. Chatel-Pelage, F., R. Varagani, P. Pranda, N. Perrin, H. Farzan, S. J. Vecci, Y. Lu, S. Chen, M. Rostam-Abadi, and A. C. Bose, "Applications of Oxygen for NO_x Control and CO_2 Capture in Coal-Fired Power Plants," *Thermal Science*, 10(3), 119–142 (2006).

54. Abu-Zahra, M. R. M., J. P. M. Niederer, P. H. M. Feron, and G. F. Versteeg, "CO_2 Capture from Power Plants: Part II. A Parametric Study of the Economical Performance based on Mono-Ethanolamine," *Interational Journal of Greenhouse Gas Control*, 1(1), 135–142 (2007)

55. Abu-Zahra, M. R. M., L. H. J. Schneiders, J. P. M. Niederer, P. H. M. Feron, and G. F. Versteeg, "CO_2 Capture from Power Plants: Part I. A Parametric Study of the Technical Performance Based on Mono-Ethanolamine," *International Journal of Greenhouse Gas Control*, 1(1), 37–46 (2007).

56. Singh, D., E. Croiset, P. L. Douglas, and M. A. Douglas, "Techno-Economic Study of CO_2 Capture from an Existing Coal-Fired Power Plant: MEA Scrubbing vs. O_2/CO_2 Recycle Combustion," *Energy Conversion and Management*, 44(19), 3073–3091 (2003).

57. Rochelle, G. T., "Amine Scrubbing for CO_2 Capture," *Science*, 325, 1652–1654 (2009).

58. Mattisson, T., A. Lyngfelt, and P. Cho, "The Use of Iron Oxide as an Oxygen Carrier in Chemical-Looping Combustion of Methane with Inherent Separation of CO_2," *Fuel*, 80(13), 1953–1962 (2001).

59. "Chilling News for Carbon Capture," *Modern Power Systems*, 26(12), 17–18 (2006).

60. Rhudy, R., "Chilled-Ammonia Post Combustion CO_2 Capture System—Laboratory and Economic Evaluation Results," EPRI, Framingham, MA (2006).

61. Electric Power Research Institute, "The Challenge of Carbon Capture," *EPRI Journal* (Spring, 2007).

62. Gray, M. A., "Carbon Capture Demonstration Projects: AEP's Perspective," http://www.westgov.org/wga/meetings/coal07/GRAY.ppt. (2007)

63. White, C. M., B. R. Strazisar, E. J. Granite, J. S. Hoffman, and H. W. Pennline, "Separation and Capture of CO_2 from Large Stationary Sources and Sequestration in Geological Formations—Coalbeds and Deep Saline Aquifers," *Journal of the Air & Waste Management Association*, 53(6), 645–715 (2003).

64. Aaron, D., and C. Tsouris, "Separation of CO_2 from Flue Gas: A Review," *Separation Science and Technology*, 40(1–3), 321–348 (2005).

65. Iyer, M., H. Gupta, B. B. Sakadjian, and L.-S. Fan, "Multicyclic Study on the Simultaneous Carbonation and Sulfation of High-Reactivity CaO," *Industrial & Engineering Chemistry Research*, 43(14), 3939–3947 (2004).

66. Fan, L.-S., and M. Iyer, "Coal Cleans Up Its Act," *The Chemical Engineers*, 36–38 (2006).

67. Sakadjian, B. B., W. Wang, D. Wong, M. Iyer, S. Ramkumar, S. Li, L.-S. Fan, and R. Statnick, "Sub-Pilot Demonstration of the CCR Process: A Calcium Oxide Based CO_2 Capture Process for Coal Fired Power Plants," Paper presented at the 33rd International Conference of Coal Utilization and Fuel Systems, Clear Water, FL (2008).

68. Lu, D. Y., R. W. Hughes, and E. J. Anthony, "*In-Situ* CO_2 Capture Using Ca-Based Sorbent Looping in Dual Fluidized Beds," Paper presented at the 9th International Conference on Circulating Fluidized Beds, Hamburg, Germany (2008).

69. Lee, K. B., M. G. Beaver, H. S. Caram, and S. Sircar, "Novel Thermal-Swing Sorption-Enhanced Reaction Process Concept for Hydrogen Production by Low-Temperature Steam-Methane Reforming," *Industrial & Engineering Chemistry Research*, 46(14), 5003–5014 (2007).

70. Energy Administration Information, "International Energy Outlook 2007," U.S. Department of Energy, Washington, D.C. (2007).

71. Buhre, B. J. P., L. K. Elliott, C. D. Sheng, R. P. Gupta, and T. F. Wall, "Oxy-fuel Combustion Technology for Coal-Fired Power Generation," *Progress in Energy and Combustion Science*, 31(4), 283–307 (2005).

72. Intergovernmental Panel on Climate Change, "Climate Change 2007: Synthesis Report," http://www.ipcc.ch/pdf/assessment-report/ar4/syr/ar4_syr.pdf. (2007).

73. Lackner, K. S., "A Guide to CO_2 Sequestration," *Science*, 300, 1677–1678 (2003).

74. Parson, E. A., and D. W. Keith, "Fossil Fuels without CO_2 Emissions," *Science*, 282(5391), 1053–1054 (1998).

75. Herzog, H. J., and E. M. Drake, "CO_2 Capture, Reuse, and Sequestration Technologies for Mitigating Global Climate Change," Proceedings of the 23rd International Technical Conference on Coal Utilization & Fuel Systems, Clearwater, Florida, March 9–13, 615–626 (1998).

76. Blunt, M., F. J. Fayers, and M. Orr Franklin, Jr., "Carbon Dioxide in Enhanced Oil Recovery," *Energy Conversion and Management*, 34(9–11), 1197–1204 (1993).

77. AGI, "Demonstrating Carbon Sequestration," Geotimes, http://www.geotimes. org/mar03/feature_demonstrating.html (2003).

78. Intergovernmental Panel or Climate Change, Carbon Dioxide Capture and Storage, Cambridge University Press, Cambridge, UK (2005).

79. Goldberg, P., "Mineral Sequestration Team Activities: Introduction, Issues & Plans," http://www.netl.doe.gov/publications/proceedings/01/minecarb/goldberg. pdf (2001).

80. Park, A.-H. A., and L.-S. Fan, "Carbon Dioxide Mineral Sequestration: Physically Activated Dissolution of Serpentine and PH Swing Process," *Chemical Engineering Science*, 59, 5241–5247 (2005).

81. Stiegel, G. J., and M. Ramezan, "Hydrogen from Coal Gasification: An Economical Pathway to a Sustainable Energy Future," *International Journal of Coal Geology*, 65(3–4), 173–190 (2006).

82. Basu, P., Combustion and Gasification in Fluidized Beds, CRC Press, Boca Paton, FL (2006).

83. Higman, C., and M. van der Burgt, Gasification, 2nd edition, Gulf Professional Publishing, Houston, TX (2008).

84. Fan, L.-S., and C. Zhu, Principles of Gas-Solids Flows, Cambridge University Press, Cambridge, UK (1998).

85. EAGLE Gasifier, "Coal Energy Application for Gas, Liquid and Electricity (EAGLE)," Project Brochure, Electric Power Development Co. (2003).

86. E-GasTM Gasifier, http://www.coptechnologysolution.com/egas/process_overview/index.htm. (2004).

87. Texaco Gasifier, "Clean Coal Technology—The Tampa Integrated Gasification-Combined Cycle Project," USDOE/Tampa Electric Company, Topical Report No. 6 (Oct., 1996).

88. Parsons Infrastructure and Technology Group Inc., "The Cost of Mercury Removal in an IGCC Power Plant," U.S. Department of Energy NETL, Washington, D.C., http://www.netl.doe.gov/technologies/coalpower/gasification/pubs/pdf/MercuryRemoval%20Final.pdf (2002).

89. Shinada, O., A. Yamada, and Y. Koyama, "The Development of Advanced Energy Technologies in Japan IGCC: A Key Technology for the 21st Century," *Energy Conversion and Management*, 43(9–12), 1221–1233 (2002).

90. Chiesa, P., T. G. Kreutz, and G. Lozza. "CO_2 Sequestration from IGCC Power Plants by Means of Metallic Membranes," *Journal of Engineering for Gas Turbines and Power-Transactions of the Asme*, 129(1), 123–134 (2007).

91. Grainger, D., and M. -B. Hagg, "Techno-Economic Evaluation of a PVAm CO_2-Selective Membrane in an IGCC Power Plant with CO_2 Capture," *Fuel*, 87(1), 14–24 (2008).

92. Carbo, M. C., D. Jansen, W. G. Haije, and A. H. M. Verkooijen, "Advanced Membrane Reactors for Fuel Decarbonisation in IGCC: H_2 or CO_2 Separation?" Paper presented at the 5th Annual Conference on Carbon Capture and Sequestration, Alexandria, Virginia (2006).

93. Kotas, T. J., The Exergy Method of Thermal Plant Analysis, Butterworth-Heinemann, London, UK (1985).

94. Tsutsumi, A., "Exergy Recuperative Gasification Technology for Hydrogen and Power Co-Production," Abstracts of Papers, 231st ACS National Meeting, Atlanta, Georgia (2006).

95. Kubo, S., H. Nakajima, S. Kasahara, S. Higashi, T. Masaki, H. Abe, and K. Onuki, "A Demonstration Study on a Closed-Cycle Hydrogen Production by the Thermochemical Water-Splitting Iodine-Sulfur Process," *Nuclear Engineering and Design*, 233(1–3), 347–354 (2004).

96. U.S. Department of Energy, "Thermochemical Cycles," Nuclear Hydrogen R&D Plan, U.S. Department of Energy Office of Nuclear Energy, Science and Technology, Washington, D.C. (2004).

97. Steinfeld, A., and R. Palumbo, "Solar Thermochemical Process Technology," *Encyclopedia of Physics Science and Technology*, 15, 237–256 (2001).

98. Kirk-Othmer Concise Encyclopedia of Chemical Technology, 3rd edition, edited by M. Grayson and D. Eckroth, Wiley-Interscience, New York (1985).

99. Contractor, R. M., "Dupont's CFB Technology for Maleic Anhydride," *Chemical Engineering Science*, 54(22), 5627–5632 (1999).

100. Dudukovic, M. P., "Frontiers in Reactor Engineering," *Science*, 325, 698–701 (2009).

101. Fan, L.-S., and H. Gupta, "Sorbent for Separation of Carbon Dioxide (CO_2) from Gas Mixtures," U.S. Patent 7,067,456 B2 (2006).

102. Ishida, M., H. Jin, and T. Okamoto, "Kinetic Behavior of Solid Particle in Chemical-Looping Combustion: Suppressing Carbon Deposition in Reduction," *Energy & Fuels*, 12(2), 223–229 (1998).

103. Mattisson, T., M. Johansson, and A. Lyngfelt, "The Use of NiO as an Oxygen Carrier in Chemical-Looping Combustion," *Fuel*, 85(5–6), 736–747 (2006).

104. Gupta, P., L. G. Velazquez-Vargas, and L.-S. Fan, "Syngas redox (SGR) Process to Produce Hydrogen from Coal Derived Syngas," *Energy & Fuels*, 21(5), 2900–2908 (2007).

105. Iyer, M., S. Ramkumar, D. Wong, and L.-S. Fan, "Enhanced Hydrogen Production with *In-Situ* CO_2 Capture in a Single Stage Reactor," Proceedings of the 23rd Annual International Pittsburgh Coal Conference, Pittsburgh, Pennsylvania, 5/1–5/16 (2006).

106. Gupta, R. P., B. S. Turk, J. W. Portzer, and D. C. Cicero, "Desulfurization of Syngas in a Transport Reactor," *Environmental Progress*, 20(3), 187–195 (2001).

107. Kwauk, M., Personal Communication (2009).

108. Forret, A., A. Hoteit, and T. Gauthier, "Chemical Looping Combustion Process Applied to Liquid Fuels," Paper presented at the AIChE Annual Meeting, Nashville, Tennessee (2009).

109. Forret, A., A. Hoteit, and T. Gauthier, "Chemcial Looping Combustion Process Applied to Liquid Fuels," Proceedings of the 4th European Combustion Meeting, Vienna, Austria (Apr. 14–17, 2009).

110. Hurst, S., "Production of Hydrogen by the Steam-Iron Method," *Journal of the American Oil Chemists' Society*, 16(2), 29–36 (1939).

111. Gasior, S. J., A. J. Forney, J. H. Field, D. Bienstock, and H. E. Benson, "Production of Synthesis Gas and Hydrogen by the Steam Iron Process—Pilot Plant Study of Fluidized and Free-Falling Beds," U.S. Department of the Interior, Bureau of Mines, Washington, D.C. (1961).

112. Institute of Gas Technology, "Development of the Steam-Iron Process for Hydrogen Production," Institute of Gas Technology, Des Plaines, IL (1979).

113. Teed, P. L., "The Chemistry and Manufacture of Hydrogen," Longmans, Green and Co. White Plains, NY (1919).

114. Lewis, W. K., and E. R. Gilliland, "Production of Pure Carbon Dioxide," U.S. Patent 2,665,972 (1954).

115. Ishida, M., D. Zheng, and T. Akehata, "Evaluation of a Chemical-Looping-Combustion Power-Generation System by Graphic Exergy Analysis," *Energy*, 12, 147–154 (1987).

116. Knoche, K. F., and H. J. Richter, "Improvement of the Reversibility of Combustion Processes," *Brennstoff-Waerme-Kraft*, 20(5), 205–210 (1968).

117. Richter, H. J., and K. F. Knoche, "Reversibility of Combustion Processes," Efficiency and Costing, edited by R. A. Gaggioli, ACS Symposium Series 235, Washington, D.C., 71–86 (1983).

118. Anheden, M., and G. Svedberg, "Exergy Analysis of Chemical-Looping Combustion Systems," *Energy Conversion and Management*, 39(16–18), 1967–1980 (1998).

119. Jin, H. G., and M. Ishida, "A New Type of Coal Gas Fueled Chemical-Looping Combustion," *Fuel*, 83(17–18), 2411–2417 (2004).

120. Johansson, E., T. Mattisson, A. Lyngfelt, and H. Thunman, "Combustion of Syngas and Natural Gas in a 300W Chemical-Looping Combustor," *Chemical Engineering Research & Design*, 84(A9), 819–827 (2006).

121. Fan, L.-S., L. G. Velazquez-Vargas, and S. Ramkumar, "Chemical Looping Gasification," Paper presented at the 9th International Conference on Circulating Fluidized Beds, Hamburg, Germany (2008).

122. Leion, H., T. Mattisson, and A. Lyngfelt, "The Use of Petroleum Coke as Fuel in Chemical-Looping Combustion," *Fuel*, 86(12–13), 1947–1958 (2007).

123. Cao, Y., Z. Cheng, L. Meng, J. Riley, and W.-P. Pan, "Reduction of Solid Oxygen Carrier (CuO) by Solid Fuel (Coal) in Chemical-Looping Combustion," Volume 50, American Chemical Society, Division of Fuel Chemistry, Washington, D.C., 99–102 (2005).

124. Gao, L., N. Paterson, D. Dugwell, and R. Kandiyoti, "Zero-Emission Carbon Concept (ZECA): Equipment Commissioning and Extents of the Reaction with Hydrogen and Steam," *Energy and Fuels*, 22(1), 463–470 (2008).

125. Slowinski, G., "Some Technical Issues of Zero-Emission Coal Technology," *International Journal of Hydrogen Energy*, 31(8), 1091–1102 (2006)

126. Fan, L.-S., P. Gupta, L. G. Velazquez-Vargas, and F. Li, "Systems and Methods of Converting Fuels," WO Patent 2,007,082,089 (2007).

127. Thomas, T., L.-S. Fan, P. Gupta, and L. G. Velazquez-Vargas, "Combustion Looping Using Composite Oxygen Carriers," U.S. Provisional Patent Series No. 11/010,648 (2004); U. S. Patent No. 7,767,191 (2010).

128. Velazquez-Vargas, L. G., T. Thomas, P. Gupta, and L.-S. Fan, "Hydrogen Production via Redox Reaction of Syngas with Metal Oxide Composite Particles," Proceedings of the AIChE Annual Meeting, Austin, Texas, November 7–12 (2004).

129. Gupta, P., L. G. Velazquez-Vargas, F. Li, and L.-S. Fan, "Chemical Looping Reforming Process for the Production of Hydrogen from Coal," Proceedings of the 23rd Annual International Pittsburgh Coal Conference, Pittsburgh, Pennsylvania, May 1–16 (2006).

130. Dobbyn, R. C., H. M. Ondik, W. A. Willard, W. S. Brower, I. J. Feinberg, T. A. Hahn, G. E. Hicho, M. E. Read, C. R. Robbins, and J. H. Smith, "Evaluation of the Performance of Materials and Components Used in the CO_2 Acceptor Process Gasification Pilot Plant," Technical Report, U.S. Department of Energy, DOE-ET-10253-T1, Washington, D.C. (1978).

131. Lin, S. Y., M. Harada, Y. Suzuki, and H. Hatano, "Process Analysis for Hydrogen Production by Reaction Integrated Novel Gasification (HyPr-RING)," *Energy Conversion and Management*, 46(6), 869–880 (2005).

132. Rizeq, R. G., J. West, A. Frydman, R. Subia, V. Zamansky, H. Loreth, L. Stonawski, T. Wiltowski, E. Hippo, and S. B. Lalvani "Fuel-Flexible Gasification-Combustion Technology for Production of H_2 and Sequestration-Ready CO_2," Annual Technical Progress Report, U.S. Department of Energy, DE-FC26-00FT40974, Washington, D.C. (2002).

133. Rizeq, R. G., R. K. Lyon, V. Zamansky, and K. Das, "Fuel-Flexible AGC Technology for Production of H_2 Power and Sequestration-Ready CO_2,"

Proceedings of the 26th International Technical Conference on Coal Utilization & Fuel System, Clearwater, Florida, March 5–9 (2001).

134. Andrus, H. A. E., Jr., G. Burns, J. H. Chiu, G. N. Liljedahl, P. T. Stromberg, and P. R. Thibeault, "Hybrid Combustion-Gasification Chemical Looping Power Technology Development," ALSTOM Technical Report, U.S. Department of Energy, DE-FC26-03NT41866, Washington, D.C. (2006).

135. Andrus, H. A. E., Jr., J. H. Chiu, G. N. Liljedahl, P. T. Stromberg, P. R. Thibeault, and S. C. Jain, "ALSTOM's Hybrid Combustion-Gasification Chemical Looping Technology Development," Proceedings of the 22nd Annual International Pittsburgh Coal Conference, Pittsburgh, Pennsylvania (2005)

136. Li, F., and L.-S. Fan, "Coal Conversion Processes—Progress and Challenges," *Energy and Environmental Science*, 1, 248–267 (2008).

CHAPTER 2

CHEMICAL LOOPING PARTICLES

P. Gupta, F. Li, L. Velázquez-Vargas, D. Sridhar, M. Iyer,
S. Ramkumar, and L.-S. Fan

2.1 Introduction

A successful operation of chemical looping processes strongly depends on the effective performance of the looping particles. Many factors are involved in the design of particles that possess desirable properties. Two types of looping reaction systems, which are characterized by two different types of looping particles, are considered in this chapter. The Type I looping system has cyclic redox reactions in two or more separate reactors, that is, the reducer and the oxidizer and/or combustor. The reducer carries out the particle reduction reaction by reducing reactants such as syngas, natural gas, and/or coal, and forms the gaseous reaction products as carbon dioxide and steam. The oxidizer and combustor carry out the particle oxidation reaction by such oxidation reactants as air, water, and/or carbon dioxide. The oxidation reaction with air is a combustion reaction that generates heat, leading to the production of electricity. The oxidation reaction with water and/or carbon dioxide is a gasification reaction that generates hydrogen and/or carbon monoxide, leading to the production of chemicals, liquid fuels and electricity. In the Type I looping system, the looping particle serves as an oxygen carrier and is commonly composed of metal oxides. The typical metals used are Ni, Fe, and Cu. The specific particle formulation depends on the intended looping applications for either combustion or gasification. The reactions in the Type I looping system take place at high temperatures.

The Type II looping system is conducted with cyclic reactions in two reactors, that is, the carbonator and the calciner. The carbonator or carbonation reactor carries out the particle carbonation reaction with carbon dioxide. The calciner or calcination reactor carries out the particle calcination reaction. In this system, the particle is a carrier of carbon dioxide with the calcination reaction taking place at a temperature higher than the carbonation reaction. The typical metal employed is calcium. The preparation of particles for the Type I looping system operation involves more complex steps than for the Type II looping system operation because of the inherent variation of the processes and products evolved from the Type I looping systems. For either type of looping particles, however, the selection of the type of materials for these particles and the determination of the method for their synthesis or regeneration require a thorough knowledge of particle characteristics such as equilibrium behavior, reaction kinetics, mechanical properties, heat effects, process configuration, and the intended reaction products. Furthermore, for looping particles to be of commercial interest, the materials and synthesis costs will have to be low, and the particles will have to be effectively recyclable during the reaction/regeneration steps of the looping process system. In this aspect, desirable processes will employ specially engineered metal oxide particles of high redox activity for the Type I looping system that can undergo many redox cycles, as opposed to metal ore or other particles of low activity that can undergo only a few redox cycles.

This chapter examines the physical and chemical characteristics of looping particles, particularly those that are recyclable. The particle properties for the Type I system and the Type II system are discussed in Sections 2.2 and 2.3, respectively. In Section 2.2, the performance criteria for the oxygen-carrier particles for Type I process applications are described. The key physical properties of various types of oxygen carrier particles and their chemical properties are discussed. These properties include melting points, oxygen-carrying capacity, reactivity, and heat of reactions. A detailed account of the thermodynamic behavior of metal oxides and their composite form is given, along with particle formulation and synthesis procedures and their relationship with particle reactivity, recyclability, physical strength, heat of reaction, process heat integration, pollutant effects, and reactor design. Section 2.3 discusses metal carbonate with an emphasis on the calcium-based carbonate. The rationale for the use of CaO as the carbon dioxide reacting sorbent is given first. Then the thermodynamic properties, reaction characteristic, synthesis procedure, and recyclability of the calcium-based sorbent particles are illustrated.

2.2   Type I Chemical Looping System

2.2.1   General Particle Characteristics

In the following, the desired properties of the looping particles for the Type I process system are given. Some of these properties also are applicable to the particles for the Type II process system.

1. Good oxygen-carrying capacity. A higher oxygen-carrying capacity for the particle gives rise to a lower particle circulation rate. The maximum oxygen-carrying capacity of the particle is determined mainly by the property of the primary metal oxide and the extent of support used for preparing the particle. The effective oxygen-carrying capacity, defined by the amount of transferable oxygen in the particle during the looping operation, reflects the extent of the solid conversion, which is affected by the gas and solid residence time in the looping reactor. For oxygen-carrier particles with multiple oxidation states, the type of the looping reactors and their gas solid contact patterns also can notably affect the effective oxygen-carrying capacity of the particles.

2. Good gas conversions in both the reduction and the oxidation reactions. Higher gas conversions result in increased energy conversion efficiencies. High gas conversions can be achieved for both oxidation and reduction reactions using properly selected primary metal oxide materials, reactor type, and gas–solid contact mode. The thermodynamic relationships among various oxidation states of metal oxide and the reactant and product gas concentrations are important factors that contribute to this selection.

3. High rates of reaction. A higher rate of reaction allows a smaller reactor to be used to achieve the same reactant conversion. A proper selection of primary metal oxides, supports, promoters, particle synthesis techniques and reaction conditions can enhance the reaction rate.

4. Satisfactory long-term recyclability and durability. Improving the recyclability and durability of the particle leads to a reduced spent particle purging rate and, hence, to a fresh particle makeup rate. The particle recyclability and durability can be improved by the use of suitable support onto primary metal oxides coupled with optimized particle synthesis procedures.

5. Good mechanical strength. Good mechanical strength can lower the rate of particle attrition or spalling. The mechanical strength of the particle is related closely to the particle's composition and method of preparation. For a specific primary metal oxide, improvement of particle mechanical strength can be achieved by the use of support or binder materials, additives, and suitable synthesis procedures.

6. Suitable heat capacity and high melting points. The heat capacity of the particle can be enhanced so that the particle also can serve as a heat transfer medium. Thus, the heat effect in the reactor caused by the heat of reaction, either endothermic or exothermic, can be moderated by the particles without resorting, to a great extent, to the use of heat exchangers. This enhancement of the heat capacity of the particle can be achieved mainly by incorporation of an inert support material within the particle. To maintain the integrity and activity of the particles, the melting temperature for metal oxides and their reduced forms, as well

as the support materials, will need to be well above the reactor operating temperatures.

7. Ability to change the heat of reaction. In some cases, it is desirable to alter the heat of reaction of the particle to minimize the heat duty required in the reactor so that the heat integration scheme for the process can be more streamlined. Altering the particle's heat of reaction can be achieved by incorporating a secondary reactive metal oxide into the particle, that participates in at least one redox reaction. The secondary metal oxide has a different heat of reaction and is often opposite in sign of the heat of reaction compared with the primary metal oxide.

8. Low cost and ease in scale-up of synthesis procedure. The raw material cost and the cost of synthesizing the particles are of important economic consideration. The particle synthesis procedures that can be scaled up easily for industrial operation are desirable for improved process economics.

9. Suitable particle size. Particles need to be sized so that they are suitable for the specific types of reactors intended for the chemical looping operation. The particle size is associated closely with the flow properties in the reactor system. It also can affect the reaction rates.

10. Resistance to contaminant and inhibition of carbon formations. Various contaminants may be present in the fuels for chemical looping processes. To obtain an oxidizer/combustor exhaust gas stream with minimal contaminants, it is desirable to minimize the interaction between the contaminants and the oxygen carrier particles in the reducer. Carbon formation in the reducer also can be detrimental to the process. Therefore, it is desirable to reduce carbon formation in order to avoid particle deactivation and contamination of gaseous products. Inhibition of these side reactions can be realized by properly tailoring the reducing gas concentrations or by addition of suitable doping agents onto the particles.

11. Pore structure. A stable pore structure is important, especially at lower temperatures where the reaction rates are determined by the diffusivity of the gaseous reactants and products. At high temperatures, the particle sintering effect can alter the pore structure and affect the reactivity.

12. Health and environmental impacts. Because of a large particle circulation rate, a large quantity of purged particles needs to be disposed of in the chemical looping process. Thus, particles with low health and environmental impacts are desirable. Considering such impacts, it is desirable that primary metal oxides and their support materials be environmentally benign and be disposable with ease.

The oxygen-carrying composite particles usually are composed of primary metal oxide, support, and promoter or doping agents. The design and preparation of these oxygen-carrying composite particles involve, to date, a complex trial-and-error procedure. Possible variables include types of metal oxide and support, weight percentage, as well as synthesis method and procedure. More than 700 different particles have been tested for reactivity, attrition behavior, and properties related to looping applications in gasification and combustion.[1,2] Several promising particles have been identified. In terms of primary metal/metal oxides, they are Ni/NiO, Cu/CuO, Fe/FeO, Fe_3O_4/Fe_2O_3, and MnO/ Mn_3O_4 with the support of Al_2O_3, TiO_2, Ni-, Co-, or Mg-Al_2O_4, bentonite, or ZrO_2. These are the particles under consideration in this chapter. Their fundamental characteristics in chemical looping combustion and gasification applications are given as follows.

The physical and chemical properties of some materials used for oxygen-carrying composite particles noted above are given in Tables 2.1–2.4. Table 2.1 presents the theoretical oxygen-carrying capacity of various metal oxides when used to carry out specific redox reactions. The oxygen-carrying capacity is a key parameter that measures the effectiveness of a chemical looping particle in performing its redox function. A higher oxygen-carrying capacity would require a lower oxygen-carrier circulation rate, leading potentially to a smaller reactor volume. As shown in Table 2.1, the oxygen-carrying capacity is described by the oxygen conversion relative to the total available oxygen in

TABLE 2.1. Oxygen-Carrying Capacity of the Metal Oxide Particles in Their Pure Form

Initial State	Reduced State	Extent of Conversion (%)	Extent of Maximum Weight Change (%)
Fe_2O_3	Fe	100.00	30.06
Fe_2O_3	FeO	33.33	10.02
Fe_2O_3	Fe_3O_4	11.11	3.34
Fe_3O_4	Fe	88.89	26.72
Fe_3O_4	FeO	22.22	6.68
FeO	Fe	66.67	20.04
NiO	Ni	100.00	21.42
CuO	Cu	100.00	20.11
CuO	Cu_2O	50.00	10.06
Cu_2O	Cu	50.00	10.06
MnO_2	Mn	100.00	36.81
MnO_2	MnO	50.00	18.40
MnO_2	Mn_3O_4	33.33	12.27
MnO_2	Mn_2O_3	25.00	9.20
Mn_2O_3	Mn	75.00	27.61
Mn_2O_3	MnO	25.00	9.20
Mn_2O_3	Mn_3O_4	8.33	3.07
Mn_3O_4	Mn	66.67	24.54
Mn_3O_4	MnO	16.67	6.13
MnO	Mn	50.00	18.40

TABLE 2.2. The Reactivity of Various Metal/Metal Oxides with Different Inert Support in Reduction Reactions with CH₄ and Oxidation Reactions with Air at Different Temperatures[3]

All cells are reported as red.-oxid. pairs. ■ = Melting or decomposition of the materials.

Inert	Metal Oxide wt%	Fe 950	Fe 1,100	Fe 1,200	Fe 1,300	Ni 950	Ni 1,100	Ni 1,200	Ni 1,300	Cu 950	Cu 1,100	Cu 1,200	Cu 1,300	Mn 950	Mn 1,100	Mn 1,200	Mn 1,300
Al₂O₃	80	a-a	a-a	b-b	b-b	a-a	a-b*	b-c*	b-c*	a-a	e-e	■	■	a-a	b-b	e-e	■
	60	a-a	b-b	b-b	b-b	a-a	a-b*	b-c*	e-e	a-a	e-e	■	■	b-b	c-c	e-e	■
	40	a-a	a-a	b-b	b-b	a-a	b-c	c-c	e-e	a-a	e-e	■	■	c-c	e-e	e-e	■
Sepiolite	80	a-a	a-a	■	■	a-a	a-a	a-a	■	a-a	■	■	■	a-a	a-a	■	■
	60	a-a	a-a	■	■	a-a	a-a	b-b	■	a-a	■	■	■	a-a	c-e	■	■
	40	a-a	a-a	■	■	a-c*	b*-a	e-e	a-b*	a-a	■	■	■	a-a	b-b	■	■
SiO₂	80	b-b	b-b	b-c*	b-c*	a-c*	a-c*	a-c*	■	a-a	■	■	■	a-a	e-e	■	■
	60	a-a	b-b	b-c*	c-c	a-c*	a-d	a-d	a-c	a-a	■	■	■	a-a	e-e	■	■
	40	a-a	a-a	b-c*	b-d	a-b*	b-d	c-d	b*-c*	a-a	■	■	■	a-a	b-b	■	■
TiO₂	80	a-a	b-b	b-c*	b-b	a-a	a-b*	a-b*	a-b*	a-a	■	■	■	b-b	e-e	■	■
	60	a-a	b-b	b-c*	b-b	a-a	a-a	a-a	a-b*	a-a	■	■	■	e-e	e-e	■	■
	40	a-a	b-c*	b-c*	b-b	a-c*	a-a	a-a	a-b*	a-a	■	■	■	e-e	e-e	■	■
ZrO₂	80	a-a	a-a	a-a	b-b	a-b*	a-c*	a-d	a-d	a-a	■	■	■	a-a	a-a	a-a	a-b*
	60	a-a	a-a	a-a	b-b	a-b*	a-b*	a-b*	a-b*	a-a	■	■	■	a-a	a-a	a-a	a-b*
	40	a-a	a-a	a-a	b-b	a-b*	a-b*	a-b*	a-b*	a-a	■	■	■	a-a	a-a	a-a	b*-b*

*Near complete conversion is reached after a long time.

[a] High reactivity and high conversion (X > 0.8 in 1 min).

[b] Conversion between 0.5 and 0.8 in 1 min.

[c] Conversion between 0.3 and 0.5 in 1 min.

[d] Conversion lower than 0.3 in 1 min, but higher than 0.3 in 20 min.

[e] Low reactivity or low conversion (X < 0.3 in 20 min).

■ Melting or decomposition of the materials.

TABLE 2.3. Physical Properties of Various Oxygen Carriers in Reduction Reactions[4]

Reduction Reaction	Enthalpy of Reaction at 1,000°C and 1 atm (kJ/mol)	Melting Point of the Reduced Metal Form (°C)	Melting Point of the Oxidized Metal Form (°C)	Specific Density of Reduced Metal Form (kg/m³)	Specific Density of Oxidized Metal Form (kg/m³)	Weight Percentage of Transferable Oxygen (%)
$2CuO + C \rightarrow 2Cu + CO_2$	−99.3	1,083	1,026	8,920	6,450	20.11
$CuO + CO \rightarrow Cu + CO_2$	−133.5					
$CuO + H_2 \rightarrow Cu + H_2O$	−101.3					
$4CuO + CH_4 \rightarrow 4Cu + CO_2 + 2H_2O$	−211.6					
$4CuO + C \rightarrow 2Cu_2O + CO_2$	−135.1	1,235	1,026	6,000	6,450	10.06
$2CuO + CO \rightarrow Cu_2O + CO_2$	−151.4					
$2CuO + H_2 \rightarrow Cu_2O + H_2O$	−119.2					
$8CuO + CH_4 \rightarrow 4Cu_2O + CO_2 + 2H_2O$	−283.3					
$2Cu_2O + C \rightarrow 4Cu + CO_2$	−63.4	1,083	1,235	8,920	6,000	11.81
$Cu_2O + CO \rightarrow 2Cu + CO_2$	−115.6					
$Cu_2O + H_2 \rightarrow 2Cu + H_2O$	−83.5					
$4Cu_2O + CH_4 \rightarrow 8Cu + CO_2 + 2H_2O$	−139.9					
$2NiO + C \rightarrow 2Ni + CO_2$	75.2	1,452	1,452	8,900	7,450	21.42
$NiO + CO \rightarrow Ni + CO_2$	−47.2					
$NiO + H_2 \rightarrow Ni + H_2O$	−15.0					
$4NiO + CH_4 \rightarrow 4Ni + CO_2 + 2H_2O$	133.5					
$2Co_3O_4 + C \rightarrow 6CoO + CO_2$	−8.6	1,480	895	8,900	6,070	6.64
$Co_3O_4 + CO \rightarrow 3CoO + CO_2$	−88.2					
$Co_3O_4 + H_2 \rightarrow 3CoO + H_2O$	−56.0					
$4Co_3O_4 + CH_4 \rightarrow 12CoO + CO_2 + 2H_2O$	−30.3					

TABLE 2.3 Continued

Reduction Reaction	Enthalpy of Reaction at 1,000°C and 1 atm (kJ/mol)	Melting Point of the Reduced Metal Form (°C)	Melting Point of the Oxidized Metal Form (°C)	Specific Density of Reduced Metal Form (kg/m³)	Specific Density of Oxidized Metal Form (kg/m³)	Weight Percentage of Transferable Oxygen (%)
$\frac{1}{2}Co_3O_4 + C \rightarrow \frac{3}{2}Co + CO_2$	53.9	1,480	895	8,900	6,070	26.58
$\frac{1}{4}Co_3O_4 + CO \rightarrow \frac{3}{4}Co + CO_2$	-58.1					
$\frac{1}{4}Co_3O_4 + H_2 \rightarrow \frac{3}{4}Co + H_2O$	-25.9					
$Co_3O_4 + CH_4 \rightarrow 3Co + CO_2 + H_2O$	90.2					
$2CoO + C \rightarrow 2Co + CO_2$	73.9	1,480	1,800	8,900	5,680	8.54
$CoO + CO \rightarrow Co + CO_2$	-48.0					
$CoO + H_2 \rightarrow Co + H_2O$	-15.8					
$4CoO + CH_4 \rightarrow 4Co + CO_2 + 2H_2O$	741.8					
$6Mn_2O_3 + C \rightarrow 4Mn_3O_4 + CO_2$	-216.6	1,564	1,080	4,856	4,810	3.38
$3Mn_2O_3 + CO \rightarrow 2Mn_3O_4 + CO_2$	-192.2					
$3Mn_2O_3 + H_2 \rightarrow 2Mn_3O_4 + H_2O$	-160.0					
$12Mn_2O_3 + CH_4 \rightarrow 8Mn_3O_4 + CO_2 + 2H_2O$	-446.3					
$2Mn_2O_3 + C \rightarrow 4MnO + CO_2$	-36.1	1,650	1,080	5,180	4,810	10.13
$Mn_2O_3 + CO \rightarrow 2MnO + CO_2$	-101.9					
$Mn_2O_3 + H_2 \rightarrow 2MnO + H_2O$	-69.7					
$4Mn_2O_3 + CH_4 \rightarrow 8MnO + CO_2 + 2H_2O$	-85.2					
$\frac{2}{3}Mn_2O_3 + C \rightarrow \frac{4}{3}Mn + CO_2$	239.6	1,260	1,080	7,200	4,810	30.40
$Mn_2O_3 + CO \rightarrow Mn + CO_2$	35.9					
$Mn_2O_3 + H_2 \rightarrow Mn + H_2O$	68.1					
$\frac{4}{3}Mn_2O_3 + CH_4 \rightarrow \frac{8}{3}Mn + CO_2 + 2H_2O$	466.2					

TABLE 2.3   Continued

Reduction Reaction	Enthalpy of Reaction at 1,000°C and 1 atm (kJ/mol)	Melting Point of the Reduced Metal Form (°C)	Melting Point of the Oxidized Metal Form (°C)	Specific Density of Reduced Metal Form (kg/m³)	Specific Density of Oxidized Metal Form (kg/m³)	Weight Percentage of Transferable Oxygen (%)
$2Mn_3O_4 + C \rightarrow 6MnO + CO_2$	54.2	1,650	1,564	5,180	4,856	6.99
$Mn_3O_4 + CO \rightarrow 3MnO + CO_2$	−56.8					
$Mn_3O_4 + H_2 \rightarrow 3MnO + H_2O$	−24.6					
$4Mn_3O_4 + CH_4 \rightarrow 12MnO + CO_2 + 2H_2O$	95.4					
$\frac{1}{2}Mn_3O_4 + C \rightarrow \frac{3}{2}Mn + CO_2$	296.7	1,260	1,564	7,200	4,856	27.97
$\frac{1}{4}Mn_3O_4 + CO \rightarrow \frac{3}{4}Mn + CO_2$	64.5					
$\frac{1}{4}Mn_3O_4 + H_2 \rightarrow \frac{3}{4}Mn + H_2O$	96.7					
$Mn_3O_4 + CH_4 \rightarrow 3Mn + CO_2 + H_2O$	580.2					
$6Fe_2O_3 + C \rightarrow 4Fe_3O_4 + CO_2$	83.7	1,538	1,560	5,200	5,120	3.34
$3Fe_2O_3 + CO \rightarrow 2Fe_3O_4 + CO_2$	−42.0					
$3Fe_2O_3 + H_2 \rightarrow 2Fe_3O_4 + H_2O$	−9.9					
$12Fe_2O_3 + CH_4 \rightarrow 8Fe_3O_4 + CO_2 + 2H_2O$	154.2					
$2Fe_2O_3 + C \rightarrow 4FeO + CO_2$	158.4	1,420	1,560	5,700	5,120	10.01
$Fe_2O_3 + CO \rightarrow 2FeO + CO_2$	−4.7					
$Fe_2O_3 + H_2 \rightarrow 2FeO + H_2O$	27.5					
$4Fe_2O_3 + CH_4 \rightarrow 8FeO + CO_2 + 2H_2O$	303.7					
$\frac{2}{3}Fe_2O_3 + C \rightarrow \frac{4}{3}Fe + CO_2$	143.8	1,275.5	1,560	7,030	5,120	30.06
$Fe_2O_3 + CO \rightarrow Fe + CO_2$	−12.0					
$Fe_2O_3 + H_2 \rightarrow Fe + H_2O$	20.2					
$\frac{1}{3}Fe_2O_3 + CH_4 \rightarrow \frac{8}{3}Fe + CO_2 + 2H_2O$	274.5					

TABLE 2.3 Continued

Reduction Reaction	Enthalpy of Reaction at 1,000°C and 1 atm (kJ/mol)	Melting Point of the Reduced Metal Form (°C)	Melting Point of the Oxidized Metal Form (°C)	Specific Density of Reduced Metal Form (kg/m³)	Specific Density of Oxidized Metal Form (kg/m³)	Weight Percentage of Transferable Oxygen (%)
$2PbO + C \rightarrow 2Pb + CO_2$	−18.5	327.5	886	11,340	8,000	7.17
$PbO + CO \rightarrow Pb + CO_2$	−93.1					
$PbO + H_2 \rightarrow Pb + H_2O$	−60.9					
$4PbO + CH_4 \rightarrow 4Pb + CO_2 + 2H_2O$	−50.1					
$2CdO + C \rightarrow 2Cd + CO_2$	124.9	320.9	900	8,650	8,150	12.46
$CdO + CO \rightarrow Cd + CO_2$	−21.4					
$CdO + H_2 \rightarrow Cd + H_2O$	10.8					
$4CdO + CH_4 \rightarrow 4Cd + CO_2 + 2H_2O$	236.8					
$SnO_2 + C \rightarrow Sn + CO_2$	175.6	232	1,630	7,285	6,950	21.23
$½SnO_2 + CO \rightarrow ½Sn + CO_2$	3.9					
$½SnO_2 + H_2 \rightarrow ½Sn + H_2O$	36.1					
$2SnO_2 + CH_4 \rightarrow 2Sn + CO_2 + 2H_2O$	338.1					
$2SnO + C \rightarrow 2Sn + CO_2$	106.7	232	1,080	7,285	6,450	11.88
$SnO + CO \rightarrow Sn + CO_2$	−30.5					
$SnO + H_2 \rightarrow Sn + H_2O$	1.7					
$4SnO + CH_4 \rightarrow 4Sn + CO_2 + 2H_2O$	200.4					

Thus, an ideal metal oxide for a chemical looping process will be able to react with various carbonaceous fuels and completely convert them to CO_2 and H_2O. If the conversion is incomplete, the unconverted feedstock will need to be combusted with oxygen in order to arrive at a stream containing primarily CO_2 and H_2O. The reduced form of metal oxides should be reactive with the oxidizing agents such as steam, air, and CO_2 in the regeneration reaction. As described, many oxides such as those of Ni, Cu, Cd, Co, Mn and Fe that have the ability of undergoing cyclic reduction and oxidation cycles have been studied in the literature for chemical looping combustion and gasification applications. Various factors that affect the properties of these particles for the chemical looping operation are illustrated further in the following sections.

TABLE 2.4. Maximum Per Pass Conversion of H₂O to H₂ in the Regeneration Reactor and the Stable Phase Obtained at 600°C for Countercurrent Gas–Solid Operation

	γ^a (%)	Oxidized Phase
Ni	0.40	NiO
Cd	1.83	CdO
Cu	0.00	Cu_2O
Co	2.27	CoO
Sn	40.82	SnO_2
MnO	0.00	Mn_3O_4
Fe	74.79	Fe_3O_4

$^a\gamma$ is the conversion of H_2O to H_2, obtained from ASPEN Plus simulation.

a fully oxidized metal oxide or the extent of the weight change of the oxygen carrier resulting from the oxygen conversion. For example, the oxygen conversion relative to the total available oxygen for Fe_2O_3 to Fe is 100%, whereas the extent of the weight change of the oxygen carrier is 30%. The weight change of the oxygen carrier provides a direct measure on the maximum amount of oxygen transferable from the given redox reactions. Table 2.1 shows that the extent of the oxygen conversion is proportional to the extent of the weight change for a fixed initial state of the metal oxide. The maximum oxygen-carrying capacity of a metal oxide varies with the type of the metal and its oxidation states. Although the oxygen-carrying capacity achieved in an actual chemical looping process is often lower than the theoretical oxygen-carrying capacity given in Table 2.1, the values given in the table provide a useful reference for the oxygen carrier selection. It is also noted that, when an inert additive is used as support to the particle for enhancing its chemical and/or physical properties, the oxygen-carrying capacity of the particle decreases. For example, with 30% by weight of inert support on Fe_2O_3, the pure Fe_2O_3 portion in the composite particle becomes 70%. Thus, the full extent of the oxygen conversion for Fe_2O_3 to Fe in the composite particle yields a weight change of 21%, instead of 30%.

Table 2.2 shows the reactivity of various metal/metal oxides with different inert support in reduction reactions with CH_4 and oxidation reactions with air at different temperatures.[3] The reactivity is described in terms of the conversion within a specific amount of time (1 minute in most cases). The metals considered include Cu, Fe, Mn, and Ni. It is observed that in general, Ni and Cu have high activities; however, Cu sinters at 950°C, which limits its application at high temperature. Fe exhibits a moderate activity with the Fe–Fe_2O_3 conversion usually being incomplete.[3] The Mn activity varies with the active metal oxide content and type of inert support.

Table 2.3 shows the physical properties of various oxygen carriers under a reduction reaction by various types of fuels.[4] The physical properties listed include melting temperatures of reduced and oxidized metals and the specific density of the reduced metals and oxidized metals. The oxygen-carrying capacity is expressed in terms of the weight change of the metal oxide. It is shown that, although Cu has a high reactivity, its melting temperature is low (1,083°C). The Mn_2O_3–Mn_3O_4 and Fe_2O_3–Fe_3O_4 pairs have the lowest oxygen-carrying capacity, whereas Mn_2O_3–Mn and Fe_2O_3-Fe have the highest oxygen-carrying capacity.[4] Table 2.3 also shows the enthalpy of the reactions. Depending on the types of metal oxides, the reduction reaction by carbon can be either endothermic (e.g., NiO–Ni and Fe_2O_3–Fe) or exothermic (CuO–Cu and Mn_2O_3–MnO).[4] Different fuels (e.g., CH_4, H_2, or CO) also would yield a different heat effect and, hence affect the heat integration scheme in the looping system. Other reducing agents that may be of interest include coal, biomass, heavy oil, and light hydrocarbons.

2.2.2 Thermodynamics and Phase Equilibrium of Metals and Metal Oxides

Coal-derived syngas, which is composed of CO and H_2, is a frequently used fuel in chemical looping processes. In these processes, as mentioned in Chapter 1, metal oxide is first reduced with syngas and then regenerated with steam and/or air. Specifically, the reduction of metal oxide with CO can be written as

$$yCO + y/zMO_z \rightarrow yCO_2 + y/zM \qquad (2.2.1)$$

The equilibrium constant is defined as

$$K_{eq1} = P_{CO_2}^y / P_{CO}^y \qquad (2.2.2)$$

P is the total pressure or the partial pressure of a gaseous species. Similarly, the reduction reaction between metal oxide and H_2 is

$$yH_2 + y/zMO_z \rightarrow yH_2O + y/zM \qquad (2.2.3)$$

The equilibrium constant is

$$K_{eq2} = P_{H_2O}^y / P_{H_2}^y \qquad (2.2.4)$$

Note that when the reducing gas is a mixture of CO and H_2, the reaction among CO, H_2, H_2O, and CO_2 also will take place:

$$CO + H_2O \rightarrow CO_2 + H_2 \qquad (2.2.5)$$

The above reaction is the water–gas shift (WGS) reaction and has the equilibrium constant:

$$K_{eq3} = \frac{P_{CO_2}P_{H_2}}{P_{CO}P_{H_2O}} \qquad (2.2.6)$$

K_{eq3} can be expressed in terms of K_{eq1} and K_{eq2} by combining Equations (2.2.2), (2.2.4) and (2.2.6) as

$$K_{eq3} = \left(\frac{K_{eq1}}{K_{eq2}}\right)^{1/y} \qquad (2.2.7)$$

Therefore, the overall thermodynamic equilibrium condition for a system composed of metal oxide, CO, H_2, and their reaction products is determined by K_{eq1} and K_{eq2}. It also can be shown that the molar fractions of CO and H_2 exiting from the reducer, x, expressed by

$$x = \frac{P_{CO} + P_{H_2}}{P_{CO} + P_{H_2} + P_{CO_2} + P_{H_2O}} \qquad (2.2.8)$$

can be derived, under the thermodynamic equilibrium, from K_{eq1}, K_{eq2}, and the initial CO fraction in the syngas f_{CO}^0 as

$$x = \frac{f_{CO}^0}{1 + K_{eq1}^{1/y}} + \frac{(1 - f_{CO}^0)}{1 + K_{eq2}^{1/y}} \qquad (2.2.9)$$

where x represents the minimum amount of CO and H_2 exiting from the reducing step and is an important parameter for the selection of particles as oxygen carriers. An ideal oxygen carrier particle should have a very small value of x.

To illustrate the thermodynamic properties of various metal oxides for syngas conversions, phase equilibrium diagrams for various metals and their oxides, including Ni, Cu, Cd, Co, Mn, Sn, and Fe, in the presence of CO/CO_2 and H_2/H_2O as a function of temperature are given in Figures 2.1–2.7. K_{eq1} and K_{eq2} can be derived readily from these figures. Based on these phase equilibrium diagrams, the maximum syngas conversion and the corresponding metal oxides in reduced form can be determined.

From Figures 2.1–2.7, the equilibrium constants are dependent on temperature, type of fuel, and the oxidation state of the metal. For example, under the equilibrium condition at 1,000°C, the reduction by H_2 of CuO to Cu, and Fe_2O_3 to Fe_3O_4 yields a H_2 conversion of ~100%, whereas that of NiO to Ni yields a H_2 conversion of 99%. Under the equilibrium condition at 1,000°C, the reduction of Fe_3O_4 to FeO and FeO to Fe by H_2 yields the H_2 conversions of 78% and 41%, respectively. Thus, the oxidation state of the metal at a given temperature is determined by the partial pressures of CO and CO_2 or H_2 and H_2O at equilibrium conditions. A high equilibrium concentration of CO and H_2 can lead to the conversion of FeO to Fe. Thus, the phase equilibrium behavior dictates the type of products formed from the syngas (CO and H_2) reaction

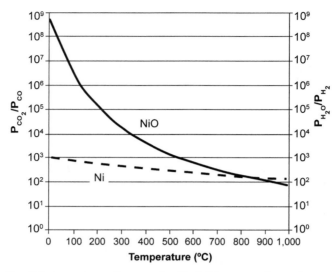

Figure 2.1. Equilibrium phase diagram for Ni–NiO system for redox reactions with CO_2/CO (—) and H_2O/H_2 (- - -).

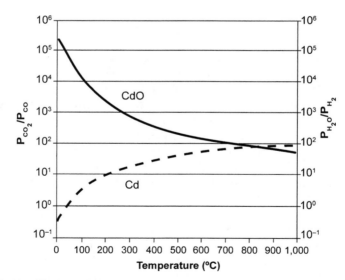

Figure 2.2. Equilibrium phase diagram for Cd–CdO system for redox reactions with CO_2/CO (—) and H_2O/H_2 (- - -).

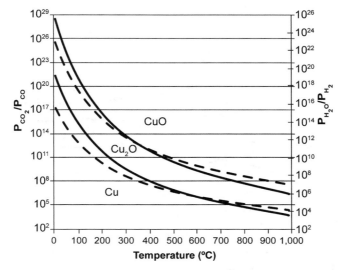

Figure 2.3. Equilibrium phase diagram for Cu–Cu$_2$O–CuO system for redox reactions with CO$_2$/CO (—) and H$_2$O/H$_2$ (- - -).

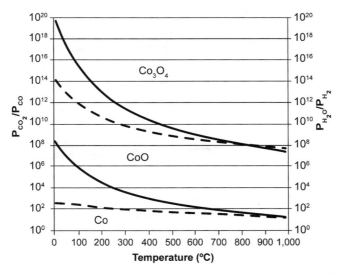

Figure 2.4. Equilibrium phase diagram for Co–CoO–Co$_3$O$_4$ system for redox reactions with CO$_2$/CO (—) and H$_2$O/H$_2$ (- - -).

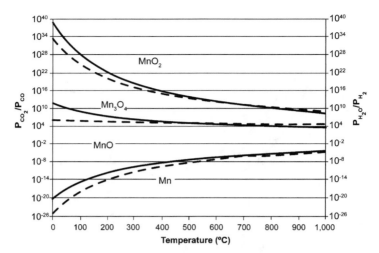

Figure 2.5. Equilibrium phase diagram for Mn–MnO–Mn₃O₄–MnO₂ system for redox reactions with CO_2/CO (—) and H_2O/H_2 (- - -).

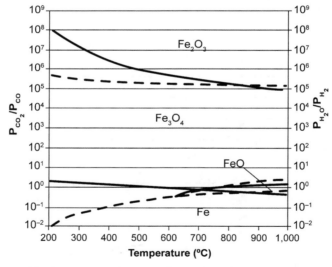

Figure 2.6. Equilibrium phase diagram for Fe–FeO–Fe₃O₄–Fe₂O₃ system for redox reactions with CO_2/CO (—) and H_2O/H_2 (- - -).

with metal oxide. When other gases are involved, the equilibrium phase diagram would change. Continued discussion of the thermodynamic equilibriums between syngas and metal oxides considering the reactor design and optimization is elaborated in Section 4.4 of Chapter 4.

Beside syngas, methane-rich fuels such as natural gas and light hydrocarbon fuel from refineries or Fischer–Tropsch synthesis also are viable feedstocks for

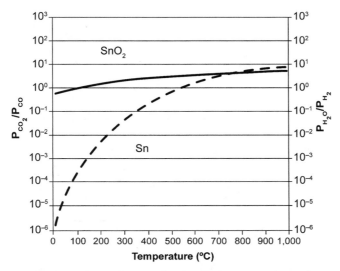

Figure 2.7. Equilibrium phase diagram for Sn–SnO$_2$ system for redox reactions for reaction with CO$_2$/CO (—) and H$_2$O/H$_2$ (- - -).

chemical looping processes. Similar to the reduction reaction with syngas, the conversion of light hydrocarbons can be carried out using metal oxides to produce carbon dioxide and steam. Metals then can be oxidized with steam and/or air. The general equation for reduction with a hydrocarbon of formula C$_x$H$_y$ can be written as

$$C_xH_y +[(2x+y)/2z]MO_z \rightarrow xCO_2 + y/2H_2O+[(2x+y)/2z]M \qquad (2.2.10)$$

The equilibrium constant is defined as

$$K_{eq} = \frac{P^x_{CO_2} P^{y/2}_{H_2O}}{P_{C_xH_y}} \qquad (2.2.11)$$

Without considering other reactions such as the Boudard reaction and reforming reactions, the maximum equilibrium hydrocarbon conversion θ at a given total pressure P_o can be expressed implicitly by the following equation:

$$K_{eq} = K =\left[x^x (y/2)^{y/2} \right]\theta^{x+y/2} P_o^{x+y/2-1}\Big/\left\{(1-\theta)[1+\theta(x+y/2-1)]^{x+y/2-1}\right\} \qquad (2.2.12)$$

Figures 2.8–2.10 show the variation of equilibrium constant K_{eq} with the temperature for methane, ethane, and butane, respectively, to reduce various metal oxides to metallic forms. Also shown are the K values required for 98% conversion of hydrocarbon to CO$_2$ and H$_2$O at a total pressure of 1 bar and 10 bar. When $K_{eq} > K$, the hydrocarbon conversion is more than 98% in a single

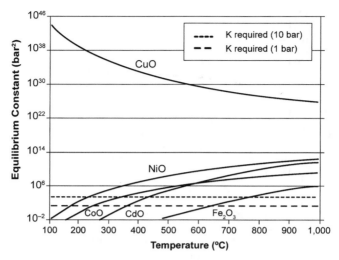

Figure 2.8. Equilibrium constant for reaction of various metal oxides with methane.

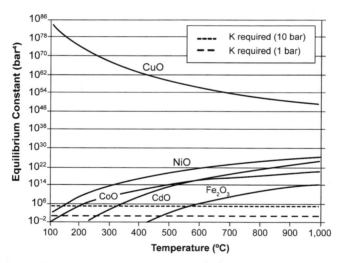

Figure 2.9. Equilibrium constant for reaction of various metal oxides with ethane.

pass reactor. Hence, the temperature at which the K_{eq} and K curves intersect is the minimum temperature at which 98% of the hydrocarbon conversion can be expected thermodynamically. It is shown that, for NiO, CoO, and CdO, the minimum temperature lies below 400 °C for reactions with methane, ethane, and butane. As gas–solid reactions in general are more favorable kinetically at temperatures higher than 400°C, the use of NiO, CoO and CdO will ensure that a 98% hydrocarbon conversion takes place at temperatures above 400°C. CuO is most favorable because it provides a complete conversion of the

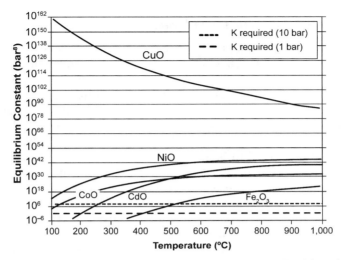

Figure 2.10. Equilibrium constant for reaction of various metal oxides with butane.

hydrocarbon at all temperatures. The conversion of Fe_2O_3 to Fe also is favorable, but at high temperatures.

Carbon formation resulting from decomposition of hydrocarbons is expected at the temperatures of interest for chemical looping processes. Higher temperatures lead to faster hydrocarbon decomposition rates, but the Boudard reaction:

$$C + CO_2 \rightarrow 2CO \qquad (2.2.13)$$

also is favored at high temperatures. Therefore, to estimate accurately the performance of the metal oxide particles under the equilibrium condition in the actual looping operations, all possible reactions such as steam/CO_2 reforming of the hydrocarbon, hydrocarbon decomposition, and Boudard reactions need to be taken into account. The thermodynamic analysis on the complex reaction network can be performed using computer simulation software such as ASPEN PLUS (Aspen Technology Inc., Houston, TX). More detailed information on thermodynamic analysis is provided in Chapter 4.

Reactions with coal and heavy oil are complicated because of the presence of long-chain hydrocarbons and asphaltenes that are not very amenable to reaction at temperatures lower than 700°C without forming significant amounts of coke. Moreover, the presence of sulfur and heavy metals such as vanadium can affect negatively the recyclability of the particles. Furthermore, ash in coal needs to be separated from looping particles recycled in the process system. Thus, an effective scheme for ash separation, particle regeneration and reactivation, and pollutant control is needed to allow particle reactivity and for recyclability to be maintained effectively in the looping process operation. Looping systems using coal as feedstock are discussed in Chapter 5.

2.2.3 Particle Regeneration with Steam

If hydrogen is the intended product, the reduced metal oxides should exhibit high per pass conversions, γ, in reaction with steam. Assuming that the reduced form of the metal oxide is used to react with pure steam, the steam-to-hydrogen conversion can be calculated under the equilibrium condition from the thermodynamic phase diagrams. Table 2.4 shows the maximum steam conversions possible to generate hydrogen from the reduced metal particles. Only Fe and Sn are found to provide reasonable conversions of steam to hydrogen. Other metals were found to be extremely unreactive with steam at the temperatures of interest. Therefore, it can be stated that for hydrogen production requiring a reasonable extent of syngas and steam conversions, the oxides of Fe and Sn are the most viable oxygen carriers among all considered in Table 2.4. In all cases, H_2O will need to be condensed from the hydrogen stream in order to produce concentrated hydrogen.

The process will, however, not be considered desirable if any of the oxide/metal materials is in liquid form or will decompose at the operating temperature of the looping process. The melting points of the various forms of iron and tin are given in Table 2.3. Assuming that the chemical looping process will operate at 600°C, it is clearly shown that all forms of iron will be in solid form and stable. Tin, however, will melt at 231°C and thus cannot be used. It also may not be feasible for tin to be used below 231°C, because the thermodynamic equilibrium does not favor the conversion of the syngas to CO_2 and H_2O at such low temperatures.

Given the high conversions for both the syngas and steam reactions, as well as the high stability of the solid particle at the reaction temperatures, oxides of iron are considered a good choice for the chemical looping process that is geared toward producing hydrogen. Fe_2O_3 is especially attractive because it also allows for near-complete conversion of syngas to CO_2 and H_2O. The successful reduction of Fe_2O_3 to Fe and FeO with complete fuel conversions, however, requires the gas–solid reducer to be in a desirable contact pattern. Further discussion on the effects of the gas–solid contact patterns on the reactant conversion are given in Section 2.2.11 and in Chapter 4.

The following works on the redox reactions for metal oxides are noteworthy. Otsuka et al.[5] reported the fast reaction of In with steam at 400°C to form In_2O_3 and H_2. Figure 2.11 shows a better maximum steam conversion for In (86%) than for Fe (~75%) at 600°C. Although In has good oxidation characteristics with steam, its very scarce supply and high cost render it unfeasible to be used as a bulk oxygen carrier. Furthermore, it has a lower oxygen-carrying capacity per weight (17%) as compared with Fe_2O_3 (30%). It may, however, be used as a promoter with iron to enhance the reaction rates, as well as steam conversion for the redox reactions. Kodama et al.[6,7] found that In(III)-ferrites react favorably with coal in chemical looping reactions. For metal oxide reactions in general, the consideration for producing carbon monoxide using CO_2 is similar to that for producing hydrogen using steam.

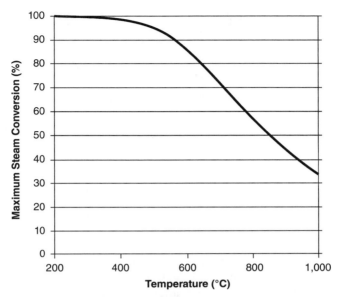

Figure 2.11. Maximum conversion of steam to hydrogen via the reaction 2In + 3H$_2$O → In$_2$O$_3$ + 3H$_2$.

2.2.4 Reaction with Oxygen and Heat of Reaction

When the oxidant for particle regeneration is air, all the reduced metal oxides can be regenerated readily to their original oxide form. The process is, however, highly exothermic, and thus caution must be exercised to ensure that the particle temperature is maintained well below its melting point. Garcia-Labiano et al.[8] provided a theoretical model to determine the heat profiles inside the particles during various chemical looping reactions when various metal oxides are used.

The exothermic heat resulting from the metal reaction with oxygen in chemical looping processes usually is used to provide the endothermic heat requirement for the reducer, to produce steam, or to produce electricity. Table 2.5 shows the heat generated when the reduced metal forms are oxidized by 1 mol of oxygen in air. As in the chemical looping process oxygen is transferred between the two reactors through the use of metal oxides, a comparison between the heat produced in the oxidizer/combustor reactor needs to be based on a similar mole of oxygen consumed rather than on a similar mole of metal used. Table 2.5 shows that the highest amount of heat is liberated when Fe is used as the oxygen carrier. The table also shows the adiabatic output temperatures when reduced metal oxides are introduced at 500°C with a stoichiometric amount of air at 25°C. When only reduced metal is considered, the highest adiabatic temperature obtained is for Cd oxidation, whereas the lowest adiabatic temperature obtained is for Cu oxidation. These temperatures all

TABLE 2.5. Heat Generated and Adiabatic Output Temperatures During Oxidation of Various Metals to Their Oxide Forms

Reaction	$-\Delta H$ (kJ) at 1,000 K	Adiabatic Output Temperature (°C)
$2Cu + O_2(g) \rightarrow 2CuO$	295.897	1,442
$2Ni + O_2(g) \rightarrow 2NiO$	468.459	2,072
$2Co + O_2(g) \rightarrow 2CoO$	466.867	2,001
$2Cd + O_2(g) \rightarrow 2CdO$	520.096	2,267
$1.333Fe + O_2(g) \rightarrow 0.667Fe_2O_3$	539.141	1,999
$4FeO + O_2 (g) \rightarrow 2Fe_2O_3$	553.568	1,357
$4Fe_3O_4 + O_2 (g) \rightarrow 6Fe_2O_3$	478.817	852
$C + O_2(g) \rightarrow CO_2 (g)$	395.165	2,140
$0.5CH_4(g) + O_2(g) \rightarrow 0.5CO_2(g) + H_2O(g)$	401.697	1,983

well exceed 1,000°C, for which the choice of the materials for the reactors requires a serious consideration. Mitigation or control of the excess temperature would require the use of excess air, inert solids in the metal oxides, and/or continuous heat removal from the oxidizer or combustor.

The heat of combustion for carbon and methane also is presented in Table 2.5 in comparison with that for metal or metal oxide. It is shown that many metals and metal oxides have higher heat of combustion than do carbon and methane, indicating that all the reduction reactions for metal or metal oxide by CH_4 or C are endothermic except those for Cu. It is noted that, from the exergy standpoint (see Section 1.6), provided that the operating temperature of the endothermic reducer is lower than that of the exothermic combustor, the low-grade heat provided to the reducer is essentially "recuperated" to the high-grade heat produced in the combustor, rendering improved exergy conversion efficiency of the process.

2.2.5 Particle Design Considering Heat of Reaction

Depending on the reactions under consideration, the temperature in the looping reactors may vary from 400°C to more than 1,000°C. To design particles that can maintain their stability and recyclability in this temperature range, it is necessary to identify the key parameters of the particles that affect such properties. For lower temperatures, pore structure is expected to play an important role. An open pore structure will allow the gaseous species to diffuse in and out of the particle so that high conversions of the particle can be reached. A high-porosity particle with a high surface area also will promote higher reaction rates than does a low-porosity particle.

At temperatures above 600°C, sintering of the pore structure may lead to complete annihilation of the mesopores and micropores, and to resultant loss of the surface area. At higher temperatures, however, the reaction rates are faster and the reaction may proceed toward completion through ionic diffusion in the particle crystal structure. Hence, additives and promoters that increase the ionic diffusion rate through the crystal structure may need to be included in the particle formulation to ensure particle reactivity and recyclability at high temperatures. The additives are mostly in the form of heteroatoms placed in the crystal lattice in such a manner that vacancies for the diffusing ions are created leading to ease in diffusion and high reaction rates of the diffusing molecules.

When a large amount of heat is generated by the reaction, the temperature of the particles may rise and may eventually reach the melting temperature. Similarly, if the reaction is endothermic, the temperature of the particles may drop and the reaction may eventually come to a halt. For heat integration, however, it may be desirable to carry the heat from one reactor where the exothermic reaction takes place to the other reactor where the endothermic reaction takes place using particles as the heat transfer medium. In all cases, it is necessary to manage the temperature variation in the particles. Particle temperature management can be achieved, as noted, by adding such inert materials as Al_2O_3, SiC, and TiO_2 to the looping particle, thereby altering the heat effect of the particle. Particle recyclability also may be improved by having the active and inert materials evenly placed in the particle. The temperature change in the particle also can be managed by introducing excessive active metal oxide materials; the unconverted metal oxides in the reaction will serve in the same role as do the inert materials for temperature moderation. If, however, the active metal oxide has multiple oxidation states, for example, iron oxides, this temperature management approach may affect the gas conversion in the reactions because the dominant phases in the partially converted iron oxides may be different from those in the completely converted metal oxides. A difference in the dominant phases of metal oxides will lead to varied equilibrium gas concentrations and, hence, to differences in optimum gas conversions. Therefore, it is necessary to carefully ensure a desired gas conversion for a specific state of the metal oxides to be reached when this temperature management approach is adopted.

2.2.6 Particle Preparation and Recyclability

As mentioned, recyclability and reactivity are two key properties required of oxygen carrier particles. Other properties that are also of importance include attrition resistance, raw material cost, and pollutant tolerance. Clearly, proper preparation or synthesis of the particle is necessary to achieve desired particle properties. This section reviews various methods that can be used for preparing the oxygen carrier particle. The advantages and disadvantages of these methods also are discussed.

The oxygen carriers used in the early chemical looping processes such as the steam-iron processes were of naturally occurring ore in powder form.[9–11] It was found that the impurities in the ore enhanced the performance of the oxygen-carrying metal oxides.[12] For particles developed today, support materials commonly are added to the primary metal oxide; as a result, the current particle preparation methods emphasize the enhancement of mixing between the support materials and oxygen-carrying metal oxides, and the role of support materials in improving the particle morphology and its chemical and physical properties.

Particle Preparation
The procedures for the modern oxygen carrier particle synthesis can be divided into three steps.

1. Formation of the metal and support matrix.
2. Bulk morphology control.
3. Rigidity and curing.

The methods for morphology control and rigidity and curing are well established. The differences among the various synthesis techniques are mainly in the methods used to form the metal and support matrix. In the following discussion, the morphology control and rigidity and curing methods will be introduced. A detailed discussion of the matrix formation methods will follow.

Morphology Control, Rigidity, and Curing Methods. This section introduces the various procedures for bulk morphology control and rigidity and curing.

Drying: The drying and calcination steps are important for all processes. The drying procedure removes moisture, water, and other solvents contained in the particles obtained from the metal-support matrix formation step. The drying of the material can be performed in many ways. Common drying practices include thermal treatment, spray-drying, and freeze drying. The thermal treatment method heats the metal-support matrix to high temperatures to remove moisture. This method is suitable when the moisture or solvent content is low. For the case when the solvent content is high, such a method may result in cluster formation. The spray drying method passes the metal-support matrix solution through a nozzle to form small droplets that are subsequently treated with either a cocurrent or a countercurrent stream of hot gas (usually air). This one-step drying procedure results in a dry, fine powder product. The freeze-drying method involves an elaborate process that yields fine spherical particles. The metal-support matrix is passed through a spray nozzle where it is made into spherical droplets by atomizing air. These spherical droplets are then frozen with liquid N_2. The water in the particle then is removed by sublimation in a freeze drier that operates at $-10°C$ with a pressure similar to the vapor pressure of ice. The particles obtained by this method are spherical as their structure is formed during the freezing step.[13]

Product Formation and Morphology Control: The product formation step processes the particles into the desired size and shape. For example, the ball mill grinding method is used to produce fine powders. The granulation method is used to obtain small granules. The pelletization and extrusion methods are used to produce large pellets.

The first method to be discussed is ball mill grinding. To obtain fine powders, the dried metal-support powder is mixed with a small amount of distilled water and sometimes dispersants like polyacrylic acid. This mixture is treated in a ball mill for an extended period of time. The duration in the ball mill depends on the size of the particle feed and that of the product. The ball mill generally uses a grinding medium that reduces the size of the powder by impaction. This technique commonly is used in industries to reach small particle size. High-energy ball mills can reduce the size to the nanometer range.[14] Such powders have a high surface-to-volume ratio and are useful for improving reactivity. The resulting powder needs to be dried and cured. Thus, the grinding step usually takes place before the drying step.

The next method to be discussed is granulation. This method is adapted to aggregate finely divided powder into small granules. There are many granulation methods, including fluidized bed granulation, mixer granulation or the single pot method, spray drying, and freeze granulation. To assist in granulation, a binder material often is added. The most common methods used to granulate chemical looping particles are spray drying, which combines the drying and granulation in one step, and freeze granulation. These two methods result in spherical particles with well-controlled sizes.

Next, the pelletization method is explored. This method produces pellets by compressing dried/moist powder under high pressure using a punch and a die. The powder obtained after drying/calcination or that obtained by granulation can be used. The strength of the pellet is dependent on the compression strength and the properties of the metal and the support. In certain cases, binding agents are added to increase the physical strength needed to withstand the compression and retain the shape.

In the extrusion method, the metal-support matrix is mixed with distilled water to obtain a paste. This paste is forced through the extrusion die using pressure and is cut into small pellets after exiting the die. The small pellets are then heat treated for removing the moisture and hardening it. The drying and calcination step often is performed after extrusion.

Calcination and Product Curing: Calcination involves heating the particle to temperatures close to or higher than the sintering temperature for an extended period of time. Such a procedure usually is performed after the drying step. The calcination procedure usually is carried out under an air atmosphere. Therefore, metal salts in the particle such as metal nitrates will be converted to metal oxides. The calcination procedure can induce certain interactions between support and metal, aiding the physical strength. Cho et al.[15] showed the increase in crushing strength with an increase in the sintering temperature.

It also was observed that a 60:40 NiO and $MgAl_2O_4$ looping medium exhibited a higher conversion when sintered at 1,300°C compared with the same medium sintered at lower temperatures. Jin et al.[16] observed the formation of a binder material $NiAl_2O_4$ during calcination, that improved particle functionality. Ishida et al.[17] observed a fall in reactivity with increasing sintering temperature for iron-based looping medium with alumina support. Generally, higher sintering temperatures result in higher physical strengths.[15,17]

The above major methods are used for bulk morphology control and rigidity and curing. The methods for metal-support matrix synthesis are given as follows.

Metal-Support Matrix Formation. Satisfactory reactivity and recyclability of the oxygen carrier particles are achieved through the interaction between the primary metal and support. This section introduces the various methods for metal-support matrix formation.

Mechanical Mixing: As the simplest method for particle mixing, mechanical mixing is economical and can yield a satisfactory result. The synthesis procedure involves the direct mixing of the metal oxide and the support in a certain ratio to form the composite looping medium. The powders of both materials first are dried. The dry powder is then mixed mechanically. Binding agents may be added in the mixing step to strengthen the composite matrix. After the formation of the metal-support matrix, the powder is processed into the desired size and shape. For example, extrusion or compression can be used to produce pellets, whereas granulation can be used to produce powders. A calcination step often is performed after obtaining particles in a desirable morphology. In some cases, when bulk pores are desirable, graphite can be mixed into the matrix initially and can then be combusted during calcination, forming pores.[18] Ishida et al.[19,20] performed tests on many Fe-, Ni- and Co-based metal oxides synthesized by this method and found them to be chemically stable over multiple cycles. Iron-based composite particles developed at The Ohio State University (OSU) were found to be reactive for more than 100 cycles.

Although cheap and effective in most applications, mechanical mixing yields a less homogeneous medium compared with other applications. The particles obtained by mechanical mixing are observed to have a metal oxide-rich phase and a support-rich phase. In particular, mechanical mixing was found not suitable for metal oxides such as copper oxide that sinter at low temperatures. The study conducted by de Diego et al.[18] indicated that copper-based oxygen carriers prepared by mechanical mixing lose activity because of sintering of the metal-rich phase.

Freeze Granulation: Freeze granulation is best suited for the synthesis of small spherical particles. The method was used extensively by Mattisson and his coworkers[13,15,21,22] for the synthesis of iron- and nickel-based metal oxides for chemical looping combustion. To prepare the particles, the metal oxide and the support are mixed in distilled water with a small amount of dispersant such as polyacrylic acid. The dispersant improves the homogeneity of the mixture. This mixture is then ground in a ball mill for an extended period of time to obtain a fine powder slurry. The slurry is then treated with a small amount of binder to achieve improved particle strength. The resulting product then is dried using the freeze-drying technique. Finally, the particles are sintered at elevated temperatures and then sieved to yield well defined sizes. Thus, freeze granulation can produce particles of well-defined morphology. After being treated at certain sintering temperatures, the spherical shape particles obtained using this method can have a very smooth surface.[21] Similar to the mechanical mixing method, freeze granulation cannot be extended to the synthesis of copper-based oxygen carriers because of the sintering of the copper-rich phase.[24] The chemical reactivity of the particles prepared by this method is highly dependent on the metal oxide and binder interaction. The crush strength of the particle improves with the number of redox cycles; the reason for this improvement is believed to be the adjustment of the structure of the particle to the best suited form.[21,25] Extensive work on various metal oxides such as Fe, Ni, and Mn was performed in this field.[13,15,21–25]

The two techniques discussed earlier involve the formation of the metal-support matrix via direct mixing of the constituents in the final form observed in the particle. The impregnation techniques, which will be discussed next, involve the generation of the matrix from a solid support material and an aqueous metal solution.

Dry Impregnation and Incipient Wet Impregnation: The dry impregnation and incipient wet impregnation techniques predominantly are used in the catalyst industry to impregnate valuable (noble metal) ingredients in the pore of a support medium. This method also can be applied to chemical looping oxygen carrier synthesis.

The method begins with the selection of the support medium. The support medium with well-defined pore volume is used in its powder form. The support is exposed to a metal salt solution with volume equal to the total pore volume. Solutions of metal nitrate salts often are used because of high solubility and availability. After being doped on the support, the oxygen carrier is calcined. The calcination step results in the decomposition of the metal nitrates to metal oxides. The process is repeated until the desired metal oxide percentage is reached. Finally, high-temperature calcination is carried out to obtain the desired physical stability. It is difficult to predict the exact amount of metal loading because of the inherent nature of the technique. Therefore, two parameters are used to quantify the metal oxide percentage, that is, theoretical metal loading (percentage) and actual metal loading (percentage). The theoretical metal loading is determined based on the amount of the metal salt solution used, whereas the actual metal loading is determined from reduction/oxidation experiments.[26]

The method is suitable for the synthesis of all types of oxygen carriers. However, because of the tedious synthesis procedure and hence the high synthesis cost, the method predominantly is used for a copper-based looping medium to reduce the effect of copper sintering. It is believed that the reduced sintering effect results from limited metal loading in the pores of the support.[27] The particles synthesized using this method showed a crushing strength similar to that of the support, thus highlighting the importance of the support selection.[26] Apart from the elaborate synthesis procedure, the metal loading of the particle obtained using such a method is limited by the pore volume of the support.

Wet Impregnation: The wet impregnation technique is more empirical compared with the previous techniques. The difference lies in the amount of metal nitrate solution used. This method involves the soaking of the support in metal nitrate solution followed by low-temperature calcination to decompose the nitrates into insoluble oxides. Such a procedure is repeated until the desired metal loading is achieved. The amount of solution used in wet impregnation in each soaking step is higher than that in dry impregnation. As a result, the synthesis procedure is simplified. Similar to the previous case, a final high-temperature sintering step is used to improve the properties of the particles.

The particles obtained by wet impregnation have properties similar to the dry impregnation technique. The study carried out by de Diego et al.[18] indicated that copper-based particles generated using this method had very good chemical stability and reactivity coupled with mechanical resistance when compared with other methods. The problem with wet impregnation is the formation of an outer shell of metal oxide, which has a weak bond with the support and tends to attrite after initial cycles.

The next two methods involve the formation of more homogenous mixtures. Raw materials in liquid form are used as the source of the metal oxide and the support.

Coprecipitation and Dissolution: Coprecipitation and dissolution are widely practiced methods in the chemical looping field to obtain an oxygen carrier that is homogenous. This method also can be used when the support material is less porous, and hence the impregnation approach is not feasible. Three precursors, that is, the metal salt solution, the support solution, and the precipitating agent, are important for this method. To prepare particles using the coprecipitation method,[18] the metal salt solution, and the support solution are first mixed together in liquid form. The precipitating agent, usually alcohol, then is added to the mixed solution, initiating the precipitation of metal and support in powder form. The dissolution method[16,20,28–32] involves the addition of the metal salt solution and the support solution to the alcohol (precipitating agent) and water simultaneously. Powders that contain metal and support are obtained in a similar manner to the coprecipitation method. Thus, the procedures followed to obtain the products are different, but the final result is the same for the two techniques.[33]

To remove the water, alcohol, and acid, and to improve the interaction between the metal and the support, multiple drying steps are carried out. For example, Jin et al.[29] used a four-step drying and calcining path where the sample initially was dried at 100°C for 12 hours to remove the water content completely. This is followed by drying at 150°C and 200°C for 24 hours and 5 hours, respectively, to heat-treat the particles to initiate interactions between support and metal and to remove the alcohol. Finally, calcination at 500°C for 3 hours in the presence of air removes the acid formed and induces physical stability by sintering.[16] The resulting powder then is formed to the final product.

The occurrence of simultaneous precipitation of metal and support is essential for the dissolution and coprecipitation method. Some governing factors for such a simultaneous precipitation include the pH of the solution and reaction rates of the indiviual precipitation reactions. This method was used extensively by Ishida and coworkers[16,17,28,30,31] with methane as the reducing gas and observed very good recyclability and reactivity for a Ni-based oxygen carrier.

Sol-Gel Synthesis: The sol-gel synthesis method provides excellent control over the physical parameters of the particle. Along with this control comes high cost and difficulty in procedures. Highly homogeneous particles can be synthesized from this method.

This method originally was defined as "the preparation of ceramic materials by preparation of a sol, gelation of the sol, and removal of the solvent."[34] Numerous particle synthesis procedures with slight modifications from the above definition also have been included in this category. This process begins with the selection of the dissolved or solution precursors of the intended metal oxide and support. The precursors can be metal alkoxides, metal salt solutions, and/or other solutions containing metal complexes. Metal alkoxides commonly are employed because of the high purity observed in the final product. These precursors then are mixed together and undergo a series of hydrolysis and condensation reactions with water to form amorphous metal oxide or oxy-hydroxide gels. These gels contain colloidal particles of metal, metal oxide, metal oxy-hydroxides, and/or other insoluble compounds, which subsequently is molded to the required shape and rigidified using dehydration, chemical cross-linking, or freezing. This rigid mass then is cured by calcinations. In some cases, the calcination step is essential for obtaining the desirable metal oxide.

This sol-gel method produces a homogenous mixture and provides good control over the microstructure. The degree of aggregation or flocculation of the colloidal precursor can control the pore size of the yield. The drawback of the sol-gel method is the high raw materials cost and the elaborate synthesis procedures.[35] Ni-based particles have been synthesized using the sol-gel method by Ishida et al.[19] showing excellent reactivity.

Solution Combustion: The solution combustion is a relatively new method adopted for the synthesis of the chemical looping medium.[36] The process involves mixing nitrates of the metal and the support in the desired ratio in water solution along with glycine (combustible agent). This solution is then heated to vaporize the water. Once most of the water is driven out, the solution is ignited when the temperature is beyond the critical self-ignition temperature. The metal-oxide-based oxygen carrier is obtained after combustion. The oxygen carrier is then ground in a mill and processed to the desired shape and calcined. This method avoids the time-consuming drying steps required for previous methods. The particles obtained from this method showed good reactivity for multiple cycles and high crushing strength.[36] The major issue was the high attrition rates (~10%) and deteriorating performance (~13% reduction in activity after 10 cycles) of the extruded particles.[36]

The advantages and disadvantages of the various particle synthesis methods discussed above are summarized in Table 2.6.

In most cases, more than one method can be used to synthesize particles with satisfactory recyclability and reactivity. The cost of the synthesis method and the effect of the resulting particles on the looping process economics will be the key to method selection. The effect of particle recyclability on the economical viability of the looping processes is discussed as follows.

Particle Recyclability
To make the looping processes economical, the metal oxide particles need to have a high oxygen-carrying capacity, good recyclability and mechanical strength to sustain multiple redox cycles, and cheap and simple production procedures.

The cost of the metal oxide particles is the key to the economic analysis of the chemical looping processes. The economics of chemical looping is given in terms of the annual cost for particle replacement, C:

$$C = 24\,sry/n \qquad (2.2.14)$$

where s is the cost (in dollars per kilogram) of the metal oxide particle, r is the metal oxide circulation rate in the reactor for a given flow rate of fuel, y is the number of days the plant is in operation per year, and n is the number of reaction/regeneration cycles the particle can undertake before full replacement must be made. Note that the cost is independent of the residence time of the particles inside the reactor system. Furthermore, parameters s, r, and y are scaling factors defined by the specifications of the particle chemistry, and the plant size and operation. Once their values are determined, the cost trend for the replacement of the particles is dependent only on n. For example, assuming the value for s as $100 per ton of metal oxide, r as 1 t/h of metal oxide, and y as 330 d/yr, the trend of cost replacement versus n can be given in Figure 2.12. It is shown in the figure that, as n increases, the replacement cost asymptotically approaches zero.

Table 2.6. Comparison of Various Synthesis Techniques[13,15–19,21,24,25,27,28,37]

Property	Technique						
	Mechanical Mixing	Freeze Granulation	Dry Impregnation or IWI	Wet Impregnation	Dissolution or Coprecipitation Method	Sol–Gel Process	Solution Combustion
Economic Feasibility	++	+	–	–	+	– –	+
Metal/Support Loading	++	++	– –	– –	–	+	–
Homogeniety	– –	– –	n/a	n/a	+	++	+
Attrition Resistance	–	–	+	–	–	+	– –
Iron Based Looping medium	++	++	+	+	++	++	++
Nickel Based Looping medium	++	++	+	+	++	++	+
Copper Based Looping medium	– –	– –	++	+	–	–	–

++ highly suitable, + suitable, – not suitable, – – highly not suitable, n/a not applicable.

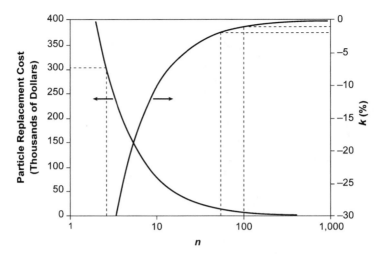

Figure 2.12. Scaled total and marginal (%) annual cost of replacement of the metal oxide particles (left scale) and k–n curve (right scale).

Although the replacement cost curve asymptotically approaches zero as the number of reaction/regeneration cycles increases, it is not clear as to how such information can be used to decide on the minimum number of reaction/regeneration cycles for a metal oxide particle in order for the process to be economical. A 10-fold increase in n causes C to decrease by 10 times. Furthermore, as n increases for a particular metal oxide particle formulation, the probability that it will perform for an even greater number of reaction cycles also increases.

To develop criteria for the selection of a metal oxide particle, the derivative of cost C is taken with respect to the number of cycles possible, n as

$$dC/dn = -24\,sry/n^2 \tag{2.2.15}$$

Combining Equations (2.2.14) and (2.2.15), the following equation is obtained:

$$(100dC/C)/dn = -100/n = k \tag{2.2.16}$$

The quantity k refers to the percentage change (decrease) in the replacement cost of the particles, if n can be improved by a small number. If $n = 50$, k is -2%, indicating that the cost reduction for improving the particle recyclability from 50 cycles to 51 cycles is 2%. As n increases, k approaches zero asymptotically. Note that k is only dependent on n and is a good criterion for use to decide on suitable values of n. The k–n curve also is shown in Figure 2.12. The actual plants may be feasible for k values lower than -1% ($n > 100$). Thus, although testing may be carried out for less cycles to compare various particle formulations, the optimized particle will need to be tested for at least 100 cycles before it can be considered commercially viable.

2.2.7   Particle Formulation and Effect of Support

The approach to finding an effective metal oxide formulation that possesses the requisite properties for chemical looping applications is complex.[3,22,24,26,37,38] Mattisson et al.[22] carried out partial reduction and oxidation of Fe_2O_3 to Fe_3O_4 and found that reaction rates for various metal-oxide particles depend on the support used. Similar findings on the reduction of Fe_2O_3 with various supports and sintering temperatures have been reported in other studies.[3,16] Urasaki et al.[39] found an approximately 22% increase in the steam oxidation rate of reaction upon adding a trace quantity (0.23 mol%) of zirconium to the Fe_2O_3. The increase was explained on the basis of reduced sintering at 500°C. Cho et al.[24] studied the relative reactivity of Fe, Ni, Cu, and Mn supported on Al_2O_3 at 950°C with alternating methane and air conditions in a fluidized-bed reactor and found Fe, Ni, and Cu to be suitable candidates. Cu and Fe oxides were found to agglomerate in the reactor, whereas Ni-based particles displayed limited mechanical strength. Takenaka et al.[40] studied the effect of Mo and Rh on the recyclability of Fe_3O_4-based particles. Rh addition was found to increase the reaction rates of reduced particles with steam by lowering the activation energy of the reaction, but increased sintering was observed leading to decreased reactivity in the long term. The addition of Mo helped prevent sintering by formation of ferrites ($Mo_xFe_{3-x}O_4$) that inhibited the contact between metallic iron particles. Montes et al.[41] studied the preparation techniques of $Ni–SiO_2$ catalysts and their cyclic oxidation and reduction characteristics at 450°C and 500°C, respectively. They found that oxidizing the initial oxygen carrier, instead of reducing it, helped increase the cyclic reaction activity. Jin et al.[16] studied many combinations of metal oxides and support materials at 600°C and found $NiO/NiAl_2O_4$ to show good recyclability. However, Ni in $NiAl_2O_4$ was not reduced completely. Lee et al.[42] studied several metal oxides and support materials for use as oxygen carrier materials. The studies conducted in a thermogravimetric analyzer (TGA) used H_2 at 600°C for reduction and air at 1,000°C for oxidation. They found Ni–Al–O to have the highest crushing strength, whereas Ni-YSZ-O had the best reactivity. Mattisson et al.[26] tested oxides of Cu, Co Mn and Ni in combination with Al_2O_3 at temperatures 750°C to 950°C in cyclic methane and air conditions, and found Ni and Cu to have high reactivity at all temperatures and cycles. Ishida et al.[17] discussed the effect of sintering temperature on hydrogen reduction activity and mechanical strength for Fe–Al–O particles. They found the particles sintered at 1,320°C contained an Al_2O_3-rich Fe_2O_3 solid solution phase, which provided a better reaction rate and a reasonable mechanical strength.

Recyclability of the Oxygen Carrier Particle in the Presence of Support
A well-formulated particle should undergo multiple reaction regeneration cycles. Because oxygen transfers in and out from the particle during reaction and regeneration, it is important to study the variation in oxygen capacity with the number of cycles. The oxygen capacity for metal oxide based particles may be defined as:

$$\text{Oxygen Capacity } (\%) = 100 \left[1 - \frac{w_0 - w}{n x w_0} \right] \qquad (2.2.17)$$

where w_0 is the fully oxidized weight of the particle, w is the instantaneous weight of the particle, x is the weight fraction of pure metal oxide in the fully oxidized particle, and n is the weight fraction of oxygen in metal oxide. For example, the oxygen capacity for iron-oxide-based particles with 30% by weight of inert material can be expressed by:

$$\text{Oxygen Capacity } (\%) = 100 \left[1 - \frac{w_0 - w}{0.3 \times (1 - 0.3) \times w_0} \right] \qquad (2.2.18)$$

Thus, a particle in a fully oxidized state initially has 100% of its oxygen capacity. A particle with its iron oxide fully reduced to Fe will have 0% of its oxygen capacity. The oxygen capacity swings. The reaction capacity, defined as the difference in oxygen capacity between the reduced and the oxidized states, is used as a measure of the useful oxygen content that can be successfully exchanged during the chemical looping operations.

Figure 2.13 shows the reaction capacity change, as measured by the TGA analysis, as a 100% Fe_2O_3 particle undergoes cyclic reduction and oxidation. With this formulation, the reaction capacity drops by as much as ~60% within a few cycles. Furthermore, the incomplete conversion of iron oxide will decrease the gaseous reactant conversions from the thermodynamics standpoint. Clearly, such a particle may be used only for a few cycles before fresh particle makeup is required. Coombs and Munir[43] studied the cyclic behavior

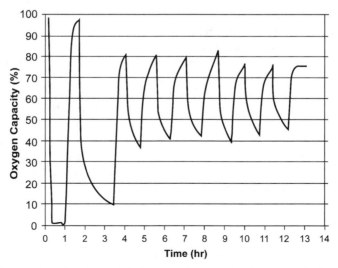

Figure 2.13. Cyclic reduction–oxidation reaction studies for pure Fe_2O_3 powders at 900°C using H_2- and O_2-rich gases.

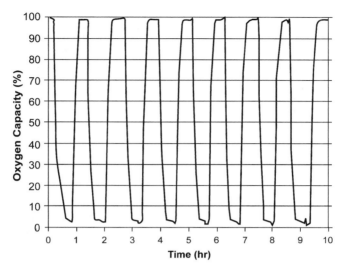

Figure 2.14. Cyclic reduction–oxidation reaction studies for Fe–Ti–O powders at 900°C using H_2- and O_2-rich gases.

of hematite powders and found that the loss in recyclability was a direct result of sintering during oxidation.

Figure 2.14 shows a similar cyclic study conducted on an Fe–Ti–O formulation. It is observed that the reaction capacity is greater than 95% and is maintained over multiple cycles. As discussed in Section 2.2.6, this particle formulation is desired for chemical looping applications. Figure 2.15 shows the variation in surface area and pore volume as determined by N_2 adsorption for both 100% Fe_2O_3 and Fe–Ti–O formulations at the reduced and oxidized states during various cycles. It is shown that in the first reduction step, the surface area and pore volume of the two formulations dropped substantially. This change resulted in near nonporous particles for both the formulations. This result may be expected because at 900°C, excessive sintering occurs, which leads to the collapse of the pore structure in the iron oxide containing particles.

Figure 2.16 shows the recyclability characteristics of an Fe–Al–O formulation. The reaction capacity increases with an increasing number of cycles, although a slowing of the reaction rate can be observed during reduction. Figure 2.17 shows the excellent recyclability that can be achieved using an Fe–YSZ–O formulation. Because YSZ is a well-known oxygen ion conductor, its incorporation into the formulation leads to near 100% reaction capacity and to very high reaction rates.

When steam (3% in N_2) is used for oxidizing Fe–Y–Zr–O particles, the oxidation reaction rates were observed to be slower than for oxidation using air. The reaction capacity also decreased because of the formation of Fe_3O_4

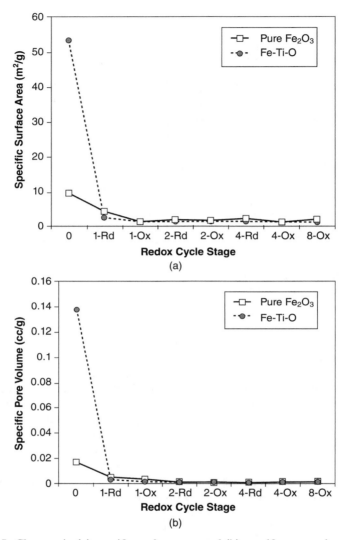

Figure 2.15. Changes in (a) specific surface area and (b) specific pore volume for pure Fe_2O_3 and Fe–Ti–O powders with multiple redox cycles at 900°C.

rather than Fe_2O_3, as indicated in Table 2.4. The particles maintain activity for numerous redox cycles tested, as shown in Figure 2.18.

Figure 2.19 shows the recyclability behavior when both H_2 and CO (syngas) were used for reducing the Fe–Y–Zr–O particles, and steam was used to oxidize the particles. A small quantity of CO_2 was added to the syngas to prevent carbon formation. It is again noted that the particles maintained activity for numerous redox cycles. Figure 2.20 shows an experiment where the

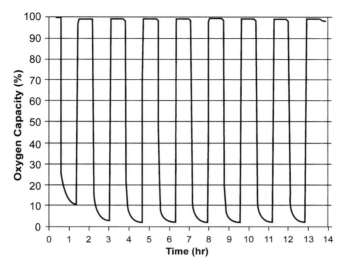

Figure 2.16. Redox cyclic studies for Fe–Al–O powders using H_2 and air for reduction and oxidation, respectively.

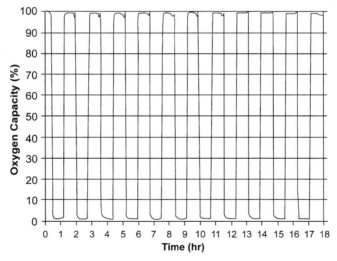

Figure 2.17. Redox cyclic studies for Fe–Y–Zr–O powders using H_2 and air for reduction and oxidation, respectively.

cyclic redox activity of the Fe–Y–Zr–O particles was tested with varying gas conditions. First, four cycles were conducted using H_2 and air, followed by two cycles using H_2 and CO_2, and then three cycles in H_2 and air, and finally four cycles using syngas and air. The oxidation with CO_2 led to the formation of Fe_3O_4, and thus full oxygen capacity could not be achieved. After reacting

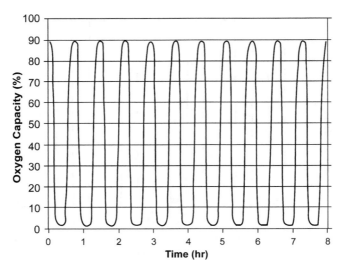

Figure 2.18. Cyclic reduction–oxidation performance of Fe–Y–Zr–O powders at 900°C in H_2 and steam conditions.

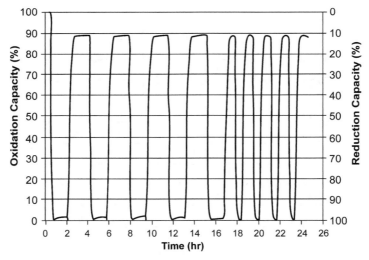

Figure 2.19. Cyclic reduction–oxidation performance of Fe–Y–Zr–O powders at 900°C in syngas and steam conditions.

with syngas, carbon formation was observed and most of the iron oxide was converted to Fe. The carbon was burned off on oxidation with air. The particle showed excellent regenerability even though varying gas conditions were imposed.

As discussed, the oxygen carrier particle is expected to maintain activity for at least 100 cycles for the process to be economical. Figure 2.21 shows a long-

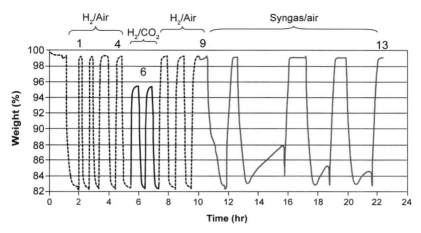

Figure 2.20. Cyclic reduction–oxidation performance of Fe–Y–Zr–O powders at 900°C under various gas conditions.

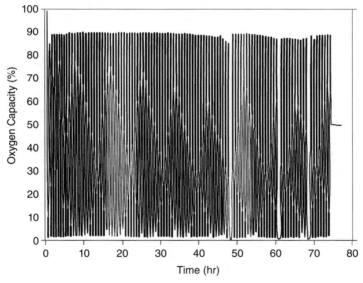

Figure 2.21. Long-term recyclability of Fe–Y–Zr–O powders at 900°C using H_2 and steam conditions for redox reactions.

term recyclability study performed on the Fe–Y–Zr–O particle formulation. Particles were reduced and oxidized with 10% hydrogen and 10% H_2O in N_2, respectively, for a total of 102 cycles in a TGA apparatus at 900°C. It was found that the particles were completely recyclable after 100 redox cycles (>75 hours). No decrease in reaction capacity was observed with time which suggests that the reaction capacity could be sustained for even more cycles.

Reactivity of the Oxygen Carrier Particle in the Presence of Support and Promoter

Particles with higher reaction rates can benefit the redox processes by reducing the size of the reactors. That is, with higher reaction rates, a larger quantity of fuel can be processed in a shorter time or using a smaller reactor. One way to improve the rate of reaction is to add support materials to the metal oxide oxygen carrier. As mentioned earlier in this chapter, studies on the support materials have been carried out extensively.[3,18,19,21,23,24,26,28,38,42,44]

Another way of modifying the rate of reaction is to introduce promoters or additives to the particle. For example, two types of Fe–Y–Zr–O formulations were synthesized by sol-gel methods with promoters, one with 0.5% CuO and the other with 1.0% NiO. These particles were subjected to several oxidation and reduction cycles using a TGA at 900°C. Particles were reduced using a gas mixture of 26.6% CO and 13.3% H_2 in N_2, and oxidized using 11% H_2O in N_2. The overall gas flow was kept similar for each cycle for all particles. Figures 2.22 and 2.23 show the cyclic TGA curves for the 0.5 wt.% CuO and 1.0 wt.% NiO, respectively. As shown in the figures, both particles were fully recyclable. The reaction rates at the initial stage of the reactions are used to estimate the effect of the promoters on the reaction rates. The reaction rates of the particles were compared with the rates of particles without any promoters.

Table 2.7 compares the initial rates for reduction and oxidation for the last redox cycle performed. The table also shows the initial rates of particles without any promoter. It can be observed that the reduction rates of the particles with promoters are lower than the particles without any promoters.

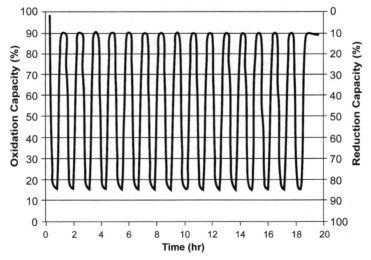

Figure 2.22. Reduction–oxidation cyclic studies at 900°C of Fe–Y–Zr–O powders containing 0.5% CuO.

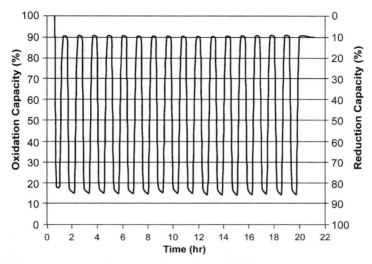

Figure 2.23. Reduction–oxidation cyclic studies at 900°C of Fe–Y–Zr–O powders containing 1.0% NiO.

TABLE 2.7. Rate of Reaction (Fractional Conversion/min) for Oxygen Carriers Containing Promoters

Type of Promoter	Relative Initial Rate of Reduction	Relative Initial Rate of Oxidation
No promoter	1.000	0.470
NiO	0.575	0.660
CuO	0.745	0.825

However, the oxidation rates are increased with the addition of promoters. In this case, the addition of 0.5% CuO to the particles increased the oxidation rate by approximately 75%, whereas the reduction rate decreased by 25%. For the case of an addition of 1% NiO, the reduction rate decreased by ~42%, whereas the oxidation rate increased by ~40%. The amount and type of promoters used affect the reaction rates. The results indicate that with a proper addition of NiO or CuO, the reaction rates can be tailored to a certain extent.

2.2.8 Effect of Particle Size and Mechanical Strength

The oxygen carrier particle needs to be processed into a certain size range for specific looping reactor applications. For chemical looping combustion systems using circulating fluidized beds (discussed in Chapter 3), particle sizes of 50–300 μm are used. For chemical looping systems using moving bed reactors (discussed in Chapters 4 and 5), pellets of sizes larger than 700 μm are used. In both cases, fine powders resulting from particle attrition can be generated and are entrained from the looping reactors, while fresh makeup particles are added to the reactors. Thus, to minimize particle attrition, good particle mechanical strength is necessary. In the following discussion, Fe–Ti–O particles operated in a moving bed and in a pneumatic conveying system are used to explore the effects of particle size and particle mechanical strength.

Pellet Strength
The strength of oxygen carrier particles needs to be sufficiently high to prevent the particle loss from thermal and mechanical stresses in the reactors. For moving bed particles, the strength can be enhanced by using techniques such as compaction to form pellets followed by calcination at high temperatures. The oxygen carriers in the looping operation will undergo densification as they proceed with multiple redox reaction cycles. This densification can lead to a further increase in the crushing strength of the pellets.

Several tests may be used to determine the strength of the pellets. The simplest one is the drop test where the pellets are dropped from a 1 m height repeatedly until they break. The number of drops before breakage is a measure of strength. Another test measures the compressive force that the pellets can take before breakage. Other tests measure the pellets' flexural strength and attrition resistance. Figure 2.24 shows the change in drop strength as an Fe–Ti–O pellet is reduced for the first time in H_2 at 900°C. The first reduction led to a significant increase in the drop strength of the pellets. Figure 2.25 shows the change in compressive strength of an Fe–Ti–O pellet after two redox

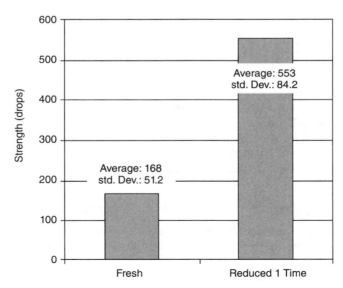

Figure 2.24. Drop strength of Fe–Ti–O pellets made by compaction before and after the first reduction reaction.

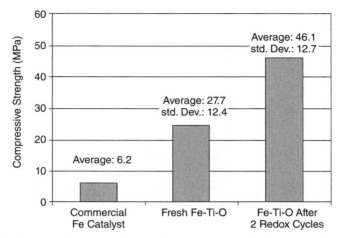

Figure 2.25. Compressive strength of Fe–Ti–O pellets made by compaction as a function of number of redox cycles.

cycles. Again, a twofold increase in the compressive strength is observed. It should be noted that the fresh pellets are still nearly four times stronger than the commercially available Fe containing catalyst of equivalent size.

The attrition rate of the pellets can be tested in an entrained flow reactor. Figure 2.26(a) shows the schematic of an entrained flow reactor that is operated under a superficial gas velocity of 18 m/s. Cylindrical pellets of 5 mm diameter and 1.5 mm height are used in the test. From Figure 2.26(b), the attrition rate of the pellets is 0.57% per cycle, which corresponds to less than a 1% pellet makeup rate.

Pellet Recyclability
Pellets that may be used in moving bed reactors are expected to be of sizes greater than 700 μm. Controlling formulation chemistry and synthesis procedures becomes important in ensuring particles with good oxygen swing capacity, as well as redox recyclability. Figure 2.27 shows the recyclability study conducted on crushed Fe–Ti–O pellets (750 and 1,000 μm) at 900°C using H_2 and air for the cyclic redox reactions. It can be shown that the pellets can be fully recyclable for 100 cycles. A similar result is obtained from Fe–Ti–O cylindrical pellets of 5-mm diameter and 4.5-mm height. These results indicate that pellets of a larger size can be prepared to maintain full oxygen transport capacities at high temperatures (see Section 2.2.10).

2.2.9 Carbon and Sulfur Formation Resistance

Carbon Formation on Oxygen Carriers
When syngas with a high CO content is used as the reducing gas, carbon deposition may occur as a result of the reverse Boudard reaction:

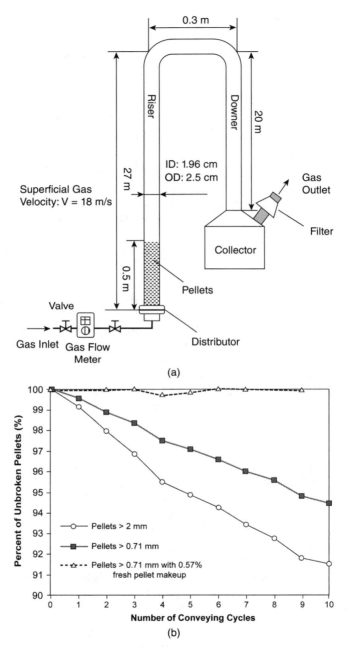

Figure 2.26. (a) Experimental setup for particle attrition test; (b) attrition test results.

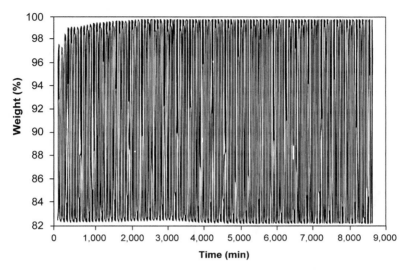

Figure 2.27. Recyclability of pellets prepared from an Fe–Ti–O formulation.

$$2CO \rightarrow C + CO_2 \qquad (2.2.19)$$

The deposited carbon can cause significant problems in the chemical looping operation. The carbon will deposit in the interstitial space between the particles, which may lead to the plugging of the reactor and to the subsequent increase in the pressure drop across the reactor. Also, the deposited carbon may affect the physical structure of the particle and may decrease its reactivity and strength. Furthermore, the deposited carbon may be carried over to the oxidizer/combustor where it will react with steam or air to form carbon monoxide and/or carbon dioxide that contaminate the gaseous products in the particle regeneration process. Some studies[20,45] discuss carbon formation and methods to avoid it over Fe containing catalysts.

Several techniques can be used to reduce carbon formation. They include incomplete reduction of metal oxide introducing steam and/or CO_2 to inhibit carbon deposition, operating at higher temperatures introducing hydrogen, and adding suitable support materials/additives to the metal oxide. In the following discussion, the Fe_2O_3-based oxygen carrier is used as an example to illustrate the control techniques for carbon deposition, recognizing that similar techniques can be applied to other types of oxygen carriers. It is noted that metal oxides such as NiO exhibit a stronger tendency for carbon formation.

Figure 2.28 shows the carbon formation during the reduction of Fe_2O_3 in a TGA in the presence of CO at 600°C. The details of the experiments involve flushing the TGA with N_2 for 30 minutes to remove trace quantities of air. The reduction is then carried out with a gas stream containing 22% CO and 11% H_2, with the balance being N_2. After the reduction, the TGA is flushed with

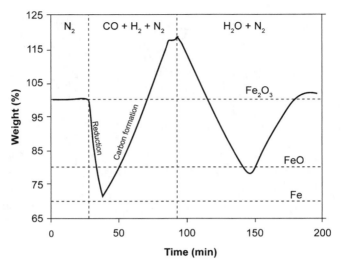

Figure 2.28. Carbon formation studies on Fe$_2$O$_3$-containing particles in a TGA with simulated syngas followed by partial regeneration with steam at 600°C.

N$_2$. Before introducing a gas stream containing water vapor in N$_2$, a rapid reduction of Fe$_2$O$_3$ takes place, leading to an almost complete conversion of Fe$_2$O$_3$ to Fe. The weight of the TGA sample, however, first decreases and then increases. The increase in weight reflects the carbon formation. The carbon formation takes place when most of the Fe$_2$O$_3$ is reduced to Fe. Fe then acts as a catalyst for the reverse Boudard reaction, and thus, carbon forms and deposits onto the surface of the reduced iron oxide. After the addition of steam, the deposited carbon is removed by the steam–carbon reaction. Thus, for the iron looping reaction, avoidance of the full reduction of iron oxide to iron could help minimize carbon deposition. A detailed discussion on general metal oxide conversions is given in Chapter 4.

Steam or CO$_2$ also can inhibit carbon deposition. Only limited steam/CO$_2$, however, can be added to the reducing gas, because the steam/CO$_2$ will impede the reduction of syngas. The effect of steam addition on the carbon disposition is shown in Figure 2.29. It is shown in the figure that an increased steam concentration leads to reduced carbon deposition. Thus, it may be possible that with the steam concentration increased to a certain level that the carbon formation may be avoided. Naturally, optimal operation of the reducer in a looping process requires consideration of the proper balance between the conversion of iron oxide and that of the carbon deposition when steam/CO$_2$ is introduced to the reducer.

An increase in temperatures favors the Boudard reaction, which is the reverse reaction of the carbon deposition reaction. The TGA experiments using the Fe$_2$O$_3$ composite particles are shown in Figure 2.30. In the experiments, the particles are first heated in N$_2$ to 900°C and are kept for 30 minutes

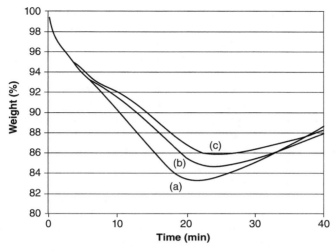

Figure 2.29. Effect of addition of H_2O on carbon formation at 600°C and atmospheric pressure. Gas composition used: 21% CO, 11.7% H_2 with (a) 0.27% H_2O, (b) 1.92% H_2O, (c) 16.5% H_2O, and the balance of N_2.

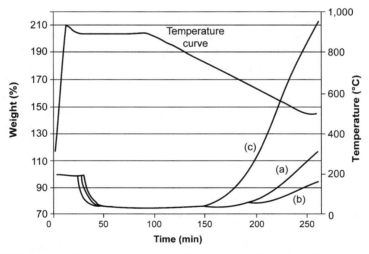

Figure 2.30. Effect of temperature and CO concentration on carbon deposition at atmospheric pressure. Gas compositions used: (a) 21% CO, 11.7% H_2, 0.27% H_2O; (b) 21% CO, 11.7% H_2, 1.92% H_2O; (c) 32.2% CO, 17.5% H_2, 0.4% H_2O; and the balance of N_2.

to flush out any traces of oxygen that may be present in the system. The particles are then reduced by a simulated syngas; no carbon formation occurs at 900°C when Fe_2O_3 is reduced completely to the Fe form. As the temperature is slowly decreased by 2.5°C/min to the point where carbon formation is observed to first occur, this temperature is marked as the critical temperature. The effects of the moisture and the partial pressure of CO on the critical temperature are shown in Figure 2.30 for three gas compositions. For the gas composition in Figure 2.30(a), the critical temperature is observed to be 702°C. The addition of water vapor to the gas [the composition in Figure 2.30(b)] leads to a decrease in the critical temperature (660°C), as well as to a decrease in the carbon formation rate. A gas composition with a higher partial pressure of CO [the composition in Figure 2.30(c)] leads to a considerable increase in the critical temperature (800°C). The syngas produced in commercial gasifiers has a high partial pressure of CO as a result the high-pressure operation of the gasifier. Thus, more significant carbon deposition can be expected when Fe_2O_3 particles are reduced fully to Fe. Higher temperatures will lead to lower carbon formation rates when syngas is used as the fuel. When natural gas or other hydrocarbons are used, however, a higher temperature can lead to more severe carbon deposition because of the tendency for the fuels to decompose.

Another method of reducing carbon deposition in an actual reactor operation is to flow H_2 through the reaction. Introducing hydrogen to the reactor can inhibit carbon deposition, because hydrogen will react with iron oxides to produce steam, which in turn reacts with carbon, and hence minimizes carbon formation. Other methods include adding more inert support or doping a certain promoter to the particle that inhibits the reverse Boudard reaction. Ishida et al.[20] studied the effect of different support, including YSZ, Al_2O_3, and TiO_2 on Fe_2O_3 and NiO based on reduced carbon formation when CO is used as a reducing agent. They reported that, for NiO particles, an addition of 40% YSZ support effectively improved the carbon deposition resistance of the particle.

Resistance to Sulfur
It is desired that a particle also possess good reactivity and recyclability even in a reactive environment where sulfur is present. This property can be important especially when gaseous fuels that contain H_2S are used (i.e., coal-derived syngas and sour natural gas). H_2S reacts with both metal oxides and metal to form metal sulfides according to:

$$MO + H_2S \leftrightarrow MS + H_2O \qquad K_{eq} = P_{H_2O}/P_{H_2S} \qquad (2.2.20)$$

$$M + H_2S \leftrightarrow MS + H_2 \qquad K_{eq} = P_{H_2}/P_{H_2S} \qquad (2.2.21)$$

Figure 2.31 shows the deactivation of an Fe–Ti–O particle when H_2S is present at 775 ppm with the reducing gas for several reduction/regeneration cycles in

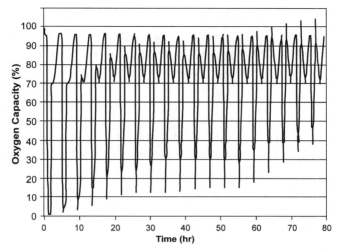

Figure 2.31. Deactivation of Fe–Ti–O particles on reduction with 775 ppm H$_2$S, 25% H$_2$, and 75% He, and regeneration with steam in N$_2$.

a TGA. Steam is used for regeneration. The experiments indicate that sulfur slowly penetrated the interior of the particle, rendering it inactive for redox reactions. Although the effect may be reduced by variation of the particle formulation, the long-term effects of sulfur can lead to particle deactivation. Note that the reactivity of the particle can be restored when the particle is regenerated by air according to the equation:

$$MS + 3/2O_2 \rightarrow MO + SO_2 \qquad (2.2.22)$$

Regeneration of metal sulfide with air, however, produces hazardous gaseous sulfur compounds. One method to prevent sulfur poisoning is to inhibit the reaction between H$_2$S and the metal oxide. As can be shown from the preceding reactions, a high hydrogen or steam partial pressure can reduce sulfidation of the metal oxide or metal particles. Because hydrogen and steam also are interconverted through a reaction with the metal oxide and metal, an excess of either hydrogen or steam can lead to substantially reduced sulfidation of the metal oxide or metal. The sulfidation behavior for metal oxide/metal is governed by the equilibrium constants of the reaction. When the H$_2$S concentration is below the equilibrium limit, there will be no sulfidation of either the metal oxide or the metal taking place. Thus, reducing the H$_2$S concentration in the feed gas stream to the subequilibrium level before the feed stream enters the chemical looping reactors allows the metal oxide or metal particles in the chemical looping reactors to be free from sulfidation. Continued discussion on the relationship between the H$_2$S concentration in the syngas and the sulfidation in the looping reactor is given in Chapter 4.

2.2.10 *Particle Reaction Mechanism*

In most studies of the reduction and oxidation of metal oxides such as iron oxides, it is assumed that the gas–solid reaction is carried out in the following steps.

Step 1. Diffusion and/or convection of gaseous reactant across the gas film to the surface of the particle.

Step 2. Diffusion of gaseous reactant through the pores of the particle.

Step 3. Adsorption of the gaseous reactant onto the reactive pore surface.

Step 4. Chemical reaction on the pore surface.

Step 5. Desorption of gaseous product from the pore surface.

Step 6. Diffusion of gaseous product out of the particle.

Step 7. Diffusion and/or convection of gaseous product across the gas film to the bulk gas phase.

Based on these steps, the pore structure and pore volume in the particle will significantly affect the reactivity of the particle. However, the experimental results shown in Figures 2.14 and 2.15 indicate that the significant decrease in pore volume after the first redox cycle does not affect the reactivity of the Fe–Ti–O particle. This may suggest a different gas–solid reaction mechanism.

The excellent recyclability of the Fe–Ti–O particles may be attributed to the increased ionic diffusivities for both the oxygen ion and the iron ion. Ionic diffusion takes place because of the defects formed in the TiO_2 crystal structure in the presence of Fe^{3+} and Fe^{2+} ions. The introduction of Fe_2O_3 into the TiO_2 can cause either lattice substitution defects or interstitial addition defects in the TiO_2 crystal structure. The atomic radius of Ti^{4+} ions is 0.61 Å, and for Fe^{3+} it is 0.65 Å.[46] Because of the similarities in the atomic radius and the close packing structure of the lattice usually encountered for larger ions such as Ti^{4+} and Fe^{3+}, it is reasonable to assume that cationic lattice substitution occurs rather than interstitial addition. To maintain electrical neutrality in the crystal structure, one oxygen ion (O^{2-}) will be vacated for every two Ti^{4+} ions substituted by two Fe^{3+} ions. The resulting oxygen ion vacancies will create point defects in the crystal structure, if the vacancy concentration is lower than a critical concentration.

When the vacancy concentration is higher than the critical concentration, other defects such as shear planes will occur. The presence of such vacancies in Fe–Ti–O particles has been confirmed by various studies. These vacancies are related closely to improved ionic diffusivity. Wang et al.[47] determined that the critical atomic concentration was 2% Fe^{3+} in TiO_2. Zhao and Shadman[48,49] identified the diffusion of atomic irons and oxygen through the ilemnite ($FeTiO_3$) structure during the reduction of ilmenite with carbon monoxide and hydrogen. Fe crystals were reported to be formed as a result of diffusion and coalescence of the Fe ions during the reduction. The formation of a 1–2 atom%

Fe containing TiO_2 in-between ilemnite crystals and Fe crystals also was identified as the cause for enhanced diffusivity of atomic iron, as well as of oxygen. Moreover, the presence of 1–5.5 atom% Ti was found in the Fe crystals. Because of the presence of these oxygen vacancies, an oxygen ion will be able to diffuse through the crystal structure. Such diffusion will be limited severely in pure Fe_2O_3 when such vacancies in the crystal lattice are absent. In pure Fe_2O_3, alternative mechanisms for oxygen to diffuse out of the particle include anionic substitution and interstitial migration. The activation energy for these diffusion mechanisms, however, will always be higher as compared with Fe–Ti–O where an oxygen diffusion path may readily exist.

Hydrogen ion diffusivity also could contribute to the reaction rates. Steinsvik et al.[50] studied the conductivity of hydrogen ions through defect containing crystal lattices of the Sr–Fe–Ti–O system and found that hydrogen ionic conductivity was an order of magnitude smaller than oxygen ion conductivity at temperatures above 700°C. Similar values were reported when steam was present. The contribution of hydrogen ion diffusion was found to increase with a decrease in temperature, reaching almost 50% of the total ionic conduction at 500°C.

Sakata et al.[51,52] studied the reduction oxidation mechanism of pure Fe_2O_3 at 300°C with the oxygen capacity varying between 98.9% and 100%. Using oxygen isotopes, they determined that bulk oxygen did not take part in the reduction and oxidation reactions, reflecting the low diffusivity of the oxygen ions to the surface. A mechanism involving diffusion of Fe ions, instead of oxygen ions, was proposed. The relative diffusivities of iron ions and oxygen ions are not expected to change at higher temperatures because iron ions are much smaller than oxygen ions. At higher temperatures, the iron will diffuse and sinter, leading to decreased reactivity over redox cycles. Urasaki et al.[53] reported cyclic redox experiments conducted on Fe_2O_3 doped with 0–0.4% ZrO_2. The oxygen capacity can be varied by 30% to 50% at 450°C when the H_2/He and H_2O/He gas mixtures were used. The presence of ZrO_2 led to improved oxidation rates compared with pure Fe_2O_3 and helped stabilize the surface area of the particle. The oxidation rates purely depend on the surface area, which decreases over several cycles, leading to decreased reactivity. Similar results were reported when Pd was the dopant.[39] The studies indicate that a stable pore structure is important for the recyclability of pure Fe_2O_3 at lower temperatures. This behavior can be explained by the low ionic diffusivity of Fe_2O_3 at low temperatures. At high temperatures, however, the enhancement of oxygen ion diffusion becomes an important factor for consideration in the development of optimal oxygen carrier particles for chemical looping processes.

2.2.11 Effect of Reactor Design and Gas–Solid Contact Modes

The optimal design of the reactors for chemical looping process applications requires consideration of such factors as reaction thermodynamics and kinetics, transport phenomena, and multiphase flows. The fundamental modes of gas–solid contact in reactor systems typically can be characterized by those represented by fluidized beds, operated at the bubbling, turbulent or fast fluidized-bed regime, and moving beds, operated at cocurrent or countercurrent flow conditions. Note that it is possible to approximate the moving bed mode by a series of fluidized beds. Two key factors that dictate the selection of gas–solid reactor are the type of metal oxide carriers employed for the looping operation and the type of products to be produced.

In chemical looping processes, it is essential to maximize the extent of the fuel conversion in the reactor. To illustrate the interplay among factors that govern the optimum process operation, two types of oxygen carriers are considered, that is, NiO and Fe_2O_3. With NiO as the looping particle in the reducer, the fuel and NiO conversions at their maximum extent or at the fuel gas–Ni/NiO equilibrium condition can be achieved with little effect due to the types of multiphase flow reactor used. This is because the oxygen carrier, NiO, has only two oxidation states. On the other hand, with Fe_2O_3 as the looping particle in the reducer, the extent of both the fuel and looping particle conversions would differ depending on the types of multiphase reactors used. This is because the oxygen carrier, Fe_2O_3, has more than two oxidation states as shown in Figure 2.6. When the fluidized bed is used as a reducer, the partial pressure ratio of CO_2 to CO is expected to be high, giving rise to a reduced product for Fe_2O_3 to be in the form of Fe_3O_4 as indicated in Figure 2.32. Thus, using a fluidized bed, the extent of the conversion of Fe_2O_3 in the reduction

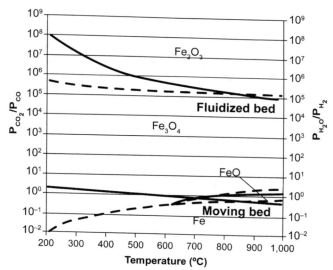

Figure 2.32. Thermodynamics of the $Fe_2O_3 + CO/H_2$ reaction in fluidized bed and countercurrent moving bed reactors.

reaction is to be low. In contrast, when a countercurrent moving bed reactor is used, the extent of the conversion of Fe_2O_3 for the reduction reaction is to be high as a result of a low partial pressure ratio of CO_2 to CO, yielding a reduced product for Fe_2O_3 to be in the form of Fe/FeO as indicated in Figure 2.32. Thus, to achieve a given extent of the fuel and Fe_2O_3 conversions in a reducer, a significantly less amount of Fe_2O_3 particles will be needed using a moving bed reactor than using a fluidized bed.

Fluidized beds have been used extensively in the industry for catalytic or noncatalytic reactor operations because of their excellent transport properties. The fluidized bed is desirable for use as an oxidizer/combustor in chemical looping combustion processes, where $Fe/FeO/Fe_3O_4$ or Ni is combusted in air to form Fe_2O_3 or NiO in the particle regeneration or heat generation process. Fluidized bed-based chemical looping combustion systems are described in detail in Chapter 3. For chemical looping gasification processes that produce hydrogen with steam, the oxidizer design requires consideration of the thermodynamics, kinetics, and metal oxide properties. The desired metal oxide particles for hydrogen production are Fe/FeO because as discussed in Section 2.2.3, Ni has a low steam-to-hydrogen conversion rate and poor reaction kinetics. The steam reaction with Fe/FeO from the reducer can take place in the countercurrent moving bed oxidizer to produce Fe_3O_4 with much higher steam-to-hydrogen conversion compared with that in a fluidized bed oxidizer. An entrained flow combustor can further oxidize Fe_3O_4 to Fe_2O_3.

To summarize, the development of an ideal looping medium requires thorough consideration of such intertwining factors as the thermodynamic properties of the particle, reaction kinetics, cost, recyclability, physical strength, ease in heat integration, resistance to contaminants, and environmental and health effects. For metal oxide particles with more than two oxidation states such as Fe_2O_3, a countercurrent fuel-particle contact mode conducted in a moving bed can achieve a maximum metal oxide reduction conversion, which is not practically achievable in a fluidized bed. For a given countercurrent moving bed design, the P_{CO_2}/P_{CO} and P_{H_2O}/P_{H_2} ratios, along with the thermodynamic phase diagrams, can be used to predict the fuel gas conversion and the oxidation state of the solid product. For metal oxides with two oxidation states such as NiO, the fuel–metal oxide contact mode yields little effect on the maximum fuel conversion reflected by fuel gas–metal/metal oxide equilibrium condition. Continued discussion on the relationship between fuel and particle conversions and gas–solid contact mode is given in Chapters 4 and 5. Although there is no omniparticle that possesses all the desired properties, as described earlier, and can be used for all looping process applications, a proper selection of a suitable primary metal oxide material, as well as its support and doping agent based on the type of the fuel to be converted and the intended product to be generated, is the key to the ultimate development of a desired looping particle. Moreover, optimized reactor design is required in order to render the process operation economically feasible. The following section illustrates the primary metal selection for one specific application.

2.2.12 Selection of Primary Metal for Chemical Looping Combustion of Coal

This section illustrates an example of the selection of a suitable primary metal oxide for chemical looping combustion applications using coal as feedstock. Table 2.8 presents the chemical and physical properties of several possible choices of metal oxides.

Among the choices of metal oxides in Table 2.8, iron is available at a low cost and has a high oxygen-carrying capacity, favorable thermodynamics, a high melting point, and good strength. Furthermore, iron induces low health and environmental effects. The reactivity of iron is, however, relatively low. NiO is often referred to as an ideal oxygen carrier as it reacts faster with CO or H_2[57] when compared with Fe_2O_3. Note that the reaction rate between metal oxides and coal char is controlled by char gasification,[58] and that the solid–solid reaction between coal char and metal oxide is slow. The oxygen carrier and the char conversion are mainly through the following reactions in the presence of CO_2 or H_2O:

$$H_2O/CO_2 + C \rightarrow CO + H_2/CO \qquad (2.2.23)$$

$$MeO + H_2/CO \rightarrow Me + H_2O/CO_2 \qquad (2.2.24)$$

The overall reaction of Reactions (2.2.23) and (2.2.24) is:

$$2MeO + C \rightarrow 2Me + CO_2 \qquad (2.2.25)$$

Therefore, CO_2 or H_2O acts as an enhancer that serves to improve the solid–solid reaction rate. For metal oxides such as NiO and Fe_2O_3, Reaction (2.2.24)

TABLE 2.8. Comparisons of the Key Properties of Different Metal Oxide Candidates[3,54–56]

	Fe_2O_3	NiO	CuO	CoO
Cost	+[a]	−	−	−
Oxygen capacity[b] (wt%)	30	21	20	21
Thermodynamics[c]	+	+	+	+
Kinetics/reactivity[d]	−	+	+	−
Melting points	+	~	−	+
Strength	+	−	~	~
Environmental and health effects	~	−	−	−

[a]+, positive; −, negative; ~, neutral.
[b]Maximum-possible oxygen-carrying capacity by weight.
[c]Capability to fully oxidize C, CO, and H_2 to CO_2 and H_2O, according to thermodynamic principles.
[d]Reactivity refers to the rates of the reactions between metal oxides and syngas (CO and H_2).

TABLE 2.9. Cost for Making Oxygen-Carrier Particles

Particle Type	OSU Iron-Based Composite Particle				NiO Based Particle
Raw Material[a]	Iron Ore	Alumina	Rutile	Silicon Carbide	Nickel Oxide
Cost ($/ton)	100	350	500	850	10,990
Raw Material Cost for Particles ($/ton)[b]			460		~10,990
Processing Cost ($/ton)[b]			40		n/a
Fresh Particle (Pellet) Cost ($/ton)			500		~10,990

[a]Market price as of January 2007. Costs of iron, alumina, rutile, and silicon carbide ores are from *Industrial Minerals*,[59] cost of NiO is obtained from Brazil Department of Mineral Production website.[60]
[b]A safety factor of 2 is given here for the raw material and processing cost estimated based on the energy consumption for processing. Electricity is assumed to be 5 cent/kWh.

is much faster than Reaction (2.2.23). Therefore, in the presence of CO_2/H_2O, the reaction rate between NiO and coal char is similar to that between Fe_2O_3 and coal char, although NiO reacts with CO/H_2 faster than does Fe_2O_3. Thus, the reactivity advantage of NiO with syngas is "silenced" when using coal in the chemical looping combustion system. Table 2.9 further compares the estimated cost for producing Fe-based particles and Ni-based particles. From the table, it is shown that the iron-based oxygen carrier particle is more economical than the nickel-based oxygen carriers. Therefore, iron is a more favorable primary metal for chemical looping combustion of coal.

Iron-oxide-based formulations have been tested for reaction with coal, with high conversions observed. Kodama et al.[61] studied reduction reactions of carbon with various metal ferrites to produce CO and then oxidation reactions of reduced metal ferrite with steam to produce hydrogen. They reported the highest reactivity and selectivity for Ni(II)-ferrite both in reduction at temperatures above 700°C and in oxidation above 500°C. Kodama et al.[7] also studied similar reactions with coal and found In(III)-ferrite to be the fastest metal ferrite. They subsequently suggested that CO_2 could be used to oxidize the reduced metal ferrites to form CO, instead of steam, to form hydrogen; In(III)-ferrite was again found to be the best performing material. Coal conversions of 60%–90% were reported; however, the oxygen-carrying capacity diminished during several cycles as a result of ash layer formation. L'vov[62] proposed a different mechanism from Reactions (2.2.23) and (2.2.24) for the carbothermal reduction of oxides of Ni, Fe, Co and Cu. L'vov[62] proposed that at high temperatures, the reaction between metal oxides and carbon is carried out by thermal decomposition of metal oxides to metal and oxygen followed by the reaction between carbon and oxygen forming CO_2 as given below:

$$2MeO \rightarrow 2Me + O_2 \quad (2.2.26)$$

$$C + O_2 \rightarrow CO_2 \qquad (2.2.27)$$

L'vov[62] further stated that carbon serves as a "buffer" in this case to maintain a low partial pressure of oxygen, which favors the decomposition or reduction of metal oxide. This reaction scheme is noted as the chemical looping oxygen uncoupling, which is discussed in Chapter 6.

2.3   Type II Chemical Looping System

2.3.1   Types of Metal Oxide

The fundamental concept underlying the Type II chemical looping system lies in the employment of metal oxide (MO) sorbents for *in situ* removal of CO_2 generated from the reactive process at high temperatures, yielding a reaction product of metal carbonate (MCO_3). The carbonation reaction can be given as

$$MO(s) + CO_2(g) \rightarrow MCO_3(s) \qquad (2.3.1)$$

The metal carbonate (MCO_3) thus formed can then be calcined to regenerate the metal oxide. The regeneration is achieved by heating the carbonate beyond the corresponding calcination or decomposition temperature. The calcination reaction is given as

$$MCO_3(s) \rightarrow MO(s) + CO_2(g) \qquad (2.3.2)$$

The key to the Type II looping process is to identify available MO-based sorbents that are cost-effective and can be used directly or engineered to arrive at desired morphological, reaction, and regeneration properties with respect to the CO_2 capture. The process that uses such high-efficiency MO-based sorbents provides the ultimate (wt%) sorption capacity of the metal oxide sorbent (quantified by kg CO_2/kg of the sorbent) and kinetics of the CO_2 sorption process. Higher sorption capacity and reactivity allow lower sorbent requirements and handling as well as lower regeneration costs leading to lower capital and operating costs. Efficient integration of this process for gasification and combustion systems mandate high operating temperatures of the sorbents in the range of 500–900°C. After regeneration, it is desirable that a pure stream of CO_2 be obtained from the sorbent calcination process. CO_2 purity plays an important role in CO_2 transportation, sequestration, and other commercial applications.

Many metal oxides undergo carbonation and calcination reactions. However, many of these metal carbonates are thermally stable and would impose huge energy penalties in the regeneration cycle. Very few metal carbonates decompose between 200°C and 900°C. Some viable metal oxide candidates include oxides/hydroxides of calcium, magnesium, zinc, copper and manganese. They react with CO_2 to form their respective carbonates. The calcination temperatures of carbonates so formed including $CaCO_3 \approx 890°C$, $MgCO_3 \approx 385°C$, $ZnCO_3 \approx 340°C$, and $MnCO_3 \approx 440°C$ are within the temperature range of interest.

Aqueous-phase MgO carbonation has been studied as a method of CO_2 sequestration,[63] but the carbonation/calcination reaction cycle for MgO has not been looked into for the purpose of CO_2 separation. Extensive literature exists on the dehydration of $Mg(OH)_2$, but carbonation studies of MgO and $Mg(OH)_2$ are rare. Butt et al.[64] conducted the carbonation of $Mg(OH)_2$ to investigate its sequestration properties. They obtained ~10% carbonate formation between 387°C and 400°C and 6% carbonate formation between 475°C and 500°C. They attributed the low conversion to the formation of a non-porous carbonate product layer from the carbonation reaction. This layer hinders the inward diffusion of CO_2 and the outward diffusion of H_2O, leading to a low conversion. They fitted the data to a contracting-sphere model and obtained an activation energy of 72 kcal/mol.

Carbonation reactions of ZnO, CuO and MnO_2 have been reported in only a few studies. Sawada et al.[65] observed a very low conversion of ZnO to $ZnCO_3$ at room temerature, but high-temperature carbonation studies have yet to be done. Okuma et al.[66] noticed a beneficial effect of moisture on the carbonation of ZnO. Shaheen and Selim[67] conducted thermal stability studies on basic $CuCO_3$ and found the decomposition temperature of $CuCO_3$ to be 225°C. They observed that $MnCO_3$ undergoes a more complex thermal degradation phenomenon. $MnCO_3$ first decomposes to MnO_2 at 300°C, which in turn changes to Mn_2O_3 at 440°C. At higher temperatures (~900°C), the final thermal decomposition product was found to be Mn_3O_4. Different oxides of manganese provide the flexibility of investigating the carbonation/calcination reaction over a wider range of temperatures.

The initial screening of the metal oxides based on their reactivity toward the carbonation reaction presents CaO as a viable candidate. The carbonation reaction of CaO and the calcination reaction of $CaCO_3$ can be, respectively, given by Equations (2.3.3) and (2.3.4) as

$$CaO + CO_2 \rightarrow CaCO_3 \qquad (2.3.3)$$

$$CaCO_3 \rightarrow CaO + CO_2 \qquad (2.3.4)$$

Figure 2.33 shows the variation in the free energy of the carbonation reaction as a function of temperature. From the figure, it can be shown that the carbonation reaction is thermodynamically favored more with a decrease in temperature. At lower temperatures, however, the kinetics of carbonation is slow. In fact, it takes geological time scales for the formation of $CaCO_3$ by the reaction between CaO and atmospheric CO_2 (at 280–360 ppm) at ambient temperatures. It also can be noted, however, that the carbonation reaction would be

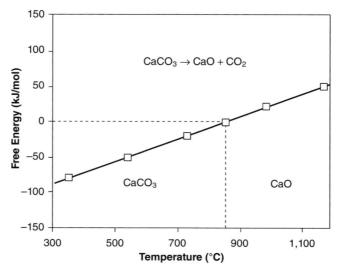

Figure 2.33. Variation in the free energy of the carbonation reaction as a function of temperature.

favored as long as the free energy is negative. This creates an upper limit of 890°C for carbonation to occur under a CO_2 partial pressure of 1 atm as observed in Figure 2.33. Theoretically, carbonation kinetics would be fastest just below the calcination temperature.

Calcium oxide has been most widely studied for its carbonation reaction. Bhatia and Perlmutter[68] studied the carbonation of medium surface area CaO (6–10 m²/g) at 400–725°C. The carbon dioxide concentration was varied between 10% and 50%. They obtained a conversion of 65% in under a minute for 81-μm particles. They found that the reaction is initially rapid (and chemically controlled) and goes through a sudden transition to a much slower regime (controlled by diffusion in the product $CaCO_3$ layer). The magnitude of the established product layer diffusivity is from 10–18 to 10–21 m²/s, and the corresponding activation energy was estimated to be 88.9 ± 3.7 kJ/mol below 400 K and 179.2 ± 7.0 kJ/mol above that temperature. The latter activation energy corresponds to a solid state diffusion controlled reaction. Barker[69,70] studied this carbonation/calcination cycle of CaO to exploit its potential for energy storage. He found that there was a decrease in surface area of CaO in the first 10 cycles. This was attributed to the loss of porosity of CaO and the sintering of $CaCO_3$. In the first cycle, the reaction occurs to a greater extent on the surface of the CaO particle compared with the conversion occurring in the interior. $CaCO_3$, being higher in molar volume, leads to the plugging of pores. Moreover, $CaCO_3$ sinters to a higher extent, exacerbating the problem.[71–73] As the number of cycles increases, the interior of the CaO particle becomes less likely to be available for carbonation. The carbonation reaction of interest also includes that of $Ca(OH)_2$ as given by

$$Ca(OH)_2 + CO_2 \rightarrow CaCO_3 + H_2O \tag{2.3.5}$$

MgO carbonation has been studied as a method of carbon dioxide sequestration,[63] but little work has been performed with respect to the carbonation/calcination reaction cycle for MgO for the purpose of CO_2 separation. The reactions of interest are

$$MgO + CO_2 \rightarrow MgCO_3 \tag{2.3.6}$$

$$Mg(OH)_2 + CO_2 \rightarrow MgCO_3 + H_2O \tag{2.3.7}$$

$$MgCO_3 \rightarrow MgO + CO_2 \tag{2.3.8}$$

Figure 2.34 shows the heat of reaction for different metal oxide/carbonate sorbents with increasing decomposition or calcination temperatures. Thus, it can be observed that most metal carbonates like those of zinc, manganese and magnesium decompose at temperatures lower than ~300–400°C. These decomposition temperatures indicate that they form carbonates below these temperatures. A low temperature for CO_2 capture is characterized by slow kinetics, and the exothermic energy from carbonation is of low-grade heat, which is not of high value within the process. However, carbonates of metals such as lithium and strontium decompose at temperatures above 1,300°C. This

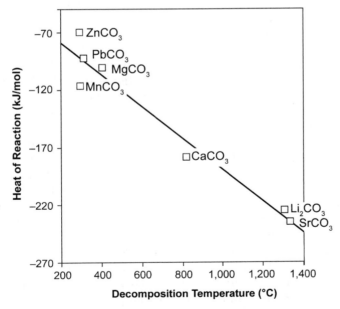

Figure 2.34. Comparison of the heat of reaction and decomposition temperature of various sorbents.

decomposition temperature mandates a severe calcination temperature, but typical gasification and combustion processes require operating temperature windows in the range of ~500–900°C. Thus, it is evident from Figure 2.34 that only calcium carbonate, which decomposes at 890°C, satisfies these operating conditions.

2.3.2 Thermodynamics and Phase Equilibrium of Metal Oxide and Metal Carbonate

Primarily, four important gas–solid reactions can occur when calcium oxide (CaO) is exposed to a fuel gas mixture obtained from coal gasification. CaO can undergo hydration, carbonation, sulfidation, and chlorination reactions with H_2O, CO_2, H_2S, and COS, and HCl, respectively. These can be stoichiometrically represented as

$$Hydration: \quad CaO + H_2O \rightarrow Ca(OH)_2 \quad\quad (2.3.9)$$

$$Carbonation: \quad CaO + CO_2 \rightarrow CaCO_3 \quad\quad (2.3.10)$$

$$Sulfur\ (H_2S)\ capture: \quad CaO + H_2S \rightarrow CaS + H_2O \quad\quad (2.3.11)$$

$$Sulfur\ (COS)\ capture: \quad CaO + COS \rightarrow CaS + CO_2 \quad\quad (2.3.12)$$

$$Halide\ (HCl)\ capture: \quad CaO + 2HCl \rightarrow CaCl_2 + H_2O \quad\quad (2.3.13)$$

All of these reactions are reversible, and the extent of each of these reactions depends on the concentration of the respective gas species and the reaction temperature. Figures 2.35–2.38 show detailed thermodynamic calculations to obtain equilibrium curves for the partial pressures of H_2O (P_{H_2O}), CO_2 (P_{CO_2}), H_2S (P_{H_2S}) and HCl (P_{HCl}) as a function of temperature for the hydration, carbonation, sulfidation, and chlorination reactions using HSC Chemistry v 5.0 (Outokumpu Research Oy, Finland). The equilibrium calculations are based on the fuel gas compositions that are typical of the different types of coal gasifiers as given in Table 2.10.

The relationship between the reaction temperatures and the equilibrium partial pressures of H_2O and CO_2 for the hydration and carbonation reactions are shown in Figure 2.35. From the data in Table 2.10, it can be inferred that the typical P_{CO_2} in the gasifiers ranges from 0.4 to 4.3 atm for entrained flow (dry) and entrained flow (slurry) gasifier systems, respectively. The equilibrium temperatures corresponding to those P_{CO_2} are 830–1,000°C, as shown in Figure 2.35. Thus, by operating below these temperatures, the condition of carbonation of CaO can be altered. Upon the addition of steam to the syngas such that the steam:CO ratio is 2:1, the moisture composition in the syngas ranges from 13.7 to 20.8 atm (P_{H_2O}) for entrained flow (dry) and entrained flow (slurry) gasifier systems, respectively. For these partial pressures of steam, the hydration of CaO occurs at all temperatures below 550–575°C, respectively. By

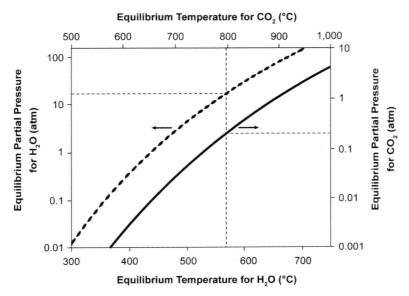

Figure 2.35. Thermodynamic data for predicting the temperature zones for hydration and carbonation of CaO in a fuel gas mixture. The dashed curve represents the equilibrium between CaO and Ca(OH)$_2$ relative to the partial pressure of H$_2$O indicated in the left axis. The solid curve represents the equilibrium between CaO and CaCO$_3$ relative to the partial pressure of CO$_2$, indicated in the right axis. ($-P_{CO_2}$, - - - P_{H_2O}).

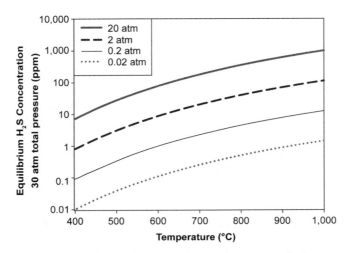

Figure 2.36. Thermodynamic data for predicting the equilibrium H$_2$S concentration for CaO sulfidation with varying steam concentration. From bottom to top curve: 0.02 atm, 0.2 atm, 2 atm, and 20 atm steam; total pressure is 30 atm.

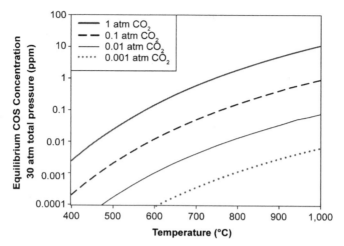

Figure 2.37. Thermodynamic data for predicting the equilibrium COS carbonyl sulfide concentration for CaO sulfidation with varying CO_2 concentration. From bottom to top curve: 0.001 atm, 0.01 atm, 0.1 atm, 1 atm CO_2; total pressure is 30 atm.

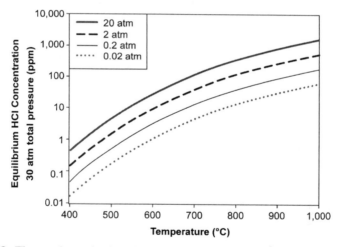

Figure 2.38. Thermodynamic data for predicting the equilibrium HCl concentration for CaO reaction with HCl with varying steam concentration. From bottom to top curve: 0.02 atm, 0.2 atm, 2 atm, 20 atm steam; total pressure is 30 atm.

operating above these temperatures, CaO-hydration can be prevented in the reactor where CO_2 is to be removed.

For the reversible sulfidation of CaO, the thermodynamic calculations depend on the concentration of moisture in the system. The thermodynamics of the sulfidation of $CaCO_3$ given below is not favorable for H_2S removal as compared with CaO:

TABLE 2.10. Typical Fuel Gas Compositions Obtained from Different Gasifiers[74]

	Moving Bed, dry	Moving Bed slagging	Fluidized Bed	Entrained Flow, slurry	Entrained Flow, dry
Oxidant	Air	Oxygen	Oxygen	Oxygen	Oxygen
Fuel	Subbituminous	Bituminous	Lignite	Bituminous	Bituminous
Pressure (psi)	295	465	145	615	365
CO	17.4	46	48.2	41	60.3
H_2	23.3	26.4	30.6	29.8	30
CO_2	14.8	2.9	8.2	10.2	1.6
H_2O	...*	16.3	9.1	17.1	2
N_2	38.5	2.8	0.7	0.8	4.7
CH_4 + HCs	5.8	4.2	2.8	0.3	—
H_2S + COS	0.2	1.1	0.4	1.1	1.3

*The composition presented is on a dry basis.

Sulfidation of $CaCO_3$: $$CaCO_3 + H_2S \rightarrow CaS + H_2O + CO_2 \qquad (2.3.14)$$

Figure 2.36 depicts the equilibrium H_2S concentrations in ppm for varying moisture concentrations (P_{H_2O}) at a total pressure of 30 atm. In the calcium looping process, the integrated water–gas shift carbonation–reactor system is operated at the near-stoichiometric steam requirement, resulting in low concentrations of steam in the reactor system. In addition, the CO_2 concentration also will be minimal because of the continuous removal of the CO_2 product by carbonation. Thus, the reactor system will favor H_2S removal using CaO at ~600–700°C. Thus, for a steam concentration of ~0.2 atm at 600°C, the equilibrium H_2S concentration corresponds to ~1 ppm, as shown in Figure 2.36. Thus, the reactor system can achieve CO_2 as well as H_2S removal while producing a pure H_2 stream. However, the typical syngas composition, with steam addition for the shift reaction, will enable H_2S removal to only 100–300 ppm. Similarly, the concepts of COS capture and HCl capture by calcium oxide in a gas mixture with minimal CO_2 and steam can be explained by Figures 2.37 and 2.38.

2.3.3 Reaction Characteristics of Ca-Based Sorbents for CO_2 Capture

The concept of utilizing lime for carbon dioxide capture has existed for well over a century. It was first introduced by DuMotay and Marechal in 1869 for enhancing the gasification of coal[75] and followed by CONSOL's CO_2 Acceptor process[76] a century later when this concept was tested in a 40-t/d plant. A variation of this process, the HyPrRing Process,[77,78] was developed in Japan for the production of hydrogen at high pressures. Gupta and Fan,[79] Iyer et al.,[80] Sun et al.,[81,82] Abanades et al.,[83,84] and Manovic et al.[85] have applied this concept to the removal of CO_2 from combustion flue gas. Fan et al.,[86] Ortiz and Harrison,[87] Han and Harrison,[88] Johnsen et al.,[89] and Balasubramanian et al.[90] have applied this concept to the removal of CO_2 and the production of hydrogen from syngas through the water–gas shift reaction and from methane through the sorption-enhanced steam methane reforming reaction.

The reaction between CaO and CO_2 occurs in two distinct stages. The first stage occurs rapidly and is kinetically controlled, while the second stage is slower and diffusion controlled. For any commercial applications, only the first stage of the reaction should be considered in order to use a compact reactor for the removal of CO_2. Abanades and Alvarez[71] studied the rate and the extent of the carbonation reaction and the variation of these parameters with multiple carbonation and calcination cycles. The calcium oxide conversion at the end of the rapid kinetically controlled regime is found to decay sharply for naturally occurring limestone with an increase in the number of cycles. Although the initial decay is smoother for dolomite and other modified sorbents, it is intrinsic to most sorbents used in the carbonation–calcination looping process. In addition to the decay in CO_2 capture capacity, dolomite and other supported sorbents also have the disadvantage of carrying more inert material in the loop, thereby increasing the parasitic energy requirement of the regeneration process. Because the cost of the supported and modified sorbents is also higher, their performance over multiple cycles needs to be significantly higher in order to maintain the economics of the process relative to natural limestone or other technologies such as the conventional amine solution for CO_2 capture. The decay in lime conversion over multiple cycles has been reported by numerous researchers including Curran et al.,[76] Shimizu et al.,[91] Silaban and Harrison,[92] Barker,[69] and Aihara et al.[93] Using these data, Abanades and Alvarez[71] concluded that the decay in conversion is dependent only on the number of cycles and independent of the reaction times and conditions. Using a simple relationship given in Equation (2.3.15), Abanades[94] related the conversion of lime for any given cycle number ($x_{c,N}$) to fitted constants (f, b) and the cycle number (N) as given by

$$x_{c,N} = f^{N+1} + b \qquad (2.3.15)$$

where the fitted parameters f and b have numerical values of 0.782 and 0.184, respectively. Taking into consideration the sorbent conversion decay over multiple cycles, the kinetics of the reaction, and mass and energy flows, Abanades[94] developed the following equation to determine the maximum capture efficiency of CO_2 in a system containing a continuous purge of solids and a make-up of fresh sorbent:

$$E_{CO_2} = \left[\frac{1 + \left(F_0 / F_R \right)}{\left(F_0 / F_R \right) + \left(F_{CO_2} / F_R \right)} \right] \left[\frac{f \left(F_0 / F_R \right)}{\left(F_0 / F_R \right) + 1 - f} + b \right] \qquad (2.3.16)$$

where E_{CO_2} is the maximum obtainable efficiency in terms of the amount of CO_2 captured from the gas stream; F_0 is the fresh feed added to the system (mol CaO/s); F_R is the total amount of sorbent required to react with the CO_2 in the system (mol CaO/s); F_{CO_2} is the flow of CO_2 (mol/s); and f and b are constants as defined in Equation (2.3.15). b is the residual carbonation conversion caused by the formation of a product layer of carbonate inside the macropores in highly sintered sorbents. This residual carbonation of the lime sorbent is beneficial as it aids in reducing the amount of fresh sorbent to be added. From an economic standpoint, it is desirable to minimize the ratios F_R/F_{CO_2} and F_0/F_R in order to minimize the energy required for calcination and the amount of fresh sorbent required.[94] For F_0 and F_R to be low, the sorbent should have a high resistance to sintering.

The $CaCO_3$ product layer formation and pore pluggage during carbonation and the sintering of CaO during calcination are both attributed to the decay and irreversibility of limestone. Abanades[94] concluded that micropores contribute to the fast stage of the carbonation reaction. The fast reaction stage ceases when the micropores connecting the crystal grains are plugged because of the increase in the molar volume during the formation of $CaCO_3$ from CaO, where $CaCO_3$ has greater than twice the molar volume of CaO. In the larger pores (mesopores and macropores), $CaCO_3$ forms a layer on the CaO wall.[95] Although the pore is sufficiently large to handle the increase in the molar volume during the formation of $CaCO_3$ from CaO, the resistance of CO_2 diffusion through the $CaCO_3$ layer dramatically increases. The increased resistance forms the boundary between the two stages of carbonation. Sintering of CaO during calcination over multiple cycles results in grain growth, which drastically reduces the CaO microporosity while increasing the mesoporosity. This leads to a reduced fast carbonation reaction zone and, therefore a decrease in CO_2 capture capacity over multiple cycles.[71]

Sun et al.[72] also investigated the sintering mechanism of limestone with an increasing number of cycles and attributed the sintering to the CO_2 released during the calcination process. They showed that the increase in the carbonation time did not have any effect on the structure of the calcine as the calcination process eliminates the changes caused by carbonation. An increase in the calcination time, however, resulted in a decrease in the pore volume for pores of size <220 nm.[72] Similar to the observation made by Abanades,[94] with the increase in the number of cycles, the pore volume decreased for pores <220 nm, and consequently increased for pores >220 nm. A sintering model has been developed by Sun et al.[72] based on the packed-bed model, the shrinking core model, and a modified sintering-kinetic model; and the average CO_2 conversion is given below:

$$X_{carb} = 1.07(n+1)^{-0.49} \qquad (2.3.17)$$

To be commercially viable, the CaO sorbent must maintain its reactivity toward CO_2 over multiple cycles. Additives and processed sorbents have been investigated, but these techniques undermine the main advantage of using natural limestone, which is its low cost. Using natural limestone also has its challenges including low reactivity and recyclability, which must be overcome.

The effect of doping calcium oxide with NaCl and Na_2CO_3 has been investigated in a TGA.[96] The addition of NaCl increased the CO_2 removal capacity of the sorbent to 40% across 13 cycles because of favorable changes in the pore structure and surface area of the sorbent, whereas the addition of Na_2CO_3 did not have any effect on the extent of carbonation. When the doped sorbents were tested in the fluidized bed, both NaCl and Na_2CO_3 caused a decrease in the CO_2 removal capacity of the calcium oxide sorbent, that might be attributed to the coating of the surface of the sorbent leading to pore blockage during the calcination stage.[96] The doping of calcium oxide with alkali metals has shown to improve the performance in the order Li < Na < K < Rb < Cs resulting from an increase in electropositivity of the alkali metal. A Cs doped CaO sorbent with 20% Cs/CaO was found to have a high sorption capacity of 50 g of CO_2/g of the sorbent.[97] Metal oxide supports like Al_2O_3, MgO, and $CaO/Ca_{12}Al_{14}O_{33}$ also have shown to improve the regenerability of the calcium oxide sorbent. But under realistic calcination conditions, the capture capacity still decreases with the number of cycles.[98–100]

2.3.4 Synthesis of the High-Reactivity PCC–CaO Sorbent

The PCC–CaO sorbent can achieve almost complete conversions (>95%) unlike those observed by Harrison and coworkers[92] for dolomite at ~50% conversion mainly from conversion of calcium that is present in dolomite. PCC refers to precipitated calcium carbonate. In addition, PCC–CaO has a CO_2 capture capacity as high as 70% by weight of the sorbent, in contrast to dolomite, which, at a reaction temperature of 600°C, has a CO_2 capture capacity of ~50% by weight of sorbent. The high CO_2 capture rate by the mesoporous PCC–CaO sorbent would ensure minimal sorbent usage and possibly smaller reactors. The PCC–$CaCO_3$ precursor[101] for PCC–CaO also has been identified as a suitable sorbent for hydrogen production with *in situ* CO_2 capture. Highly reactive PCC is obtained by bubbling CO_2 gas in a $Ca(OH)_2$ slurry. The surface properties of this novel calcium sorbent are tailored by using anionic surfactants, positively charged precipitated $CaCO_3$ particles with a positive zeta potential.[102–105] The system reaches an optimum only when the zeta potential equals zero, depicting the maximum in the surface area as shown in Figure 2.39. The sorbent optimization gives rise to sorbents of a surface area of 60 m^2/g and a pore volume of 0.18 cm^3/g.

The structurally altered PCC has a unique mesoporous structure (5–30 nm) with a maximum pore size distribution occurring at 15 nm. In contrast, the pores of the naturally occurring or commercial calcium minerals were predominantly microporous (<5 nm), as shown in Figure 2.40. The other CaO precursors are Linwood calcium carbonate (LC) and dolomite (DL). The

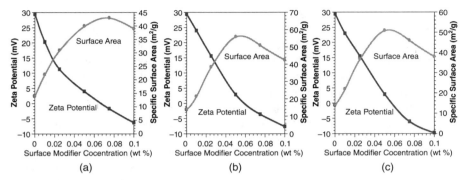

Figure 2.39. Effect of concentrations of different surface modifiers on the zeta potential and specific surface area of the sorbent (a) ligno-sulfonate, (b) Dispex N40V, and (c) Dispex A40. Adapted from Agnihotri et al.[102]

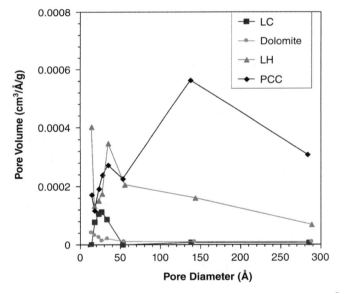

Figure 2.40. Pore size distributions for various calcium sorbents.[79]

mesoporous pores make the sorbent less susceptible to pore pluggage and filling, a phenomenon observed because of the presence of microspores, as observed by Silaban and Harrison.[92] This structure alteration now leads to almost 100% sorbent conversions. CaO obtained from PCC was found to have extraordinarily high reactivity toward SO_2, H_2S, and CO_2,[103,104,106] giving very high conversion.

Calcium carbonate primarily occurs in three different polymorphs, each of which may have multiple morphologies depending on the arrangement of the atoms and ions in the crystal structure. These polymorphs are all present in nature, as well as in synthesized PCC, and can be classified as calcite, aragonite, and vaterite. Calcite is the most stable polymorph and typically occurs in the triagonal-rhombohedral (acute to obtuse), scalenohedral, tabular, and prismatic morphologies. Calcite crystals also display intergrowth or twinning to form fibrous, granular, lamellar, and compact structures. The rhombohedral and prismatic forms find applications in paper coating and in polymer strength enhancing agents, whereas the scalenohedral form is used in paper filling because of its light scattering ability.[107] Calcite exhibits a unique property such that its solubility in water decreases with increasing temperature. The aragonite polymorph has an orthorhombic morphology with needle-shaped or acicular crystals. Twinning of these crystals results in the formation of pseudohexagonal structures that could be in a columnar or fibrous matrix. Aragonite is unstable at standard temperatures and pressures and is eventually converted to calcite over geological timescales. Aragonite also exhibits a higher density and solubility than calcite. The needle-shaped morphology of aragonite is beneficial for high-gloss paper-coating applications as well as for strength-enhancing additives in polymeric materials. Vaterite is the most unstable form of calcium carbonate at ambient conditions and is readily converted to calcite (at lower temperatures) and aragonite (at higher temperatures of 60°C) on exposure to water. Vaterite is usually spherical in shape and has a higher solubility in water than the other polymorphs. The transformation of aragonite and vaterite to calcite is accelerated with temperature.[108]

Although PCC predominantly contains calcite, various factors in the synthesis procedure such as the extent of saturation of the calcium hydroxide solution, pH of the solution, and concentration of CO_2 dictate the type and size of its morphology. For example, PCC synthesized from highly saturated aqueous calcium hydroxide solutions contains aragonite at 70°C and vaterite at 30°C.[109] Cizer et al.[110] have shown that rhombohedral calcite crystals formed by the exposure of calcium hydroxide to 100% CO_2 are micrometer-sized, whereas those precipitated with 20% CO_2 are submicrometer-sized. In addition, it also was found that, during the initial stages of carbonation, when the concentration of Ca^{2+} ions in the solution is greater than the concentration of CO_3^{2-} ions, a scalenohedral calcite is precipitated. The scalenohedral morphology is transformed into the rhombohedral form during the later stages of precipitation, when the CO_3^{2-} concentration in the solution is high.[110]

2.3.5 Reactivity of Calcium Sorbents

Fan and Jadhav[111] conducted a pilot demonstration of the OSCAR process for SO_2 removal from the flue gas stream in coal combustion using PCC sorbent. The performance of CaO obtained from different precursors such as PCC, LH, LC, and dolomite, for the carbonation reaction in a pure CO_2 stream is com-

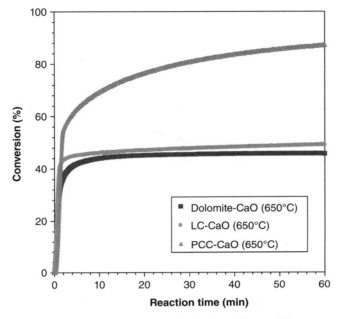

Figure 2.41. Carbonation reactions of CaO obtained from different precursors at 650°C. Adapted from Gupta and Fan.[104]

pared, as given in Figure 2.41. These experiments indicate that the reactivity of PCC–CaO is higher than those of other calcium-based sorbents.[104] It can be shown that the reaction has initially fast kinetics, followed by slow kinetics. Unlike other sorbents, the PCC–CaO does not taper off after the first several minutes of reaction. This result can be further confirmed at different temperatures, as shown in Figure 2.42.

The extent of carbonation and sulfation can be calculated by

$$X_{CO_2} = \frac{\left(\dfrac{m_C - m_B}{44}\right)}{\left(\dfrac{m_B}{56}\right)} \qquad (2.3.18)$$

$$X_{SO_2} = \frac{\left(\dfrac{m_D - m_B}{80}\right)}{\left(\dfrac{m_B}{56}\right)} \qquad (2.3.19)$$

where X_{CO_2} is the moles of CO_2 captured per mole of CaO sorbent, X_{SO_2} is the moles of SO_2 captured per mole of CaO sorbent; m_B is the initial mass of CaO in the sample, m_C is the final mass of the sample after carbonation only,

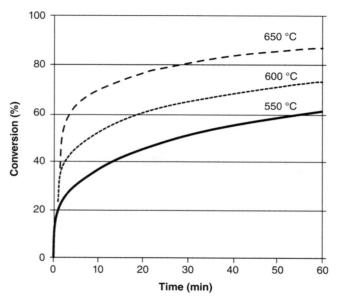

Figure 2.42. Carbonation reactions of PCC–CaO at different temperatures.[104]

m_D is the final mass of the sample after sulfation only, $m_B/56$ is the initial number of moles of CaO in the sample,

$$\frac{m_C - m_B}{44}$$

is the number of moles of $CaCO_3$ formed, and

$$\frac{m_D - m_B}{80}$$

is the number of moles of $CaSO_4$ formed.

   The extent of the sulfidation reaction with three different CaO sorbents,[106] namely, CaO obtained from Aldrich chemicals, PCC, and LC at the reaction temperature of 800°C and a total pressure of 1 MPa and P_{H_2S} of 3 kPa (0.3%) are shown in Figure 2.43. The figure indicates the high reactivity of PCC–CaO as compared with the other CaO sorbents. This result again can be attributed to the refined sorbent morphology of PCC.

2.3.6 Recyclability of Calcium Oxides

The recyclability property of calcium-based sorbents dictates the long-term viability of the sorbent for successful process applications. In the following

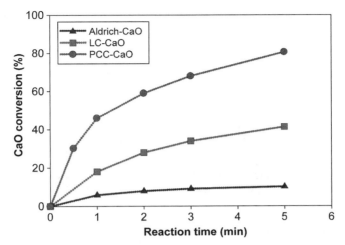

Figure 2.43. Reactivity of different CaO sorbents to H$_2$S removal.

discussion, factors that affect recyclability characteristics of various sorbents and experimental data are described.

Cyclic Calcination and Carbonation Behavior of Ca-Based Sorbents
Cyclical calcination–carbonation studies with PCC at 700°C exhibited sustained reactivity, whereas those with commercial Aldrich CaCO$_3$ could exhibit a loss in reactivity over two cycles.[104] Extended isothermal life-cycle testing of naturally occurring limestone powder (LC) and PCC sorbent at 700°C is shown in Figures 2.44 and 2.45. Figure 2.44 presents data for 50 cycles with the LC sorbent, whereas Figure 2.45 presents data for 100 CCR cycles with the PCC sorbent. The carbonation was conducted in a 10% CO$_2$ stream, whereas pure N$_2$ was used for calcination. Each of the carbonation–calcination steps was conducted for 30 minutes.

In these figures, the sorption capacity of the sorbent is quantified in kg CO$_2$ captured/kg sorbent. Theoretically, 56 g of unsupported CaO sorbent would react with 44 g of CO$_2$ corresponding to a maximum CO$_2$ sorption capacity of 78.6 wt.% at 100% conversion. In reality, 100% conversion of CaO does not occur in any cycle, because CaCO$_3$ has a molar volume (36.9 cm^3/mol) higher than that of the reactant CaO (16.9 cm^3/mol), and thus, when carbonation of CaO takes place, pore plugging in CaO occurs. Figure 2.44 shows that the wt.% capacity of the LC-based sorbent for CO$_2$ capture reduces from 58% in the first cycle to 20% at the end of the 50th cycle caused by sintering or grain growth in the LC precursor in the course of cyclic reactions, rendering the structure susceptible to pore pluggage and pore mouth closure.[104,112,113] In contrast, it is shown in Figure 2.45 that the conversion of PCC–CaO during 100 cycles is distinctly higher than that of the LC-based sorbent. The capacity

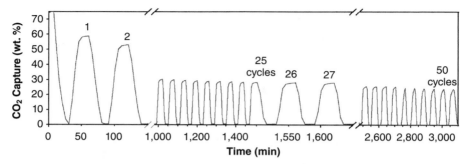

Figure 2.44. Extended calcination–carbonation cycles with Linwood carbonate (LC) fines at 700°C in a TGA in a 10% CO_2 stream.[113]

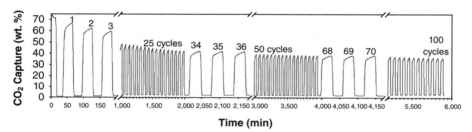

Figure 2.45. Extended calcination–carbonation cycles with precipitated calcium (PCC) fines at 700°C in a TGA in a 10% CO_2 stream.[113]

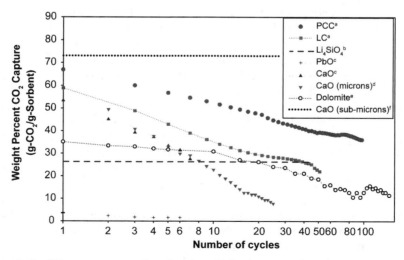

Figure 2.46. CO_2 capture capacity of various high-temperature sorbents over multiple carbonation–regeneration cycles:[113] a. Iyer et al.;[113] b. White et al.[114] and Kato et al.;[117] c. Kato et al.;[119] d. Barker;[69] e. Ortiz and Harrison;[87] and f. Barker.[70]

shown in Figure 2.45 is ~68 wt.% in the first cycle, which decreases to 40 wt.% in the 50th and slightly lower than 36 wt.% by the 100th cycle over ~6,000 minutes on stream. The higher reactivity over multiple cycles can be attributed to the predominant mesoporous structure of PCC, which allows the reactant gases to access more internal surface area of the particle through the larger pores.

Figure 2.46 depicts graphically the wt.% CO_2 capture attained by LC, PCC, and other high-temperature MO sorbents reported in the literature for multiple CCR cycles.[113] The experimental conditions used in the studies presented in Figure 2.46 are detailed in Table 2.11. They include carbonation and calcination temperatures, solid residence times, number of cycles, sorption capacities (wt.%), and the CO_2 concentration in the gas mixture during the reaction and regeneration steps. Although several studies have been conducted on a variety of Ca-based sorbents, a sorbent that exhibits a consistently high reactivity and sorption capacity over multiple cycles remains an area of considerable interest. Hydrated lime has shown consistency in reactivity over five cycles in a pilot-scale demonstration of the CCR process (see Sections 2.3.7 and 6.3). PCC–CaO attains a 66.8 wt.% increase in 30 minutes and a 71.5 wt.% after 120 minutes at the end of the first cycle. In contrast, earlier studies[104] indicated a sorption capacity of approximately 71 wt.% (90% conversion) in a pure CO_2 stream after 60 minutes at 650°C. Thus, factors such as types of calcium-based sorbent, CO_2 concentrations, temperatures, and cycle time play a significant role in determining the sorption capacity for the sorbent.

The experiments conducted by Barker on 10-μm CaO powders show a decrease in the sorption capacity from ~59 wt.% in the first carbonation cycle to 8 wt.% at the end of the 25th cycle.[69] The work suggests that because of the formation of a 22-nm-thick product layer, particles smaller than 22 nm in diameter should be able to reach stoichiometric conversion. Barker[70] later proved this hypothesis by obtaining repeated 93% conversion (73% weight capture) of 10-nm CaO particles across 30 cycles with a carbonation time of 24 hours under 100% CO_2 at 577°C. In a PbO–CaO-based chemical heat pump process, PbO attained 3.6 wt.% CO_2 capture in the first cycle, decreasing to 1.6 wt.% by the sixth cycle, and CaO showed a decrease in CO_2 capture from 53 wt.% in the first cycle to 27.5 wt.% by the fifth cycle.[115] A lithium zirconate (Li_2ZrO_3)-based sorbent provided a 20-wt.% capacity across two cycles.[116] In another study, researchers at Toshiba Corp. observed that the reactivity of lithium orthosilicate was better than that of lithium zirconate.[114,117] Extended cyclical studies performed on lithium orthosilicate samples revealed a consistent 26.5-wt.% capacity across 25 cycles.[118] As noted, Ortiz and Harrison[87] developed an enhanced hydrogen production process from the water–gas shift reaction by removing CO_2 from the gas mixture through the carbonation of CaO from dolomite. The dolomitic limestone based CCR process yielded a 35-wt.% capacity (gram of CO_2/gram of dolomite) (or 83% CaO conversion) in the first cycle that fell to 11.4-wt.%

TABLE 2.11. Comparison of Different Metal Oxide Sorbent Systems for High-Temperature CO_2 Capture for Multiple Carbonation–Regeneration Cycles

Sorbent	PCC-CaO[a]	LC-CaO[a]	Li_4SiO_4[b]	PbO[c]	CaO[c]	CaO Micron Level (10 μm)[d]	CaO Submicron Level (10 nm)[e]	Dolomite[f]	Li_2ZrO_3[g]
Carbonation time per cycle	30 min	30 min	1.3 hr	4 hr	2 hr	24 hr	24 hr	5 min	250 min
Calcination time per cycle	120 min	30 min	1.3 hr	30 min	15 min	—	—	5 min	100 min
Carbonation Temperature (°C)	700	700	600	300	880	866	577	800	500
Calcination Temperature (°C)	700	700	820	450	860	866	629	950	780
Carbonation P_{CO_2} (atm)	0.1	0.1	0.2	0.4	1.0	1.0	1.0	1.0	0.5
Calcination P_{CO_2} (atm)	0 (N_2)	0 (N_2)	0.2	1.0	0.4	0 (N_2)	0 (N_2)	1.0	0.5/dry air
Number of cycles	40	50	50	6	7	25	30	148	2
Approximate Ultimate CO_2 Capture by Weight (g-CO_2/100 g-sorbent)	45.3	22	26.5	1.6	27.5	7.9	73	11.4	20

[a]Iyer et al.[113]
[b]White et al.[114] and Kato et al.[117]
[c]Kato et al.[119]
[d]Barker.[69]
[e]Barker.[70]
[f]Ortiz and Harrison.[87]
[g]Ida and Lin.[116]

(or 27% CaO conversion) by the 148th cycle, when the carbonation experiments were performed in pure CO_2 at 800°C and calcination was conducted at 950°C.[87]

Calcium Oxide Regeneration by Vacuum/Steam/CO$_2$ Calcinations and Hydration

Steam calcination or vacuum calcination can lead to regeneration of $CaCO_3$ at lower enough temperatures that sintering can be minimized. Prior investigation has focused on vacuum calcination, which results in a pure stream of CO_2. It was reported in the literature that CaO obtained from calcination of limestone under vacuum has a higher reactivity.[120–122] Repeated calcination in N_2 leads to a loss in the surface area. Vacuum calcination of PCC followed by carbonation of PCC–CaO was repeated over two cycles. In the first cycle, PCC was vacuum calcined at 750°C and then carbonated at 700°C in pure CO_2. During the second cycle, the carbonate formed in the first cycle was vacuum-calcined and -carbonated again under the same conditions as the first cycle. The values of surface area (SA) and pore volume (PV) of the sorbent at various stages of the reactions are given in Table 2.12. The extent of carbonation was beyond 90% for every vacuum calcination–carbonation cycle. The results given in the table indicate that there is no systematic decline in SA and PV of the vacuum calcined CaO sorbent over two cycles. It is noted that this observation of reactivity over two cycles is important because the highest extent of sintering usually occurs during the first two cycles. The vacuum calcination and carbonation procedure could lead to a reasonable conversion over many cycles because of effective retention of the sorbent morphology. Steam calcination also is suitable for the production of a pure CO_2 stream for sequestration, as the steam can be condensed out from the CO_2-steam mixture at the exit of the calciner. Sintering is caused by high calcination temperatures as well as by the presence of steam, although the calcination temperature has a predominant effect on sintering. Hence, by lowering the calcination temperature, steam calcination is capable of reducing the overall sintering.

TABLE 2.12. Structural Properties of Calcium-Based Sorbents Undergoing Vacuum Calcination at 750°C and Carbonation at 700°C

Materials	Surface Area (m²/g)	Pore Volume (cm³/g)
PCC Starting Materials (Cycle 1)	38.3	0.1416
CaO after Vacuum Calcination (Cycle 1)	12.63	0.0241
CaCO₃ after Carbonation at 700°C (Cycle 2)	6.5	0.0103
CaO after Vacuum Calcination (Cycle 2)	15.93	0.0401
CaCO₃ after Carbonation at 700°C (Cycle 3)	2.361	0.00448

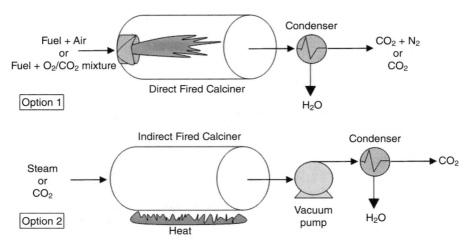

Figure 2.47. Direct- and indirect-fired calcination options for designing a calciner.

Figure 2.47 presents the direct and indirect firing calcination options for regenerating the $CaCO_3$ sorbent. The fuel used in direct calcination can be hydrogen generated in the plant. Carbon dioxide can be separated out from the hydrogen combustion stream by steam condensation.

As discussed, CaO sorbents are prone to sintering during the calcination step. Although steam and vacuum calcination aid in reducing the extent of sintering by lowering the calcination temperature, over multiple cycles, sintering progressively increases and reduces the CO_2 capture capacity of the sorbent. Extensive research has been conducted to develop methods to reduce the sintering of the sorbent. Pretreatment methods[85,123–125] have been developed that involve grinding the sorbent, preheating it in a nitrogen atmosphere, and hydration with water and steam at 100–300°C.

A method of complete reactivation through hydration of the calcined sorbent has been developed at The Ohio State University.[126] The complete reactivation of the sorbent during every cycle reverses the effect of sintering, and the history of the number of cycles is lost. Hence, this process minimizes the amount of solids circulation in the system. Two modes of hydration have been investigated: (1) ambient hydration with water and (2) high-temperature, high-pressure hydration. High-temperature, high-pressure hydration does not require the cooling and reheating of the sorbent, thereby significantly reducing the parasitic energy consumption of the process. These results are shown in Figure 2.48. The results for ambient hydration with water are shown in Figure 2.48(a). These bench-scale hydration results show that the capture capacity of the sorbent, calcined under realistic conditions, increases from 30% (6.8 mol/ kg CaO) to >55% (12.5 mol/kg CaO) on hydration with water. Sorbent reactivation by hydration can be integrated into the CCR process to alleviate

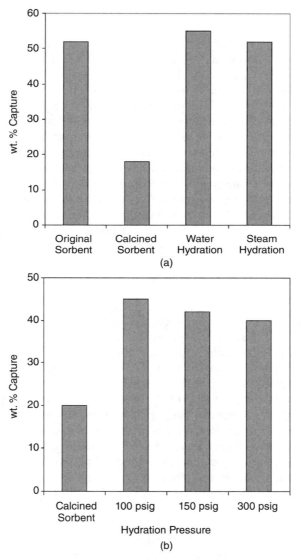

Figure 2.48. (a) Ambient pressure hydration; (b) high-pressure hydration.

sorbent decay over multiple cycles. Hydration at atmospheric pressure, in the presence of steam at 150°C, yielded a sorbent with 52 wt.% (11.8 mol/kg CaO) capture. Figure 2.48(b) shows the effect of pressure hydration at 600°C for pressures ranging from 100 psig (~8 atm) to 300 psig (~21 atm). It was found that the reactivity of the sorbent increases from 18% to 45% by pressure hydration at 600°C and 100 psig (~8 atm).

2.4 Concluding Remarks

The properties of the metal oxide particles used in the reduction and oxidation reactions or carbonation and calcination reactions play a key role to the successful operation of chemical looping processes. Syntheses of effective metal oxides that can be used to achieve a desired conversion efficiency of carbonaceous feedstock in a looping reactor system require consideration of intricate interrelationships among various reaction and process factors. For the Type I looping system, the reaction factors of importance include types of metal oxides and support materials, oxygen transfer capacity, gas and solid conversions, rates in both reduction and oxidation reactions, heat capacity and heat of reactions, melting points, mechanical strength, long-term recyclability, ease in scaleup, health and environmental effects, and particle cost. The process factors of importance include types of the products intended such as hydrogen, chemicals, liquid fuels, and electricity, reactor types, heat integration, and process intensification strategies, and overall process efficiency and economics. The key metals under the Type I looping system considered are Ni, Cu, and Fe supported by such metal oxides as Al_2O_3, TiO_2, ZrO_2, and SiO_2. Among these three metals, nickel-based particles exhibit superior thermal stability and reactivity with gaseous fuels, but it is costly. Copper-based particles also possess good reactivity, but they have a high tendency to sinter and agglomerate. Iron-based particles are cheap and are suitable for hydrogen production, but iron's reactivity with gaseous fuels is relatively low. It is expected that, through the improvements in particle formulation and in the metal-support matrix synthesis techniques, the thermal stability and reactivity of the metal oxide particles can be enhanced further. It is unlikely, however, that a universal particle will be developed that possesses all the requisite properties for all types of process applications. Thus, for optimization of a specific chemical looping process, a suitable primary oxygen carrier based on the type of fuels to be converted and the intended products to be generated needs to be selected.

Calcium oxide or calcium hydroxide is a typical metal oxide sorbent used for carbonation and calcination reactions in a Type II looping system. Among various reaction and process factors that are of importance to the Type II looping system are the recyclability of the calcium-based sorbent and the heat integration scheme for the endothermic calcination. The nature of the morphological variation caused by sintering for calcium oxide/calcium carbonate during the reactions could pose a severe limitation in the capability of sorbent recyclability in the looping applications. The use of high-surface-area reengineered sorbents like PCC aids in improving the reactivity and recyclability within the looping system. Alterations in the calcination process by the addition of steam or vacuum to lower the calcination temperature also can reduce sorbent sintering and improve recyclability. Moreover, the incorporation of a sorbent reactivation step such as hydration can completely reverse the effect of sintering and, thus, maintain sorbent reactivity a constant over multiple cycles. Further understanding of the fundamental properties of particle sintering and the particle reaction mechanism at high temperatures, is essential to the syntheses of the optimal looping particles for a long-term, economical looping process operation.

References

1. Lyngfelt, A., M. Johansson, and T. Mattisson, "Chemical Looping Combustion—Status of Development," Paper presented at the 9th International Conference on Circulating Fluidized Beds, Hamburg, Germany (2008).

2. Li, F., and L.-S. Fan, "Clean Coal Conversion Processes–Progress and Challenges," *Energy and Environmental Science*, 1, 248–267 (2008).

3. Adanez, J., L. F. de Diego, F. Garcia-Labiano, P. Gayan, A. Abad, and J. M. Palacios, "Selection of Oxygen Carriers for Chemical-Looping Combustion," *Energy & Fuels*, 18(2), 371–377 (2004).

4. Cao, Y., and W. P. Pan, "Investigation of Chemical Looping Combustion by Solid Fuels. 1. Process Analysis," *Energy & Fuels*, 20(5), 1836–1844 (2006).

5. Otsuka, K., A. Mito, S. Takenaka, and I. Yamanaka, "Production of Hydrogen from Methane without CO_2-Emission Mediated by Indium Oxide and Iron Oxide," *International Journal of Hydrogen Energy*, 26(3), 191–194 (2001).

6. Kodama, T., A. Aoki, T. Shimizu, Y. Kitayama, and S. Komarneni, "Efficient Thermochemical Cycle for CO_2 Gasification of Coal Using a Redox System of Reactive Iron-Based Oxide," *Energy & Fuels*, 12(4), 775–781 (1998).

7. Kodama, T., S. Miura, T. Shimizu, and Y. Kitayama, "Thermochemical Conversion of Coal and Water to CO and H_2 by a Two-Step Redox Cycle of Ferrite," *Energy*, 22(11), 1019–1027 (1997).

8. Garcia-Labiano, F., L. F. de Diego, J. Adanez, A. Abad, and P. Gayan, "Temperature Variations in the Oxygen Carrier Particles During Their Reduction and Oxidation in a Chemical-Looping Combustion System," *Chemical Engineering Science*, 60(3), 851–862 (2005).

9. Lane, H., "Process for Production of Hydrogen," U.S. Patent 1,078,686 (1913).

10. Messerschmitt, A., "Process for Producing Hydrogen," U.S. Patent 971,206 (1910).

11. Teed, P. L., "The Chemistry and Manufacture of Hydrogen," Longmans, Green and Co., White Plains, NY (1919).

12. Hurst, S., "Production of Hydrogen by the Steam-Iron Method," *Journal of the American Oil Chemists' Society*, 16(2), 29–36 (1939).

13. Johansson, M., T. Mattisson, and A. Lyngfelt, "Creating a Synergy Effect by Using Mixed Oxides of Iron- and Nickel Oxides in the Combustion of Methane in a Chemical-Looping Combustion Reactor," *Energy & Fuels*, 20(6), 2399–2407 (2006).

14. Fecht, H. J., E. Hellstern, Z. Fu, and W. L. Johnson, "Nanocrystalline Metals Prepared by High-Energy Ball Milling," *Metallurgical Transactions a-Physical Metallurgy and Materials Science*, 21(9), 2333–2337 (1990).

15. Cho, P., T. Mattisson, and A. Lyngfelt, "Defluidization Conditions for a Fluidized Bed of Iron Oxide-, Nickel Oxide-, and Manganese Oxide-Containing Oxygen

Carriers for Chemical-Looping Combustion," *Industrial & Engineering Chemistry Research*, 45(3), 968–977 (2006).

16. Jin, H. G., T. Okamoto, and M. Ishida, "Development of a Novel Chemical-Looping Combustion: Synthesis of a Solid Looping Material of $NiO/NiAl_2O_4$," *Industrial & Engineering Chemistry Research*, 38(1), 126–132 (1999).

17. Ishida, M., K. Takeshita, K. Suzuki, and T. Ohba, "Application of Fe_2O_3-Al_2O_3 Composite Particles as Solid Looping Material of the Chemical-Loop Combustor," *Energy & Fuels*, 19(6), 2514–2518 (2005).

18. de Diego, L. F., F. Garcia-Labiano, J. Adanez, P. Gayan, A. Abad, B. M. Corbella, and J. M. Palacios, "Development of Cu-Based Oxygen Carriers for Chemical-Looping Combustion," *Fuel*, 83(13), 1749–1757 (2004).

19. Ishida, M., H. G. Jin, and T. Okamoto, "A Fundamental Study of a New Kind of Medium Material for Chemical-Looping Combustion," *Energy & Fuels*, 10(4), 958–963 (1996).

20. Ishida, M., H. G. Jin, and T. Okamoto, "Kinetic Behavior of Solid Particle in Chemical-Looping Combustion: Suppressing Carbon Deposition in Reduction," *Energy & Fuels*, 12(2), 223–229 (1998).

21. Johansson, M., T. Mattisson, and A. Lyngfelt, "Investigation of Fe_2O_3 with $MgAl_2O_4$ for Chemical-Looping Combustion," *Industrial & Engineering Chemistry Research*, 43(22), 6978–6987 (2004).

22. Mattisson, T., M. Johansson, and A. Lyngfelt, "Multicycle Reduction and Oxidation of Different Types of Iron Oxide Particles—Application to Chemical-Looping Combustion," *Energy & Fuels*, 18(3), 628–637 (2004).

23. Liu, T., T. Simonyi, T. Sanders, R. V. Siriwardane, and G. Veser, "Nanocomposite oxygen carriers for chemical looping combustion," Abstracts of Papers, 232nd ACS National Meeting, San Francisco, CA (2006).

24. Cho, P., T. Mattisson, and A. Lyngfelt, "Comparison of Iron-, Nickel-, Copper- and Manganese-Based Oxygen Carriers for Chemical-Looping Combustion," *Fuel*, 83(9), 1215–1225 (2004).

25. Johansson, E., A. Lyngfelt, T. Mattisson, and F. Johnsson, "Gas Leakage Measurements in a Cold Model of an Interconnected Fluidized Bed for Chemical-Looping Combustion," *Powder Technology*, 134(3), 210–217 (2003).

26. Mattisson, T., A. Jardnas, and A. Lyngfelt, "Reactivity of Some Metal Oxides Supported on Alumina with Alternating Methane and Oxygen-Application for Chemical-Looping Combustion," *Energy & Fuels*, 17(3), 643–651 (2003).

27. de Diego, L. F., P. Gayan, F. Garcia-Labiano, J. Celaya, A. Abad, and J. Adanez, "Impregnated CuO/Al_2O_3 Oxygen Carriers for Chemical-Looping Combustion: Avoiding Fluidized Bed Agglomeration," *Energy & Fuels*, 19(5), 1850–1856 (2005).

28. Ishida, M., M. Yamamoto, and T. Ohba, "Experimental Results of Chemical-Looping Combustion with $NiO/NiAl_2O_4$ Particle Circulation at 1200°C," *Energy Conversion and Management*, 43(9–12), 1469–1478 (2002).

29. Jin, H., T. Okamoto, and M. Ishida, "Development of a Novel Chemical-Looping Combustion: Synthesis of a Looping Material with a Double Metal Oxide of CoO-NiO," *Energy & Fuels*, 12(6), 1272–1277 (1998).

30. Jin, H. G., and M. Ishida, "Reactivity Study on a Novel Hydrogen Fueled Chemical-Looping Combustion," *International Journal of Hydrogen Energy*, 26(8), 889–894 (2001).

31. Jin, H. G., and M. Ishida, "Reactivity Study on Natural-Gas-Fueled Chemical-Looping Combustion by a Fixed-Bed Reactor," *Industrial & Engineering Chemistry Research*, 41(16), 4004–4007 (2002).

32. Jin, H. G., and M. Ishida, "A New Type of Coal Gas Fueled Chemical-Looping Combustion," *Fuel*, 83(17–18), 2411–2417 (2004).

33. Haber, J., J. H. Block, and B. Delmon, "Manual of Methods and Procedures for Catalyst Characterization," *Pure and Applied Chemistry*, 67(8–9), 1257–1306 (1995).

34. Brinker, C. J., and G. W. Scherer, Sol-Gel Science: The Physics and Chemistry of Sol-Gel Processing, Academic Press, New York, (1990).

35. Sakka, S., L. C. Klein, and E. J. A. Pope, Sol-Gel Science and Technology, American Ceramic Society, Westerville, OH (1995).

36. Erri, P., and A. Varma, "Solution Combustion Synthesized Oxygen Carriers for Chemical Looping Combustion," *Chemical Engineering Science*, 62(18–20), 5682–5687 (2007).

37. Corbella, B. M., L. de Diego, F. Garcia-Labiano, J. Adanez, and J. M. Palacios, "The Performance in a Fixed Bed Reactor of Copper-Based Oxides on Titania as Oxygen Carriers for Chemical Looping Combustion of Methane," *Energy & Fuels*, 19(2), 433–441 (2005).

38. Mattisson, T., M. Johansson, and A. Lyngfelt, "The Use of NiO as an Oxygen Carrier in Chemical-Looping Combustion," *Fuel*, 85(5–6), 736–747 (2006).

39. Urasaki, K., N. Tanimoto, T. Hayashi, Y. Sekine, E. Kikuchi, and M. Matsukata, "Hydrogen Production Via Steam-Iron Reaction Using Iron Oxide Modified with Very Small Amounts of Palladium and Zirconia," *Applied Catalysis A-General*, 288(1–2), 143–148 (2005).

40. Takenaka, S., T. Kaburagi, C. Yamada, K. Nomura, and K. Otsuka, "Storage and Supply of Hydrogen by Means of the Redox of the Iron Oxides Modified with Mo and Rh Species," *Journal of Catalysis*, 228(1), 66–74 (2004).

41. Montes, M., J. B. Soupart, M. Desaedeleer, B. K. Hodnett, and B. Delmon, "Influence of Metal Support Interactions on the Stability of Ni/SiO$_2$ Catalysts During Cyclic Oxidation Reduction Treatments," *Journal of the Chemical Society-Faraday Transactions I*, 80, 3209–3220 (1984).

42. Lee, J. B., C. S. Park, S. Choi, Y. W. Song, Y. H. Kim, and H. S. Yang, "Redox Characteristics of Various Kinds of Oxygen Carriers for Hydrogen Fueled Chemical-Looping Combustion," *Journal of Industrial and Engineering Chemistry*, 11(1), 96–102 (2005).

43. Coombs, P. G., and Z. A. Munir, "Cyclic Reduction Oxidation of Hematite Powders," *Journal of Materials Science*, 24(11), 3913–3923 (1989).

44. Abad, A., T. Mattisson, A. Lyngfelt, and M. Johansson, "The Use of Iron Oxide as Oxygen Carrier in a Chemical-Looping Reactor," *Fuel*, 86(7–8), 1021–1035 (2007).

45. Cho, P., T. Mattisson, and A. Lyngfelt, "Carbon Formation on Nickel and Iron Oxide-Containing Oxygen Carriers for Chemical-Looping Combustion," *Industrial & Engineering Chemistry Research*, 44(4), 668–676 (2005).

46. Chiang, Y.-M., D. P. Birnie III, and W. D. Kingery, Physical Ceramics–Principles for Ceramic Science & Engineering, 2nd edition, John Wiley & Sons Inc., New York (1997).

47. Wang, X. H., J. G. Li, H. Kamiyama, M. Katada, N. Ohashi, Y. Moriyoshi, and T. Ishigaki, "Pyrogenic Iron(III)-Doped TiO_2 Nanopowders Synthesized in RF Thermal Plasma: Phase Formation, Defect Structure, Band Gap, and Magnetic Properties," *Journal of the American Chemical Society*, 127(31), 10982–10990 (2005).

48. Zhao, Y., and F. Shadman, "Kinetics and Mechanism of Ilmenite Reduction with Carbon-Monoxide," *AIChE Journal*, 36(9), 1433–1438 (1990).

49. Zhao, Y., and F. Shadman, "Reduction of Ilmenite with Hydrogen," *Industrial & Engineering Chemistry Research*, 30(9), 2080–2087 (1991).

50. Steinsvik, S., Y. Larring, and T. Norby, "Hydrogen Ion Conduction in Iron-Substituted Strontium Titanate, $SrTi_{1-x}Fe_xO_{3-x/2}$ $(0 \le x \le 0.8)$," *Solid State Ionics*, 143(1), 103–116 (2001).

51. Sakata, K., T. Nakamura, M. Misono, and Y. Yoneda, "Reduction-Oxidation Mechanism of Oxide Catalysts–Oxygen Diffusion During Redox Cycles," *Chemistry Letters*, (3), 273–276 (1979).

52. Sakata, K., F. Ueda, M. Misono, and Y. Yoneda, "Catalytic Properties of Iron-Oxides: 2. Isotopic Exchange of Oxygen, Oxidation of Carbon-Monoxide, and Reduction-Oxidation Mechanism," *Bulletin of the Chemical Society of Japan*, 53(2), 324–329 (1980).

53. Urasaki, K., Y. Sekine, N. Tanimoto, E. Tamura, E. Kikuchi, and M. Matsukata, "Effect of a Small Amount of Zirconia Additive on the Activity and Stability of Iron Oxide During Repeated Redox Cycles," *Chemistry Letters*, 34(2), 230–231 (2005).

54. Garcia-Labiano, F., L. F. de Diego, J. Adanez, A. Abad, and P. Gayan, "Temperature Variations in the Oxygen Carrier Particles During Their Reduction and Oxidation in a Chemical-Looping Combustion System," *Chemical Engineering Science*, 60(3), 851–862 (2005).

55. Gupta, P., L. G. Velazquez-Vargas, and L.-S. Fan, "Syngas Redox (SGR) Process to Produce Hydrogen from Coal Derived Syngas," *Energy & Fuels*, 21(5), 2900–2908 (2007).

56. Perry, R. H., and D. W. Green, Perry's Chemical Engineers' Handbook, 8th edition, McGraw-Hill (2008).

57. Zhao, H., L. Liu, B. Wang, D. Xu, L. Jiang, and C. Zheng, "Sol–Gel-Derived NiO/$NiAl_2O_4$ Oxygen Carriers for Chemical-Looping Combustion by Coal Char," *Energy & Fuels*, 22(2), 898–905 (2008).

58. Scott, S. A., J. S. Dennis, A. N. Hayhurst, and T. Brown, "In Situ Gasification of a Solid Fuel and CO_2 Separation Using Chemical Looping," *AIChE Journal*, 52(9), 3325–3328 (2006).

59. Factiva, "IM Prices," Metal Bulletin Plc. (2007).

60. Da Silva, C. S., "Nicke—National Department of Mineral Production (Brazil)," http://www.dnpm.gov.br/enportal/conteudo.asp?IDSecao=170&IDPagina=1093 (2007).

61. Kodama, T., Y. Watanabe, S. Miura, M. Sato, and Y. Kitayama, "Reactive and Selective Redox System of Ni(II)-Ferrite for a Two-Step CO and H_2 Production Cycle from Carbon and Water," *Energy*, 21(12), 1147–1156 (1996).

62. L'vov, B. V., "Mechanism of Carbothermal Reduction of Iron, Cobalt, Nickel and Copper Oxides," *Thermochimica Acta*, 360(2), 109–120 (2000).

63. Park, A.-H. A., R. A. Jadhav, and L.-S. Fan, "CO_2 Mineral Sequestration: Chemically Enhanced Aqueous Carbonation of Serpentine," *Canadian Journal of Chemical Engineering*, 81(3–4), 885–890 (2003).

64. Butt, D. P., K. S. Lackner, C. H. Wendt, S. D. Conzone, H. Kung, Y. C. Lu, and J. K. Bremser, "Kinetics of Thermal Dehydroxylation and Carbonation of Magnesium Hydroxide," *Journal of the American Ceramic Society*, 79(7), 1892–1898 (1996).

65. Sawada, Y., M. Murakami, and T. Nishide, "Thermal Analysis of Basic Zinc Carbonate: 1. Carbonation Process of Zinc Oxide Powders at 8 and 13 Degrees C," *Thermochimica Acta*, 273, 95–102 (1996).

66. Okuma, N., Y. Funayama, H. Ito, N. Mizutani, and M. Kato, "Adsorption and Reaction of CO_2 Gas on the Surface of Zinc Oxide Fine Particles in the Atmosphere," *Hyomen Kagaku* 9(6), 452–458 (1988).

67. Shaheen, W. M., and M. M. Selim, "Effect of Thermal Treatment on Physicochemical Properties of Pure and Mixed Manganese Carbonate and Basic Copper Carbonate," *Thermochimica Acta*, 322(2), 117–128 (1998).

68. Bhatia, S. K., and D. D. Perlmutter, "Effect of the Product Layer on the Kinetics of the CO_2-Lime Reaction," *AIChE Journal*, 29(1), 79–86 (1983).

69. Barker, R., "Reversibility of Reaction $CaCO_3$ Reversible CaO + CO_2," *Journal of Applied Chemistry and Biotechnology*, 23(10), 733–742 (1973).

70. Barker, R., "Reactivity of Calcium-Oxide Towards Carbon-Dioxide and Its Use for Energy-Storage," *Journal of Applied Chemistry and Biotechnology*, 24(4–5), 221–227 (1974).

71. Abanades, J. C., and D. Alvarez, "Conversion Limits in the Reaction of CO_2 with Lime," *Energy & Fuels*, 17(2), 308–315 (2003).

72. Sun, P., J. R. Grace, C. J. Lim, and E. J. Anthony, "The Effect of CaO Sintering on Cyclic CO_2 Capture in Energy Systems," *AIChE Journal*, 53(9), 2432–2442 (2007).

73. Wang, J. S., and E. J. Anthony, "On the Decay Behavior of the CO_2 Absorption Capacity of CaO-Based Sorbents," *Industrial & Engineering Chemistry Research*, 44(3), 627–629 (2005).

74. Stultz, S. C., and J. B. Kitto, Steam, Its Generation and Use, 40th edition, Babcock & Wilcox, Lynchburg, VA (1992).

75. Squires, A. M., "Cyclic Use of Calcined Dolomite to Desulfurize Fuels Undergoing Gasification," *Advances in Chemistry Series*, 69, 205–209 (1967).

76. Curran, G. P., C. E. Fink, and E. Gorin, "CO_2 Acceptor Gasification Process—Studies of Acceptor Properties," *Advances in Chemistry Series*, 69, 141–165 (1967).

77. Lin, S. Y., M. Harada, Y. Suzuki, and H. Hatano, "Process Analysis for Hydrogen Production by Reaction Integrated Novel Gasification (Hypr-Ring)," *Energy Conversion and Management*, 46(6), 869–880 (2005).

78. Lin, S. Y., Y. Suzuki, H. Hatano, and M. Harada, "Developing an Innovative Method, Hypr-Ring, to Produce Hydrogen from Hydrocarbons," *Energy Conversion and Management*, 43(9–12), 1283–1290 (2002).

79. Gupta, H., and L.-S. Fan, "Carbonation-Calcination Cycle Using High Reactivity Calcium Oxide for Carbon Dioxide Separation from Flue Gas," *Industrial & Engineering Chemistry Research*, 41, 4035–4042 (2002).

80. Iyer, M., H. Gupta, B. B. Sakadjian, and L.-S. Fan, "Multicyclic Study on the Simultaneous Carbonation and Sulfation of High-Reactivity CaO," *Industrial & Engineering Chemistry Research*, 43(14), 3939–3947 (2004).

81. Sun, P., J. R. Grace, C. J. Lim, and E. J. Anthony, "Removal of CO₂ by Calcium-Based Sorbents in the Presence of SO₂," *Energy & Fuels*, 21(1), 163–170 (2007).

82. Sun, P., C. J. Lim, and J. R. Grace, "Cyclic CO₂ Capture by Limestone-Derived Sorbent During Prolonged Calcination/Carbonation Cycling," *AIChE Journal*, 54(6), 1668–1677 (2008).

83. Abanades, J. C., E. J. Anthony, J. S. Wang, and J. E. Oakey, "Fluidized Bed Combustion Systems Integrating CO₂ Capture with CaO," *Environmental Science & Technology*, 39(8), 2861–2866 (2005).

84. Abanades, J. C., G. S. Grasa, M. Alonso, N. Rodriguez, E. J. Anthony, and L. M. Romeo, "Cost Structure of a Postcombustion CO₂ Capture System Using CaO," *Environmental Science & Technology*, 41(15), 5523–5527 (2007).

85. Manovic, V., D. Lu, and E. J. Anthony, "Steam Hydration of Sorbents from a Dual Fluidized Bed CO₂ Looping Cycle Reactor," *Fuel*, 87(15–16), 3344–3352 (2008).

86. Fan, L.-S., S. Ramkumar, and M. Iyer, "Calcium Looping Process for High Purity Hydrogen Production," International Patent Application PCT/US2007/079432 (2007).

87. Ortiz, A. L., and D. P. Harrison, "Hydrogen Production Using Sorption-Enhanced Reaction," *Industrial & Engineering Chemistry Research*, 40(23), 5102–5109 (2001).

88. Han, C., and D. P. Harrison, "Simultaneous Shift Reaction and Carbon Dioxide Separation for the Direct Production of Hydrogen," *Chemical Engineering Science*, 49(24B), 5875–5883 (1994).

89. Johnsen, K., H. J. Ryu, J. R. Grace, and C. J. Lim, "Sorption-Enhanced Steam Reforming of Methane in a Fluidized Bed Reactor with Dolomite as CO₂-Acceptor," *Chemical Engineering Science*, 61(4), 1195–1202 (2006).

90. Balasubramanian, B., A. L. Ortiz, S. Kaytakoglu, and D. P. Harrison, "Hydrogen from Methane in a Single-Step Process," *Chemical Engineering Science*, 54(15–16), 3543–3552 (1999).

91. Shimizu, T., T. Hirama, H. Hosoda, K. Kitano, M. Inagaki, and K. Tejima, "A Twin Fluid-Bed Reactor for Removal of CO₂ from Combustion Processes," *Chemical Engineering Research & Design*, 77(A1), 62–68 (1999).

92. Silaban, A., and D. P. Harrison, "High Temperature Capture of Carbon Dioxide: Characteristics of the Reversible Reaction between CaO(s) and CO₂(g)," *Chemical Engineering Communications*, 137, 177–190 (1995).

93. Aihara, M., T. Nagai, J. Matsushita, Y. Negishi, and H. Ohya, "Development of Porous Solid Reactant for Thermal-Energy Storage and Temperature Upgrade Using Carbonation/Decarbonation Reaction," *Applied Energy*, 69(3), 225–238 (2001).

94. Abanades, J. C., "The Maximum Capture Efficiency of CO₂ Using a Carbonation/Calcination Cycle of CaO/CaCO₃," *Chemical Engineering Journal*, 90(3), 303–306 (2002).

95. Alvarez, D., and J. C. Abanades, "Determination of the Critical Product Layer Thickness in the Reaction of CaO with CO₂," *Industrial & Engineering Chemistry Research*, 44(15), 5608–5615 (2005).

96. Salvador, C., D. Lu, E. J. Anthony, and J. C. Abanades, "Enhancement of CaO for CO_2 Capture in an FBC Environment," *Chemical Engineering Journal*, 96(1–3), 187–195 (2003).

97. Reddy, E. P., and P. G. Smirniotis, "High-Temperature Sorbents for CO_2 Made of Alkali Metals Doped on CaO Supports," *Journal of Physical Chemistry B*, 108(23), 7794–7800 (2004).

98. Fang, F., Z. S. Li, and N. S. Cai, "Experiment and Modeling of CO_2 Capture from Flue Gases at High Temperature in a Fluidized Bed Reactor with Ca-Based Sorbents," *Energy & Fuels*, 23(1), 207–216 (2009).

99. Li, Z. S., N. S. Cai, and Y. Y. Huang, "Effect of Preparation Temperature on Cyclic CO_2 Capture and Multiple Carbonation-Calcination Cycles for a New Ca-Based CO_2 Sorbent," *Industrial & Engineering Chemistry Research*, 45(6), 1911–1917 (2006).

100. Stevens, J. F., B. Krishnamurthy, P. Atanassova, and K. Spilker, "Development of 50KW Fuel Processor for Stationary Fuel Cell Applications," Final Report, DOE/GO/13102-1 (2007).

101. Fan, L.-S., A. Ghosh-Dastidar, and S. Mahuli, "Calcium Carbonate Sorbent and Methods of Making and Using Same," U.S. Patent 5,779,464 (1998).

102. Agnihotri, R., S. K. Mahuli, S. S. Chauk, and L.-S. Fan, "Influence of Surface Modifiers on the Structure of Precipitated Calcium Carbonate," *Industrial & Engineering Chemistry Research*, 38, 2283–2291 (1999).

103. Ghosh-Dastidar, A., S. K. Mahuli, R. Agnihotri, and L.-S. Fan, "Investigation of High-Reactivity Calcium Carbonate Sorbent for Enhanced SO_2 Capture," *Industrial & Engineering Chemistry Research*, 35, 598–606 (1996).

104. Gupta, H., and L.-S. Fan, "Carbonation-Calcination Cycle Using High Reactivity Calcium Oxide for Carbon Dioxide Separation from Flue Gas," *Industrial & Engineering Chemistry Research*, 41, 4035–4042 (2002).

105. Wei, S.-H., S. K. Mahuli, R. Agnihotri, and L.-S. Fan, "High Surface Area Calcium Carbonate: Pore Structural Properties and Sulfation Characteristics," *Industrial & Engineering Chemistry Research*, 36, 2141–2148 (1997).

106. Chauk, S. S., R. Agnihotri, R. A. Jadhav, S. K. Misro, and L.-S. Fan, "Kinetics of High-Pressure Removal of Hydrogen Sulfide Using Calcium Oxide Powder," *AIChE Journal*, 46(6), 1157–1167 (2000).

107. Sohara, J., Personal Communication (2006).

108. Yamaguchi, T., and K. Murakawa, "Preparation of Spherical Calcium Carbonate (Vaterite) Powder Transition to Calcite in Water," *Zairyo*, 30, 856–860 (1981).

109. Wray, J. L., and F. Daniels, "Precipitation of Calcite and Aragonite," *Journal of American Chemical Society*, 70, 2031–2034 (1957).

110. Cizer, O., K. Van Balen, J. Elsen, and D. Van Gemert, "Crystal Morphology of Precipitated Calcite Crystals from Accelerated Carbonation of Lime Binders," Paper presented at the 2nd International Conference on Accelerated Carbonation for Environmental and Materials Engineering, 149–158, Rome, Italy, October 1–3 (2008).

111. Fan, L.-S., and R. A. Jadhav, "Clean Coal Technologies: Oscar and Carbonox Commercial Demonstrations," *AIChE Journal*, 48(10), 2115–2123 (2002).

112. Wu, S., M. A. Uddin, C. L. Su, S. Nagamine, and E. Sasaoka, "Effect of the Pore-Size Distribution of Lime on the Reactivity for the Removal of SO_2 in the Presence of High-Concentration CO_2 at High Temperature," *Industrial & Engineering Chemistry Research*, 41(22), 5455–5458 (2002).

113. Iyer, M., H. Gupta, B. B. Sakadjian, and L.-S. Fan, "Multicyclic Study on the Simultaneous Carbonation and Sulfation of High-Reactivity CaO," *Industrial & Engineering Chemistry Research*, 43(14), 3939–3947 (2004).

114. White, C. M., B. R. Strazisar, E. J. Granite, J. S. Hoffman, and H. W. Pennline, "Separation and Capture of CO_2 from Large Stationary Sources and Sequestration in Geological Formations–Coalbeds and Deep Saline Aquifers," *Journal of the Air & Waste Management Association*, 53(6), 645–715 (2003).

115. Kato, Y., D. Saku, N. Harada, and Y. Yoshizawa, "Utilization of High Temperature Heat from Nuclear Reactor Using Inorganic Chemical Heat Pump," *Progress in Nuclear Energy*, 32(3–4), 563–570 (1998).

116. Ida, J., and Y. S. Lin, "Mechanism of High-Temperature CO_2 Sorption on Lithium Zirconate," *Environmental Science & Technology*, 37(9), 1999–2004 (2003).

117. Kato, M., S. Yoshikawa, and K. Nakagawa, "Carbon Dioxide Absorption by Lithium Orthosilicate in a Wide Range of Temperature and Carbon Dioxide Concentrations," *Journal of Materials Science Letters*, 21(6), 485–487 (2002).

118. Nakagawa, K., *Silicates for the Separation of CO_2 from Flue Gas*, Proceedings of the Carbon Dioxide Capture Workshop at NETL, Pittsburgh, Pennsylvania (2003).

119. Kato, Y., N. Harada, and Y. Yoshizawa, "Kinetic Feasibility of a Chemical Heat Pump for Heat Utilization of High-Temperature Processes," *Applied Thermal Engineering*, 19(3), 239–254 (1999).

120. Beruto, D., L. Barco, A. W. Searcy, and G. Spinolo, "Characterization of the Porous CaO Particles Formed by Decomposition of $CaCO_3$ and $Ca(OH)_2$ in Vacuum," *Journal of the American Ceramic Society*, 63(7–8), 439–443 (1980).

121. Beruto, D., and A. W. Searcy, "Calcium Oxides of High Reactivity," *Nature*, 263(5574), 221–222 (1976).

122. Dash, S., M. Kamruddin, P. K. Ajikumar, A. K. Tyagi, and B. Raj, "Nanocrystalline and Metastable Phase Formation in Vacuum Thermal Decomposition of Calcium Carbonate," *Thermochimica Acta*, 363(1–2), 129–135 (2000).

123. Laursen, K., W. Duo, J. R. Grace, and C. J. Lim, "Cyclic Steam Reactivation of Spent Limestone," *Industrial & Engineering Chemistry Research*, 43(18), 5715–5720 (2004).

124. Manovic, V., and E. J. Anthony, "Thermal Activation of CaO-Based Sorbent and Self-Reactivation During CO_2 Capture Looping Cycles," *Environmental Science & Technology*, 42(11), 4170–4174 (2008).

125. Zeman, F., "Effect of Steam Hydration on Performance of Lime Sorbent for CO_2 Capture," *International Journal of Greenhouse Gas Control*, 2(2), 203–209 (2008).

126. Fan, L.-S., S. Ramkumar, W. Wang, and R. Statnick, "Separation of Carbon Dioxide from Gas Mixtures by Calcium Based Reaction Separation Process," U.S. Provisional Patent Application 61/116172 (2008).

CHAPTER 3

CHEMICAL LOOPING COMBUSTION

F. Li, F. Wang, D. Sridhar, H. R. Kim, L. G. Velazquez-Vargas, and L.-S. Fan

3.1 Introduction

Combustion is characterized by a rapid chemical reaction of oxidants such as oxygen with fuels, including biomass, coal, natural gas, and oil.[1] During combustion of carbonaceous fuels, the fuel reacts with oxidants to produce heat and combustion flue gases consisting mainly of carbon dioxide and steam:

$$C_xH_{2y} + (x + y/2)O_2 \rightarrow xCO_2 + yH_2O + \text{Heat} \qquad (3.1.1)$$

Since the discovery of fire, humankind has been relying on combustion of carbonaceous fuels for heating, transportation, and electricity generation. Approximately 66% of the electricity consumed worldwide is produced from combustion of either coal or natural gas.[2] Fossil fuels are the basis for most electricity production, and fossil fuel combustion power plants, especially those burning coal, are a key source of CO_2 generation. For example, pulverized coal (PC) combustion power plants accounted for 23% of worldwide anthropogenic CO_2 emissions in 2006.[3] The CO_2 concentration continues to increase in the atmosphere, and so it is becoming increasingly important to develop and implement CO_2 control strategies to capture and store CO_2 from fossil fuel combustion power plants. Implementing a CO_2 control strategy for fossil fuel conversion plants will result in a significant decrease in the overall plant efficiency and in a significant increase in the cost of plant operation.

This chapter discusses a CO_2 control technique for fossil fuel combustion through the employment of the chemical looping combustion (CLC) approach.

Section 3.2 describes the characteristics of conventional power plants for PC combustion, as well as traditional and advanced CO_2 capture methods as applied to postcombustion or precombustion in fossil fuel power plants. In Section 3.3, CLC-based power generation systems are discussed in detail, along with the effects of the CLC operational variables, including gas flow rates, solids circulation rates, solids inventory, particle reactivity properties, and operating pressure. An analysis is provided on the feasibility of reactor scale-up for a given metal oxide, gas–solid contact mode, and operating capacity in a CLC process. The reactor configuration, operational conditions, and test results for various CLC systems at a small pilot scale of $10 kW_{th}$ and greater are also given in detail. Section 3.3 ends with a discussion on the effects of fuel type, particle performance, reactor design, and operating conditions on the fuel conversion and CO_2 capture efficiency.

3.2 CO_2 Capture Strategies for Fossil Fuel Combustion Power Plants

Because coal is more carbon intensive than crude oil and natural gas per unit energy output, carbon management for coal is more critical. This section introduces various CO_2 capture techniques for coal-based power plants. These techniques, however, generally are applicable to power plants using other types of fossil fuels.

3.2.1   Pulverized Coal Combustion Power Plants

Approximately 55% of the coal produced worldwide is combusted in PC combustion power plants for electricity generation, accounting for nearly 40% of the electricity generated worldwide.[4] When CO_2 emission control is required, the components, configuration and operating condition of the coal combustion plant directly affect the design and performance of the CO_2 emission control systems, which must be integrated to the plant. A representative process flow diagram for a PC power plant without CO_2 emission control devices is given in Figure 3.1. The figure shows that a PC power plant consists of five subsystems. They are as follows: (1) the coal handling and preparation subsystem, (2) the coal combustion and steam generation subsystem, (3) the flue gas treatment and pollutant control subsystem, (4) the turbine–generator subsystem, and (5) the heat rejection subsystem.[1] In a PC power plant, coal first is transported from the storage facility to the pulverizer where it is ground into fine powders of a mean particle size less than 74 μm. The pulverized coal powders are conveyed pneumatically through the burner nozzle to the boiler, which is a large enclosed combustion chamber. This pneumatic air stream, referred to as the primary air, is usually preheated to 130–200°F (54–93°C). The flow rate of the air and coal mixture is higher than 15 m/s to prevent sedimentation of the coal powder. To ensure good mixing between the coal powder and the air, a deflector is placed near the exit of the burner nozzle to induce turbulence

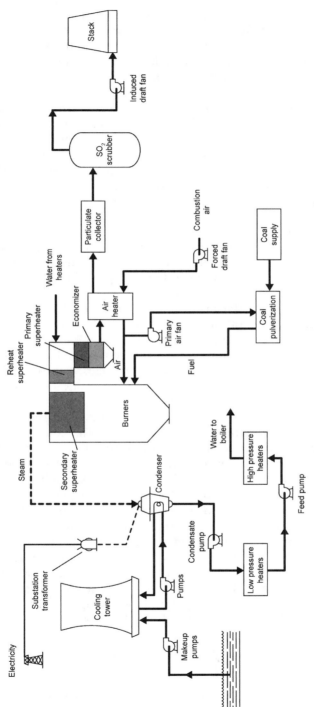

Figure 3.1. Process flow diagram of a coal-fired power plant.[1]

145

or radially pitched swirl. A stream of secondary air, preheated to approximately 600°F (316°C), usually is introduced from the outside of the burner nozzle to enhance mixing. Upon entering the boiler, the air–fuel mixture is ignited and combusted. Because of the large amount of heat generated from coal combustion, the temperature within the furnace can exceed 2,800°F (1,538°C).

The heat carried by high-temperature effluent gases from the boiler then is extracted by the secondary superheater, the primary superheater, the reheat superheater, and the economizer, in that order. The economizer, the primary superheater, and the secondary superheater are connected in series to produce high-pressure superheated steam for high-temperature steam turbines. The intermediate pressure steam discharged from these high-pressure steam turbines is superheated in the reheater superheater. The reheated intermediate pressure steam then is introduced to the intermediate-pressure and low-pressure steam turbines for subsequent generation of additional electricity. The low-grade steam discharged from the low-pressure steam turbines is condensed by cooling water prior to entering the next steam-generation cycle. The heat removed in the condensation step eventually is released to the environment through one or more cooling towers.

The steam generation scheme of the PC power plant is illustrated further in Figure 3.2. Because of the highly irreversible nature of the process, especially in the fuel combustion and steam-generation steps, a conventional subcritical PC (Sub-CPC) power plant can convert only 30%–36% of the energy in coal into electricity. Higher efficiencies can be achieved with increased steam temperature and pressure in supercritical (SCPC) and ultra-supercritical PC (USCPC) power plants.

The combustion of coal in the boiler generates flue gases that contain several by-products harmful to the environment. The control strategies for traditional pollutants such as sulfur oxides, nitrogen oxides, fine particulates, and mercury are discussed in Chapter 1. This chapter focuses on the CO_2 control strategies for fossil fuel-based power plants. A coal-based power plant again is used as an example to illustrate these strategies, noting that these strategies also can be applied to processing other types of fossil fuels, such as natural gas. As discussed, CO_2 management in a fossil fuel power plant can be costly. Table 3.1 provides cost information on carbon capture and sequestration (CCS) based on existing technologies for a PC power plant. As shown, the cost for carbon capture including CO_2 separation and compression is significantly higher than that for CO_2 transportation and storage.[5] Thus, the development of cost-effective CO_2 capture technologies is the key to affordable CO_2 emission control in fossil fuel-based power plants.

3.2.2 CO₂ Capture Strategies

Several carbon-capture techniques can be or potentially can be used in fossil fuel-based power plants. They can be categorized generally into postcom-

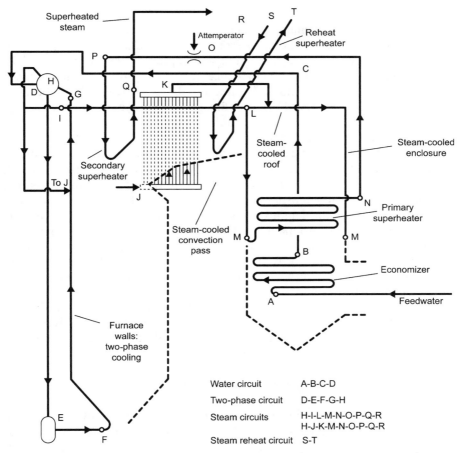

Figure 3.2. Steam- and power-generation scheme of a PC power plant.[1]

TABLE 3.1. Cost of CCS Technology[5]

CCS Steps	Cost ($/t of CO_2 Processed)
Capture (separation and compression)	15–75
Transportation	1–8
Storage (geological sequestration)	0.1–0.3

bustion capture, precombustion capture, and oxycombustion capture. Postcombustion capture directly removes CO_2 from the combustion flue gas. In precombustion capture, CO_2 is removed in the fuel conversion process prior to combustion. In oxycombustion capture, fuel combustion takes place using oxygen instead of air, generating a concentrated CO_2 stream. These methods are explored in the following sections.

Postcombustion Capture

Figure 3.3 shows the schematic flow diagram of postcombustion capture techniques. Typical coal combustion processes generate a CO_2-containing flue gas stream at ambient pressure. Postcombustion capture seeks to separate CO_2 from the flue gas stream to obtain a concentrated CO_2 stream, which is compressed for transportation and then sequestered. Because the postcombustion capture techniques separate CO_2 from flue gas after combustion, they readily can be retrofitted to existing power plants.

The CO_2 separation techniques that are considered for postcombustion applications include chemical/physical absorption solvents, low-temperature adsorbents, high-temperature sorbents, CO_2 separation membranes, metal organic frameworks, ionic liquids, molecular filtration, and enzyme-based systems.[6] Several postcombustion capture technologies that are deployed commercially or under pilot demonstration, such as monoethanolamine (MEA), chilled ammonia, and carbonation–calcination reaction (CCR) systems, were introduced in Chapter 1. In the following discussion, the MEA solvent-based chemical absorption system, which is one of the most widely applied CO_2 separation technologies in the industry, is described further. The CCR system is discussed in Chapter 6.

Figure 3.4 is a more detailed flow diagram of the MEA system in Figure 1.6(a). Because of the relatively strong chemical interaction between the solvent and CO_2, MEA can remove the low partial pressure CO_2 in the flue gas. The chemical reaction can be given by

$$C_2H_4OHNH_2 + H_2O + CO_2 \leftrightarrow C_2H_4OHNH_3^+ + HCO_3^- \tag{3.2.1}$$

In the absorber, CO_2 in the flue gas reacts with the MEA solvent to form water-soluble compounds, and thus CO_2 can be separated from the flue gas. Because absorption of CO_2 is favored at low temperatures, the absorption step is carried out at ~40°C. After absorption, the CO_2-rich solvent is introduced to the desorber, which is operated at 100–120°C and 1–2 atm. Steam is used to heat up the spent solvent. In the regenerator/desorber, the solvent–CO_2 com-

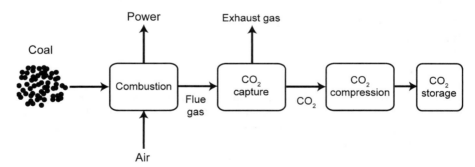

Figure 3.3. General postcombustion capture flow diagram.

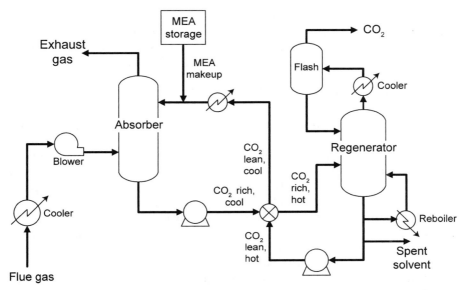

Figure 3.4. Flow diagram of the MEA scrubbing system for CO_2 capture.[7]

pound decomposes. As a result, the solvent is regenerated, producing a concentrated CO_2 stream. The MEA system is suitable for CO_2 separation from power plant flue gas, which has a CO_2 concentration of 10–15 vol.% or a CO_2 partial pressure of 0.10–0.15 atm. Note that current progress in the development of solvent-based CO_2 capture technology is expected to follow a similar path to the historical progress of the development of the flue gas desulfurization (FGD) technology.[8] FGD technology is a commercially dominant technology for SO_2 capture from flue gases in the coal boiler industry today. An industrial demonstration by Mitsubishi Heavy Industries in Tokyo, Japan, indicated that the MEA system is capable of separating ~90% of the CO_2 in the flue gas stream, producing a concentrated CO_2 stream that contains up to 99.8 wt.% CO_2.[9] Such a CO_2 stream is suitable for further compression, transportation, and sequestration.

Although considerable improvement is being made on the MEA system, it is in general costly for deployment in PC power plants. This is because the regeneration of the spent MEA solvent consumes a significant amount steam, which reduces the overall power output and, hence, the efficiency of the power plant.[7] The heat requirement of the reboiler alone can amount to 4 GJ/t of CO_2.[7,10] Furthermore, the regenerated solvent needs to be cooled before entering the absorber. A large amount of cooling water is required in this step. Although the heat exchange between the spent solvent and the regenerated solvent can alleviate the heating or cooling duty for the desorber or absorber, a large amount of parasitic energy, however, is required. In addition, the CO_2 generated from the desorber is at low pressure (1–2 atm). Further compression of CO_2 to ~150 atm is required before it can be transported and sequestered. Also, the MEA solvent has a tendency to foul. For instance, NO_x and SO_2 react with the MEA solvent to form a heat-stable salt, which reduces the absorption capacity of the solvent.[7] Because these salts are not regenerable, 1 mol of NO_x or SO_2 in the flue gas will lead to a loss of ~2 mol of MEA solvent. Thus, the presence of SO_2 and NO_x in the flue gas can increase significantly the operating cost of the MEA system. SO_2 and NO_x removal prior to the MEA system is thus necessary. The preferred levels of SO_2 and NO_x are 10 ppmv and 20 ppmv in the MEA system, respectively.[7,11]

The energy-intensive nature of the MEA system leads to a large parasitic energy penalty for CO_2 capture in a PC power plant. When retrofitted to an existing PC power plant, the MEA system reduces or derates the overall plant efficiency by 21%–42%.[12–18] Table 3.2 gives the general energy penalties for 90% CO_2 capture on various PC power plants, including Sub-CPC, SCPC, and USCPC power plants using MEA systems. The reduction in the plant efficiency is accompanied by an increase in the cost of electricity. Various studies indicate that a 90% CO_2 capture with MEA will lead to an increase in the cost of electricity of 40%–70%.[16,18,19] Table 3.3 summarizes the change in the cost of electricity using the MEA system.

To reduce the energy and cost penalties for CO_2 capture in PC power plants, the following properties are desirable for solvent-based technologies:

1. CO_2-scrubbing temperature to be comparable with ambient temperature with the pressure close to ambient pressure,
2. low heat requirement for solvent regeneration,
3. capable of regenerating CO_2 at elevated pressure,
4. high solvent tolerance to SO_x and NO_x
5. low solvent cost, and
6. low operating pressure drop in the absorber and desorber.

TABLE 3.2. Energy Conversion Efficiencies of Various Pulverized Coal Combustion Power Plants and Energy Penalties for CO_2 Capture Using MEA[12–18]

Technology	Subcritical PC	Supercritical PC	Ultra-Supercritical PC
Base Case Efficiency (%) HHV	33–37	37–40	40–45
MEA Retrofit Derating (%)[a]	30–42	24–34	21–30

[a]Percentage decrease in energy conversion efficiency when a retrofit MEA system is used to capture 90% of the CO_2 in the flue gas.

TABLE 3.3. Increase in Cost of Electricity for CO_2 Capture Using MEA[16,18,19]

	MIT, 2007	Davison, 2007	Rubin, 2007
Cost of Electricity Increase (%)	66.5	42.2	66.0

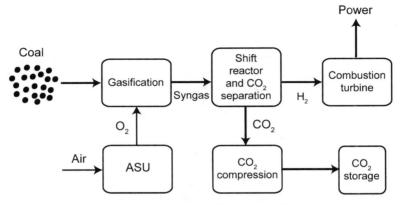

Figure 3.5. General precombustion capture flow diagram.

For instance, the chilled ammonia system discussed in Chapter 1 can regenerate CO_2 at 20–40 atm, and thus, the energy requirement for CO_2 compression for transport is reduced. Moreover, the heat required for the regeneration of ammonia carbonate and bicarbonate solvent is lower than that required for MEA solvent regeneration. Comparing the MEA and chilled ammonia-based systems, the energy requirement of the chilled ammonia system can be 50% of the MEA system.[20]

Precombustion Capture
Precombustion capture is exemplified by the coal-based Integrated Gasification Combined Cycle (IGCC) Process shown in Figure 3.5. In this process, coal first reacts with oxygen, coming from the air separation unit (ASU), in the presence of steam to produce syngas composed mainly of CO and H_2. To capture carbon in the syngas prior to combustion, the CO in the syngas stream is converted to CO_2 while forming H_2 through the water–gas shift (WGS) reaction:

$$CO + H_2O \rightarrow CO_2 + H_2 \qquad (3.2.2)$$

Thus, the resulting gas stream contains a high CO_2 concentration (up to ~40 vol.% on a dry basis). The CO_2 (and H_2S) can be captured using either chemical absorption-based acid gas removal processes, such as monoethanolamine (MEA) and methyldiethanolamine (MDEA), or physical absorption-based processes, such as Selexol (UOP, Des Plaines, IL) and Rectisol (Lurgi, Frankfurt, Germany), yielding concentrated H_2.[21] H_2 then is used to generate electricity through a combined cycle power generation system with minimal carbon emissions. Unlike PC combustion processes, which fully oxidize carbonaceous fuels to generate heat, modern coal gasifiers in the IGCC Process convert coal to syngas via partial oxidation reactions with oxygen and/or steam under elevated pressures.[1,22] The high-pressure syngas stream, undiluted by N_2

from the air, has a much lower volumetric flow rate when compared with that of the flue gas from coal-fired power plants. For instance, the volumetric flow rate of syngas generated from a dry feed, oxygen blown gasifier can be two orders of magnitude smaller than that from a PC boiler with a similar coal-processing capacity (dry basis). Meanwhile, the partial pressure of CO_2 in the syngas after the WGS reaction can be 80 times higher than that in the PC boiler flue gas (dry basis). The significantly reduced gas flow rate and increased gas partial pressures render the pollutant and CO_2 control relatively easy for gasification processes.

Despite the advantages in product versatility and pollutant controllability, gasification is capital intensive. It is estimated that an IGCC system requires 6%–10% more capital investment when compared with an USCPC plant.[23] Both plants have similar energy conversion efficiencies. Although CO_2 capture from the gasification process is relatively easier than for the combustion process, CO_2 capture also is energy and capital intensive. CO_2 capture can derate the energy conversion efficiency of the IGCC system by 13%–24%, increasing the cost of electricity by 25%–45%.[13,16,24–26] The efficiency loss mainly results from the large steam consumption for the WGS reaction, heating and cooling of the CO_2 separation solvent, and power consumption for CO_2 compression.

Oxycombustion Capture
Oxycombustion capture first was proposed by Horn and Steinberg in 1982.[27] In the oxycombustion scheme, coal is combusted with almost pure oxygen (~95%) instead of air. A fraction of the flue gas is recycled to the coal boiler to moderate the combustion temperature because current materials are not capable of withstanding the high combustion temperature with pure oxygen.[6] Unlike the conventional PC power plant where CO_2 is diluted by N_2 in air, the flue gas from oxycombustion contains mainly CO_2 and steam. Steam easily can be condensed out, resulting in a concentrated CO_2 stream. Figure 3.6 shows

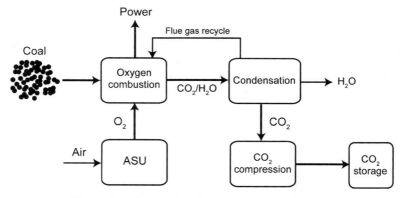

Figure 3.6. Oxycombustion capture flow diagram.

the Oxycombustion Capture Process. An air-separation unit (ASU), necessary for oxycombustion, is one of the key factors for efficiency loss and capital needs. For instance, ~230 kWh of electricity is consumed to produce 1 t of oxygen when a cryogenic ASU is used.[28] Compared with the conventional PC boiler, the decreased concentration of inert gas (N_2) in the oxycombustion boiler leads to enhanced combustion temperature and, hence, more efficient combustion. However, the overall efficiency loss for oxycombustion capture is estimated to be 20%–35%.[15,18,29] Nearly three quarters of that energy loss comes from the operation of the ASU for oxygen generation, and the increase in the cost of electricity generated is estimated to be ~46%.[18]

In addition to the various demonstration activities mentioned in Chapter 1, an oxy-fuel demonstration plant (30 MW) has been operated in Schwarze Pumpe, Germany. A larger scale demonstration (200–300 MW) is planned to be in service by 2015.[28] The Oxycombustion Process has the potential to be retrofitted to existing power plants, but the necessity to increase the ASU efficiency and reduce the ASU cost poses several significant challenges for the commercialization of the oxycombustion system. Moreover, improvements in furnace temperature control and safe oxygen handling are crucial for reliable oxycombustion plant operation. Novel oxygen-production technologies are being developed that will reduce the capital and operating costs for oxygen production. These technologies include ionic transport membranes (ITMs) or oxygen transport membranes (OTMs), where oxygen molecules are ionized under a high temperature and are separated from air using ceramic membranes; as well as the ceramic autothermal recovery (CAR) process, where oxygen is adsorbed under pressures and temperatures using ceramic solid particles and is separated by desorption using stripping gases. The typical class of materials used for membranes or particles is perovskites.[6]

CLC Process for CO_2 Capture

CLC processes are categorized as both precombustion capture and oxycombustion capture because the carbon in the fuel is separated prior to combustion, and the fuel is converted by an oxygen carrier rather than by air. This carbon-capture system has the potential to achieve high-energy conversion efficiency while CO_2 is captured, as illustrated in the next section.

3.3 Chemical Looping Combustion

CLC characterizes an indirect fuel combustion strategy in which metal oxide is used, which serves as an oxygen carrier and a combustion intermediate between the air and the fuel. As described in Chapter 1, the CLC concept can be traced back to the pioneering study of Lewis and Gilliland[30] in the 1950s when they proposed to use the reaction of copper oxide with syngas to produce carbon dioxide. In their process, carbon dioxide was the desired product from the fuel conversion. In the late 1960s, the CLC process was proposed as a

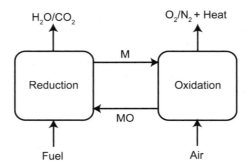

Figure 3.7. Schematic process diagram of CLC.[44]

novel fuel conversion route that reduces the irreversibility of the fuel combustion for heat and power generation.[31] Verified to be advantageous by thermodynamic analysis,[31–33] the chemical looping concept has been explored extensively during the last two decades. Studies carried out in the 1980s and 1990s focused on the development of the cyclic chemical intermediates[34,35] and on their applications in CLC processes for power generation using gaseous fuels such as methane and syngas.[36,37] From the beginning of this century, the possibilities of using liquid and solid fuels such as coal and biomass in a CLC system also have been explored.[38–43]

The reaction scheme for CLC generally involves two main reactions that are carried out in two separate reactors. The simplified schematic of CLC is given in Figure 3.7.

In the reducer, the oxygen carrier particle is reduced by the fuel, producing CO_2 and H_2O:

$$(2x + y/2)\,\text{MeO} + C_xH_y \rightarrow (2x + y/2)\,\text{Me} + x\,CO_2 + y/2\,H_2O \qquad (3.3.1)$$

As discussed in Chapter 2, this fuel conversion step can be either endothermic or exothermic depending on the types of fuel and the particles. The reduced oxygen carriers then are moved to the oxidizer where combustion takes place with air, and the reduced oxygen carrier is converted back to its original oxidation state:

$$\text{Me} + \text{Air} \rightarrow \text{MeO} + \text{Oxygen-depleted Air} \qquad (3.3.2)$$

During the oxidation step, a large amount of heat is generated, which is used for electricity generation. Because air and fuels are converted in separate reactors, the CO_2 generated from the reducer is only mixed with steam and is readily separable. The capital- and energy-intensive CO_2 separation step thus can be avoided. Moreover, the indirect fuel conversion in two separate reactors can be advantageous from the thermodynamics point of view. The heat integration between the two reactors can reduce the exergy loss by recuperat-

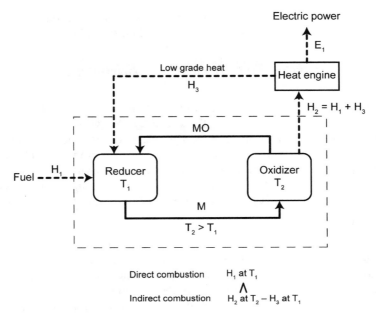

Figure 3.8. Potential heat integration scheme for CLC that reduces the exergy loss for combustion.

ing the low-grade heat while producing a larger amount of high-grade heat. Figure 3.8 illustrates such an exergy recuperative integration scheme.

As mentioned, both gaseous and solid fuels can be used for CLC. Gaseous fuels are easier to convert compared with solid fuels. However, the direct-solid fuel chemical looping systems are attractive because of the potential economic impact. The desired CLC systems using gaseous fuels require enhanced fuel and metal oxide conversion efficiencies in the reducer as well as minimum carbon deposition, solids circulation rates, and reactor sizes. The comprehensive information on oxygen carrier particle performance, reactor design and operating behavior also is required.

The conversion of solid fuels needs further consideration of issues, including ash removal, tar conversion, solids feeding and mixing, solid fuel conversion, and *in situ* oxygen carrier interaction with such gaseous pollutants as sulfur and mercury. Therefore, a CLC system using solid fuels requires the utilization of an oxygen carrier particle that is robust toward pollutants and ash. With the physical and chemical properties of oxygen carrier particles elaborated in Chapter 2, the next section describes the performance of oxygen carrier particles from the perspective of the CLC process operation. This is followed by discussions on the CLC reactor design and operating behavior. The operational results of several small pilot-scale CLC systems using gaseous or solid fuels also are presented. Novel schemes and applications of CLC such as chemical looping oxygen uncoupling and chemical looping enhanced steam methane reforming systems are presented in Chapter 6.

3.3.1 Particle Reactive Properties and Their Relationship with CLC Operation

The performance of the oxygen carrier governs the feasibility of the CLC process. It is essential to examine the particle reactive properties from the TGA test and a laboratory-scale reactor to access the reaction characteristics, thereby aiding large-scale CLC system design and operations. The particle performance evaluation methods discussed in this section are based on CLC systems utilizing a circulating fluidized bed (CFB) design. Other types of chemical looping reactor systems such as moving beds followed by an entrained bed are described in Chapters 4 and 5.

Methodology for Estimating Solids Inventory and Particle Circulation Rate
A goal to achieve for the CLC is its application as a retrofit or a stand-alone technology. The CLC currently is operated at the lab to small pilot scales. Considerable operational data on the effects of operational variables including solids inventory, gas–solid contact modes, solids circulation rate, gas velocity, solids holdup, mixing, heat and mass transfer, and gas and solid conversions are essential to ascertain the best design of CLC reactor systems. In this section, the methodology for estimating such properties as solids inventory and particle circulation rate based on that proposed by researchers at Chalmers University of Technology is discussed.[44]

To start with, several parameters must be defined to calculate the solids inventory and solids circulation rate. The first parameter is the oxygen-carrying capacity of the particle. Different metal oxides have different oxygen-carrying capacity (R_o), which is defined as the weight fraction of the total transferable oxygen in the oxygen carrier when it is at its fully oxidized form:

$$R_o = (m_{ox} - m_{red})/m_{ox} \qquad (3.3.3)$$

where m_{ox} is the weight of a fully oxidized sample, whereas m_{red} is the weight of the same sample after being fully reduced. Another parameter that characterizes the extent of oxidation of the oxygen carrier after CLC reactions is the oxygen carrier conversion (X), defined as the ratio of the transferable oxygen in the current oxygen carrier sample to that in the fully oxidized oxygen carrier, as given by

$$X = \frac{m - m_{red}}{m_{ox} - m_{red}} \qquad (3.3.4)$$

where m is the weight of the oxygen carrier sample after reaction. A mass conversion (ω) also is defined to characterize the oxidation state of the oxygen carrier:

$$\omega = m/m_{ox} = 1 + R_o(X - 1) \qquad (3.3.5)$$

The mass conversion rate of the oxygen carrier can be expressed as $d\omega/dt$. With the aforementioned parameters measured on laboratory-scale instruments, the amount of the bed material or the solids inventory required for each reactor (m_{bed}) and the solids circulation rate (m_{sol}) in a CLC process can be estimated by

$$m_{bed} = \frac{\omega \cdot m_o}{d\omega/dt} \qquad (3.3.6)$$

$$m_{sol} = \frac{\omega \cdot m_o}{\Delta\omega} \qquad (3.3.7)$$

where m_o is the stoichiometric oxygen mass flow rate required for a complete fuel conversion. $\Delta\omega$ is the difference between the average oxygen carrier mass conversion in the oxidizer (ω_{ox}) and that in the reducer (ω_{red}). $\Delta\omega$ accounts for the extent of oxygen carrier conversion in the CLC process. A higher $\Delta\omega$ corresponds to a more complete utilization of the oxygen-carrying capacity of the oxygen carrier in the redox reactions. Note that the bed mass is inversely proportional to the mass conversion rate, and the solids circulation rate is inversely proportional to the extent of mass conversion ($\Delta\omega$).

The extent of the fuel conversion is also important to the overall CLC performance. Such information can be obtained directly from the laboratory tests. When methane is used as the fuel, the CO_2 yield of the reducer (γ_{red}) is defined as the molar fraction of CO_2 among the carbon-containing products of the reducer:

$$\gamma_{red} = \frac{p_{CO_2,out}}{p_{CH_4,out} + p_{CO,out} + p_{CO_2,out}} \qquad (3.3.8)$$

This is a direct measure of the CO_2 content in the reducer exhaust gas stream. In most cases, the CO_2 yield (γ_{red}) alone is not adequate to describe the fuel conversion because the hydrogen content in the reducer exhaust gas is not taken into account. Another parameter, the fuel conversion based on the heating value (γ_{heat}), often is used along with γ_{red}[45]:

$$\gamma_{heat} = \frac{p_{CH_4,in} H_{CH_4} - \dfrac{1}{1 - p_{H_2,out}} \left(p_{H_2,out} H_{H_2} + p_{CO,out} H_{CO} + p_{CH_4,out} H_{CH_4} \right)}{p_{CH_4,in} H_{CH_4}} \qquad (3.3.9)$$

where P_i is the partial pressure of gaseous species I, when the gas mixture is set at 1 atm after condensing out the steam. H_i is the heating value of gaseous species i. The air conversion in the oxidizer, which is less important compared

Carrier properties

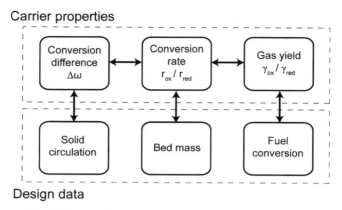

Design data

Figure 3.9. Relationship of carrier properties to the design variables.[46]

with the fuel conversion, can be defined in a manner similar to fuel conversion.

The relationships among oxygen carrier reactive properties and the key operational parameters discussed are described in Figure 3.9. It is evident from the figure that three factors—oxygen carrier mass conversion rate ($d\omega/dt$), extent of oxygen carrier mass conversion ($\Delta\omega$), and extent of fuel conversion (γ_{red})—significantly affect the CFB design parameters. All three factors are closely related to the reactive properties of the oxygen carrier.

Reaction curves of a nickel-based oxygen carrier, shown in Figure 3.10, are used to exemplify the aforementioned methods. These curves were obtained by Mattisson et al. from TGA tests.[44] As is shown, the mass conversion rates of the oxygen carrier are the highest when the oxygen carrier conversion is within the intermediate range, that is, when the oxygen carrier is neither fully oxidized nor fully reduced. Furthermore, note that a higher temperature leads to a higher mass conversion rate. Similar trends are found in other types of oxygen carriers. To determine the solids inventories and solids circulation rate, the mass conversions of the oxygen carrier in the reducer and the oxidizer need to be determined first. Researchers at Chalmers University of Technology (Gothenburg, Sweden) proposed a CFB operating strategy to be under the conditions in which the mass conversion rates, $d\omega/dt$, of the oxygen carrier are high.

Figure 3.10 also shows that the high mass conversion rates of the oxygen carrier in both reducer and oxidizer occur over a relatively narrow range of the carrier conversion, ω. Limiting ω to a small range or keeping $\Delta\omega$ small also can reduce the temperature drop of the oxygen carriers in the reducer with the exception of CuO. Note that a large temperature drop in the reducer can lead to a decreased mass conversion rate of the oxygen carrier, and thus, most CFB-based CLC systems are operated at a small $\Delta\omega$. Such an operating strategy can lead to a reduced solids inventory resulting from the fast oxygen

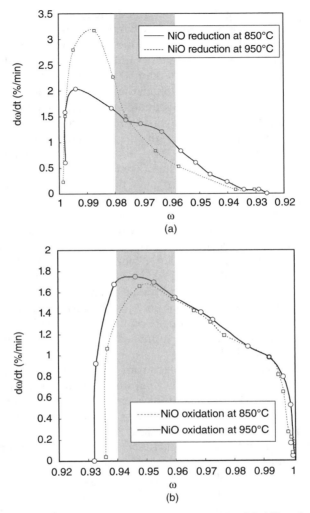

Figure 3.10. Relationship between the mass conversion rate ($d\omega/dt$) and oxygen carrier mass conversion (ω) of Al_2O_3 (67 wt.%) supported NiO (33 wt.%) for: (a) reduction and (b) oxidation. The shaded band represents the oxygen carrier conversion range where the mass conversion rates in both the reducer and the oxidizer are high.[44]

carrier reaction rates. The challenge, however, lies in the high solids circulation rate requirement because of a small $\Delta\omega$. For example, for ω between 0.98 and 0.99 ($\Delta\omega = 0.01$), Mattisson et al.[44] estimated that the solids inventories will be 230 kg/MW and 390 kg/MW for the reducer and the oxidizer, respectively. The corresponding solids circulation rate will be 8 kg/MW·s or 28,800 t/h for a power plant of 1,000 MW_{th} capacity.

Based on the operating strategy described, the laboratory-scale experimental data are used to predict the solids handling requirements of the CFB system at a commercial scale. Iron oxide modified with various supports was tested in a fluidized bed experimental setup.[47] The results indicated that a reducer solids inventory can be less than $500 \, kg/MW_{th}$ for required solids circulation rates of $6–8 \, kg/s/MW_{th}$. Johansson et al.[48] investigated $MgAl_2O_4$-supported iron oxide particles and estimated the solids inventory to be $150 \, kg/MW_{th}$ in the reducer. The mass conversion rates during the oxidation of iron oxide-based oxygen carriers are high in all cases. Mattisson et al.[44] also investigated alumina supported Ni-, Cu-, Mn-, and Co-based oxygen carriers using TGA and concluded that alumina supported Mn and Co are not suitable for CLC operation. For Al_2O_3-supported CuO, the solids inventory in the reducer is estimated as $70 \, kg/MW_{th}$, where as while the solids inventory in the oxidizer is estimated as $390 \, kg/MW_{th}$. For Al_2O_3-supported NiO, the solids inventory in the reducer is estimated as $230 \, kg/MW_{th}$, whereas the solids inventory in the oxidizer is estimated as $390 \, kg/MW_{th}$. The solids circulation rates for Al_2O_3 supported NiO are $1–8 \, kg/s/MW_{th}$. The experimental results obtained by Cho et al.[49] using a laboratory-scale fluidized bed indicate that the solids inventory of NiO-based oxygen carriers in the reducer can be lower than $80 \, kg/MW_{th}$. The corresponding solids circulation rate is $4–8 \, kg/s/MW_{th}$. Leion et al.[50] tested the possibility of using petroleum coke as the fuel, instead of methane, with an iron oxide carrier and $MgAl_2O_4$ support and found that the bed inventory requirement is $\sim 2,000 \, kg/MW_{th}$.

The results presented are obtained from the laboratory experiments performed under batch conditions. The operating conditions of these units are different from those in continuous CFB systems. Therefore, the results obtained from these laboratory experiments only can be used for estimating purposes for the parameters of the CFB operation. Several small CFB systems have been constructed to study particle reactivity behavior under continuous conditions. Johansson et al.[51] studied the combustion of natural gas using a nickel-based oxygen carrier in a 300-W CFB. The CO_2 yield of the reducer (γ_{red}) of between 97% and 99.5% was achieved with a total solids inventory of $\sim 1,000 \, kg/MW_{th}$. Abad et al.[52] used an iron-based oxygen carrier in a 300-W CFB. The researchers found that at $1,223 \, K$ a solids inventory of $2,500 \, kg/MW_{th}$ was required for the reducer to achieve the 100%-CH_4 conversion and of $1,200 \, kg/MW_{th}$ was required to achieve the reducer for the 99%-CH_4 conversion. When syngas was used as the fuel, solids inventories of $1,100 \, kg/MW_{th}$ and $650 \, kg/MW_{th}$ for the reducer were required for the 100% and 99% fuel conversion, respectively. The solids inventory in the oxidizer ranged from 320 to 380 kg/MW_{th}. The corresponding solids circulation rates ranged from 6.7 to $10 \, kg/s \cdot MW_{th}$. de Diego et al.[53] investigated the possibility of using a copper-based oxygen carrier in a $10-kW_{th}$-CLC system with methane as the fuel. The outlet methane concentration is negligible if the oxygen carrier to fuel ratio is greater than 1.5, corresponding to a solids circulation rate of $4.2 \, kg/s \cdot MW_{th}$ or higher. The total solids inventory was $2,100 \, kg/MW_{th}$. Ryu et al.[54] demonstrated a bentonite-supported NiO oxygen carrier in a 50-kW$_{th}$ CLC system with 99.7% conversion of methane using ~720 kg/MW$_{th}$-bed material and a solids circulation rate of 12 kg/m^2 s. Note that the solids inventory and solids flux required for a continuous, large CLC system are usually higher than those of a laboratory unit as a result of better reactant contact in a laboratory unit compared with that in a large unit, leading to a higher overall reaction rate.

The information presented indicates that both the solids inventory and solids circulation rates are large for a commercial CLC using CFB. For a 1,000-MW$_{th}$-CLC plant using a CFB, the total solids inventory is on the order of 400–2,500 t. The corresponding solids circulation rate is 3,600–30,000 t/h. The large solids handling requirements in the commercial CLC system represent a major solids flow challenge to CFB operation.

3.3.2 Key Design and Operational Parameters for a CFB-Based CLC System

It is evident that the design of a CFB-based CLC system methodically should consider the solids circulation rate and solids inventory requirements; these requirements can be estimated based on the reactivity of the particles. Other important properties besides reactivity include density, size, attrition, and agglomeration. The CFB system consists of two or more reactors, with at least one being the riser operated at dilute or dense fluidization conditions for particle pneumatic conveying and/or reaction. The other reactor(s) can be operated at different fluidization conditions including bubbling or turbulent fluidization. Considering the fluidization properties of the particles in the CLC system, the size of the oxygen carriers cannot be too large or too small. The size of a single particle typically would be between 75 and 250 μm. Furthermore, attrition and agglomeration of particles need to be avoided because attrition results in particle loss, whereas agglomeration leads to bed defluidization.

In this section, the key design parameters with a focus on the hydrodynamic properties of the riser in a conceptual 1,000-MW$_{th}$ CFB-based CLC system using methane as the feed are discussed. Particular attention is placed on the simultaneous handling of a large amount of gas and solid reactants during the riser operation. For a given processing capacity of the fuel—methane for example—the consumption rate of the methane fuel is fixed irrespective of the oxygen carrier used:

$$F_{CH4} = \frac{C_{plant}}{H_{CH4}} \qquad (3.3.10)$$

where F_{CH4} is the molar flow rate (mol/s) of methane to the reducer; C_{plant} is the fuel processing capacity of the plant, that is, 1,000 MW$_{th}$ or 10^6 kW$_{th}$ higher heating value (HHV); and H_{CH4} is the heating value of methane (889 kJ/mol HHV). Because the CLC plant converts nearly all methane into CO_2 and H_2O, the stoichmetric requirement of the molar flow rate of air to convert methane is

$$F_{air,sto} = \frac{2F_{CH_4}}{0.21} \qquad (3.3.11)$$

With 10% excess air for full combustion of the reduced oxygen carrier,[55] the minimum molar flow rate of air entering the CFB riser is

$$F_{air,riser} = \frac{2F_{CH_4}}{0.21} \times 1.1 \qquad (3.3.12)$$

The relationship between the cross-sectional area of the riser and the superficial gas velocity at the riser inlet thus can be obtained based on the operating pressure and temperature of the riser as

$$A_{riser} = \frac{F_{air,riser} \times 22.4 \times T_{oxidizer}}{P \times 273.15 \times U_g \times 1,000} \qquad (3.3.13)$$

where A_{riser} is the cross-sectional area of the riser; P is the pressure in the riser; and U_g is the superficial gas velocity of the riser inlet. Note that the oxygen concentration in the riser decreases continuously along the riser, as does the gas velocity. For complete combustion of the oxygen carrier, the molar flow rate of air at the riser outlet is 82% of that at the riser inlet.

Most reported experiments to date use a riser superficial velocity of 5–10 m/s. Some parametric relationships for riser flows are given here. For a riser flow with an operating superficial gas velocity above U_{tr}, defined as the transport velocity of gas–solid fluidization,[56,57] the riser can be operated in the fast fluidization regime. The transport velocity can be estimated from the following empirical relation[58]:

$$Re_{tr} = 2.28 Ar^{0.419} \qquad (3.3.14)$$

where $Re_{tr} = \dfrac{\rho U_{tr} d_p}{\mu}$, and Ar is Archimedes number, defined as:

$$Ar = \frac{\rho(\rho_p - \rho)g d_p^3}{\mu} \qquad (3.3.15)$$

where ρ, ρ_p, d_p, μ, and g are the gas density, the density of the particles, the diameter of the particles, the gas viscosity, and the gravitational constant, respectively. The relationship among the solids flux, J_s (kg/m^2) defined as the mass flow rate of solids flowing through a unit area of the riser cross section, the operating superficial gas velocity, U_g (m/s) and the solids holdup ε_s in the riser can be given by

$$J_s = \left[\frac{U_g}{1 - \varepsilon_s} - Max(U_{se}, U_t) \right] \times \rho_p \times \varepsilon_s \qquad (3.3.16)$$

where the term $\dfrac{U_g}{1-\varepsilon_s} - \text{Max}\,(U_{se}, U_t)$ gives the average particle velocity in the riser[59]; U_t is the terminal velocity of individual particles; and U_{se} is a critical velocity defined as the velocity beyond which the significant entrainment of the particles occurs.[60,61] U_{se} can be calculated based on the following correlation equation:[62]

$$\text{Re}_{se} = 1.53\text{Ar}^{0.50} \tag{3.3.17}$$

When the residence time in the riser is too short to allow the oxygen carrier to be oxidized completely, a dense bed operated in the turbulent fluidization regime is used, which is placed below the riser to provide the complete required particle reaction time. The superficial gas velocity in the turbulent fluidization regime is lower than U_{se} and higher than U_c. U_c marks the onset of the transition to the turbulent fluidization regime and is characterized by the maximum amplitude of the pressure fluctuation variation with the gas velocity in the bed. The hydrodynamic properties for several types of particles used in CLC processes were examined based on the correlation equations proposed by Bi et al.[58,62,63] It is found that U_{tr} ranges from 6.1–8.2 m/s; U_{se} ranges from 4.9 to 7.5 m/s; and U_c ranges from 3.6 to 4.9 m/s. In the following discussion, the superficial velocity for the riser is set at 7 m/s, and the superficial gas velocity U_g for the turbulent bed below the riser is set at 4.9 m/s. Based on these velocities, the cross-sectional area required for the fast fluidization section and the turbulent fluidization section is calculated.

The solids flux and solids holdup in the CLC system can be determined from the aforementioned equations coupled with CLC operating parameters. The solids flux, J_s (kg/m^2·s) is calculated by normalizing the solids circulation rate, m_{sol} (kg/s) with the cross-sectional area of the riser, A_{riser} as follows:

$$J_s = \frac{m_{sol}}{A_{riser}} \tag{3.3.18}$$

The solids holdup, ε_s, then can be calculated using Equation (3.3.16). These calculations provide essential parametric information to account for the riser and CFB flow properties.

The analyses on the CFB design parameters are performed on Ni-, Cu- and Fe-based oxygen carrier particles. Experimental results on the reactive properties of particles from both the laboratory and the small pilot systems are used. These results are given in Table 3.4. Because most CLC systems currently are operated under atmospheric pressure conditions, the behavior of the riser operated at 1 atm and 900°C first is examined. The values for U_t, U_c, and U_{se} reported in the table are based on this operating condition. It is found that the airflow rate required is so large that an extremely high (162 m^2) riser cross section is needed to maintain the superficial gas velocity of the riser at 7 m/s.

TABLE 3.4. Key Design Parameters for Different Oxygen Carriers in a 1,000-MW$_{th}$CLC System Using Methane as the Fuel

Metal Type	Ni		Cu		Fe		
Unit Scale for Basic Data	Lab Scale[44,49]	CFB 120 kW[55]	Lab Scale[44]	CFB 10 kW	Lab Scale[47,48]	CFB 300 W[52]	Moving Bed, H$_2$ 25 kW
Particle Type	NiO/MgAl$_2$O$_4$	NiO/MgAl$_2$O$_4$	CuO/Al$_2$O$_3$	CuO/Al$_2$O$_3$	Fe$_2$O$_3$/MgAl$_2$O$_4$	Fe$_2$O$_3$/Al$_2$O$_3$	Composite Fe$_2$O$_3$
Airflow Rate @ 1,000 MW$_{th}$ and 10% Excess (mol/s)	11,784						1,309
Volumetric Airflow at 1 atm and 900°C (m^3/s)	1,134						126
Particle Circulation Rate @ 1,000 MW$_{th}$ (kg/s)	4,000	10,000	3,000	6,000	8,000	10,000	800
Reducer Solids Inventory (t)	230	160	70	total 2,100	500	1,200	1,500 Total
Oxidizer Solids Inventory (t)	390	80	390	n/a	n/a	350	
Medium Particle Size (μm)	153	120	300	200	153	151	2,000
Particle Density (g/cm^3)	1.9	5	2.5	2.5	4.1	2.15	2.5
U_t (m/s) (900°C, 1 atm)	2	0.8	2	1.2	1.1	0.6	11
U_c (m/s) (900°C, 1 atm)	4	4.8	4.9	4.2	4.8	3.6	4
U_{se} (m/s) (900°C, 1 atm)	6	6.7	7.5	6.1	6.9	4.9	9.7
Typical Riser Superficial Gas Velocity (m/s)	7.00						12
Area of Turbulent Bed Section (if Required) at 1 atm (m^2)	231.47						25.18

164

TABLE 3.4 Continued

Metal Type	Ni	Cu	Fe
Area of Riser Section at 1 atm (m²)	162.03		10.49
Corresponding Riser Diameter (m)	14.37		3.66
Solids Flux at 1 atm (kg/m²·s)	24.69	18.52 37.03 49.37	76.23
Number of Beds Needed Given 8 m ID Riser	3.23		<1
Number of Beds Needed Given 1.5 m ID Riser	91.73		5.94
U_g for a Single 1.5 m ID Riser at 1 atm (m/s)	642.14		71.29
U_g for a Single 8 m ID Riser at 1 atm (m/s)	22.58		2.51 ($U_g < U_t$; Not applicable)
Required Pressure for a Single 1.5 m ID Riser at $U_g = 7$ m/s (atm)	91.73		10.00
Corresponding Solids Flux for a Single Pressurized 1.5 m ID Riser (kg/m²·s)	2,264.69	1,698.51 3,397.03 4,529.37	452.88
Required Pressure for a Single 8 m ID Riser at $U_g = 7$ m/s (atm)	3.23	5,661.71	$U_g < U_t$; Not Applicable
Corresponding Solids Flux for a Single Pressurized 8 m ID Riser (kg/m²·s)	79.62	59.71 119.43 159.24	199.04

165

Risers of such a size pose design and operational challenges and are deemed undesirable; the use of multiple risers for CLC will be more feasible. Considering a riser inner diameter (ID) of either 1.5 m, a common size of fluid catalytic cracking (FCC) riser, or 8 m, a common size of a circulating-fluidized-bed combustor (CFBC) riser,[56] a required number of risers to carry out the CLC operation can be calculated. Table 3.4 presents various design conditions including the superficial gas velocity in the riser for a single riser operated at 1 atm and the required operating pressure for a single riser operated at a superficial gas velocity of 7 m/s.

As can be seen from Table 3.4, both the solids circulation rates and the airflow rates for the CFB-based CLC systems are large for all cases. Although a multiriser design is technically feasible, construction of many high-temperature risers is not practical economically. A reduction in riser cross-sectional area leads to an increase in the riser superficial velocity, which could enhance particle attrition. One approach to reduce the riser cross-sectional area is to increase the operating pressure. For example, an increase of the operating pressure in the riser to 90 atm leads to a single riser with a diameter of 1.5 m. Operating the riser at such a high pressure and high temperature with a very high solids flux yields a riser flow of high densities or high solids holdups. Figures 3.11 and 3.12 describe the relationships between riser diameter and such variables as solids flux, solids holdup, and riser pressure when a single riser is used at a constant superficial gas velocity of 7 m/s. It is seen from the figures that the riser either can be designed at a low pressure with a large cross-sectional area and a low solids flux and holdup, or at an elevated pressure with a small cross-sectional area and a high solids flux and holdup.

Note from the figures that the solids holdup decreases and is followed by a slight increase when the riser diameter increases. With an increasing riser diameter, the decrease in the solids holdup is mainly from the decrease in the solids flux, whereas the slight increase is mainly from the increase in the par-

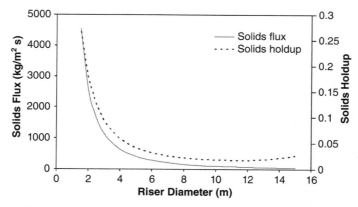

Figure 3.11. The relationship among the riser diameter and the solids flux and the solids holdup for a 1,000-MW$_{th}$ CLC with a riser superficial gas velocity of 7 m/s.

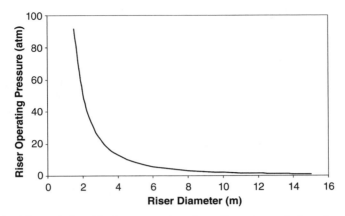

Figure 3.12. The relationship between the riser diameter and the riser operating pressure for a 1,000-MW$_{th}$ CLC with a riser superficial gas velocity of 7 m/s.

ticle terminal velocity caused by the decrease in the operating pressure of the riser. The selection of the optimum operating pressure and cross-sectional area of the riser, however, needs to take into account the incremental capital and operating cost variations with the pressure and bed cross-sectional area. It is further noted from Table 3.4 that for a moving bed chemical looping gasification system using iron oxide as the oxygen carrier, both the solids circulation rate and the airflow rate are low. For the moving-bed operation, the low solids circulation rate results from the high oxygen carrier conversion in the reducer, whereas the low airflow rate results from the significantly reduced oxygen consumption to oxidize the partially regenerated metal oxide state of Fe_3O_4 to the original state of Fe_2O_3.

3.3.3   CLC Reactor System Design

The successful design of an effective CLC system requires ensuring that no gas leak occurs between the two reactors. Because of its ability to handle solids at high circulation rates and to provide good gas and solid contact, the CFB system has been considered as a primary system for carrying out the CLC operation. The design criteria of a CFB system and its application in CLC process operation are given in the following sections.

Circulating-Fluidized-Bed System for the CLC Process
The CFB system has been used extensively in the petrochemical industry for FCC, and in the utility industry for coal combustion.[64] A circulating fluidized bed system consists of four main parts: the riser, gas–solid separator, down-comer/standpipe, and solids flow control system. Figure 3.13(a) shows the schematic diagram of a CFB system. Gases and solids are introduced at the bottom of the riser and flow cocurrently upward. Solids are separated by

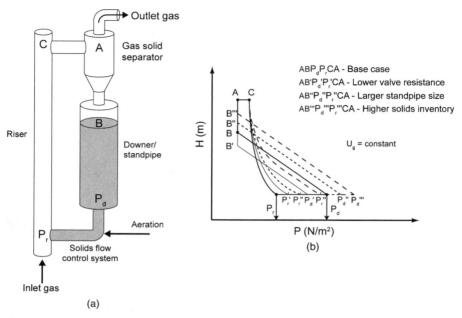

Figure 3.13. Flow diagram and pressure profile in a circulating fluidized bed loop.[56,65]

the gas–solid separator, that is, the cyclone at the top of the riser. The down-comer/standpipe serves two functions: (1) it holds solids as a solids depository used in the circulation of the riser system; (2) it builds up the pressure at the bottom of the downcomer/standpipe, thereby providing the desired solids flow rate in the riser and preventing reverse solids flow and gas leaks from the riser inlet to the downcomer/standpipe. Chemical reactions also may take place in the downcomer.

The hydrodynamics of a CFB system is determined by two operation variables—gas velocity and solids circulation rate. For general chemical looping process applications, two types of the CFB system, low-flux circulating fluidized bed (LFCFB), which usually is accompanied by low-density flow, and high-flux circulating fluidized bed (HFCFB), which usually is accompanied by high-density flow, can be used depending on the solids circulation rate required. The key characteristics of the LFCFB and the HFCFB are described as follows[65,66]:

1. LFCFB[56,65,67,68]:
   a. Solids circulation rate is smaller than $200 \, kg/m^2 \cdot s$ in the riser.
   b. Flow structure consists of core and annulus regions, with the dilute core region surrounded by a dense annulus region.
   c. Axial dispersion is high.
   d. Downward solids flux occurs at the wall.

e. Moderate pressure is required at the bottom of downcomer to drive the solids flow.

2. HFCFB[65–69]:

   a. Solids circulation rate is larger than $200 \, kg/m^2 \cdot s$ in the riser.

   b. Flow structure consists of segregated core and annulus regions that are less segregated compared with LFCFB.

   c. Axial dispersion is low.

   d. Upward solids flux can occur at the wall.

   e. High pressure is required at the bottom of the downcomer to drive the solids flow.

Most CFB systems in industry use nonmechanical solids flow control valves, and the system design is based on the pressure balance in a CFB loop. The pressure balance in both LFCFB and HFCFB and the designs of a gas–solid separator, downcomer/standpipe, and solids flow control system are discussed in subsequent sections.

Pressure Balance

Low-Flux/Low-Density Circulating Fluidized Bed (LF/LDCFB). In an LFCFB, the pressure drop in the riser is small compared with the full range of pressure variation in the CFB system.[56] Figure 3.13 shows a typical flow diagram and the corresponding pressure balance in a CFB loop. In Figure 3.13(a), P_r is the pressure at the bottom of the riser and P_d is the pressure at the bottom of the downcomer/standpipe. Point A refers to the location after the gas–solid separator, and Points B and C refer to the locations at the bed level of the downcomer/standpipe and at the top of the riser, respectively. The base case (ABP$_d$P$_r$CA) in Figure 3.13(b) represents the pressure balance in an LFCFB. The lowest pressure in a CFB loop is located after the gas–solid separator, A. As the solid particles are recovered from the gas–solid separator and return to the downcomer/standpipe, the pressure is almost unchanged from the gas–solid separator, A, to the surface of the downcomer/standpipe, B. In the downcomer/standpipe, the solid particles are under the fluidization condition. The pressure drop along the downcomer/standpipe is proportional to the bed height, and the pressure increases linearly from B to the bottom of the standpipe, P_d. Because a pressure drop needs to be overcome when the solid particles are transferred from the downcomer/standpipe to the riser through a solids flow control valve, the pressure decreases from P_d to the riser bottom, P_r. The pressure decreases gradually in the riser from P_r to the top riser, C. The decrease of pressure from C to A is caused by the pressure drop across the gas–solid separator. From the pressure balance in a CFB loop, it is noted that the key to smooth circulation of the solid particles in the CFB system is to provide a sufficient pressure head on the downcomer/standpipe.

High-Flux/High-Density Circulating Fluidized Bed (HF/HDCFB). In an HFCFB CLC, the pressure drop in the riser is higher than that in an LFCFB. To provide a sufficient pressure head for the solids circulation in a CFB unit with nonmechanical solids flow control valves, three options usually are considered: reducing the valve resistance, increasing the downcomer/standpipe size, and increasing the solids inventory in the downcomer/standpipe.[65] For example, a larger standpipe size and a higher solids inventory allows more solid particles to be stored in the CFB system, thereby maintaining a higher, relatively stable solids level and, hence, a higher-pressure head at the bottom of the downcomer/standpipe. As shown in Figure 3.13(b), all three options increase the pressure at the bottom of the riser, P_r, giving rise to a sufficient pressure head to support the high-density solids suspension in the riser. Another option to achieve a high-flux and a high-density condition in the CFB is to use two or more loop systems. A sufficient pressure head to the downcomer/standpipe also can be achieved by using a pressurized fluidized bed as a solids feeder[70,71] or by using a mechanical solids flow control valve such as a screw feeder to break up the pressure balance in the CFB loop.[72–74] However,

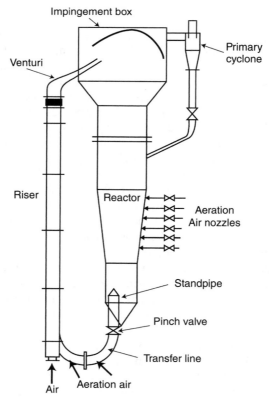

Figure 3.14. Single-loop, high-density CFB system.[75]

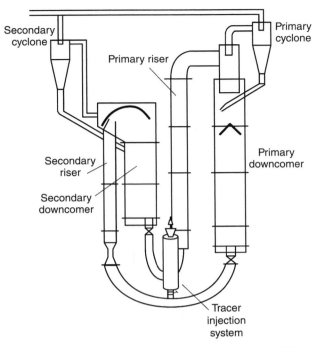

Figure 3.15. Two-loop, high-density CFB system.[76]

both the pressurized fluidized bed solids feeder and screw feeder are commonly applicable only to small units. The downcomer/standpipe with a non-mechanical solids flow control valve serves to provide desired solids feeding functions for industrial CFB CLC systems.

Figures 3.14 and 3.15 show the schematic diagrams of one-loop and two-loop CFB systems, respectively. The one-loop CFB system with a tall downcomer/standpipe and a large inventory shown in Figure 3.14 provides a high solids flux of $350 \, kg/m^2 \cdot s$,[69,75] with a high flux up to $550 \, kg/m^2 \cdot s$ in a one-loop HFCFB system. Knowlton et al.[68] achieved a higher flux up to $782 \, kg/m^2 \cdot s$. For the two-loop CFB system in Figure 3.15, the separated solids from the short downcomer/standpipe of the first riser are transported to a long downcomer/standpipe through a second riser. The long downcomer/standpipe creates sufficient pressure head for the first riser to be operated at the high solids flux condition up to $450 \, kg/m^2 \cdot s$.[76]

High solids flux also can be achieved for a shallow CFB system by a series of downcomers in such a manner that the pressure of the last downcomer is built up by all the preceding downcomers, leading to a pressure that is sufficiently high to provide a high solids flux to the riser. For both the one-loop and the two-loop CFB systems, the solids flow control valve and gas–solid separator are designed under the minimum pressure loss condition. Figure 3.16 shows the schematic diagram of the CFB system with one downcomer loop

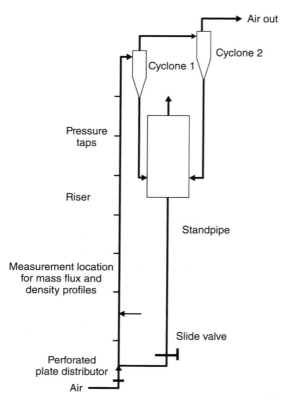

Figure 3.16. Schematic diagram of a CFB test facility at Particulate Solid Research Inc. (PSRI) (Chicago, IL) with solids flux up to 782 kg/m^2·s.[68]

that achieved the solids flux of 782 kg/m^2·s indicated earlier.[68] In the system, a fluidized bed is used to collect the solids recovered from the two cyclones at the top of the riser. The structure of a long standpipe underneath the fluidized bed is used to provide adequate pressure head for the riser to be operated under such high solids flux conditions. Figure 3.17 shows the radial profiles of the solids mass flux at low overall solids flux and high overall solids flux in this system. The figure indicates downward and upward solids flux near the wall at low and high solids fluxes, respectively. Figure 3.18 shows the radial profiles of bed density and solids mass flux at various overall solids fluxes from 49 kg/m^2·s to 782 kg/m^2·s in this system. At an overall solids flux of 782 kg/m^2·s, the local upward solids flux first increases and then decreases along the radius, and can reach a maximum of ~880 kg/m^2·s at $r/R = 0.8$, where r is the radial coordinate and R is the radius of the riser.

Gas–Solid Separator
Cyclones are used widely as the gas–solid separator in a CFB system. The cyclone separates solid particles from a fluid stream by exerting centri-

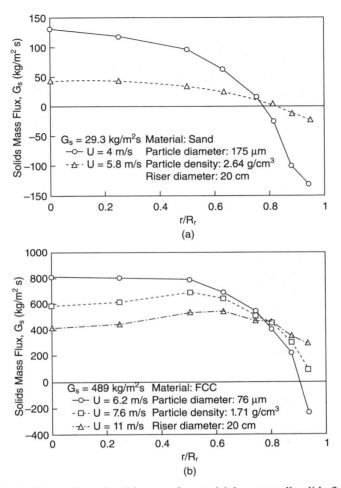

Figure 3.17. Radial profiles of solids mass flux at (a) low overall solids flux and (b) high overall solids flux.[68]

fugal force on the particles. The advantages of the cyclone over other gas–solid separators such as wet scrubber, electrostatic precipitator, and filter include[77]:

1. no moving parts,
2. relatively inexpensive to construct,
3. relatively low pressure drop,
4. low maintenance cost, and
5. suitable for high-temperature operation.

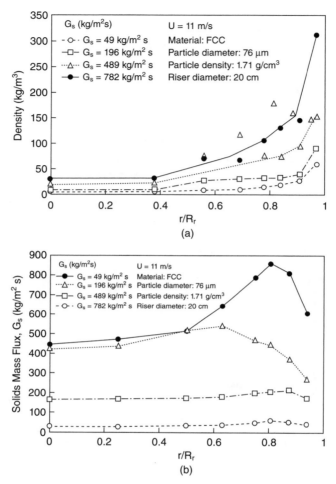

Figure 3.18. Radial profiles of (a) bed density and (b) solids mass flux at various solids fluxes.[68]

The cyclone efficiency, however, is largely reduced when separating particles smaller than 10 μm. There are two major types of cyclones used in industry, namely, the reverse-flow cyclone and the uniflow cyclone.[77]

In a reverse-flow cyclone, the gas and solids enter the cyclone from the top through tangential inlet vanes or axial inlet vanes. With the presence of the gravitational and centrifugal forces, the solids move downward along the wall of the barrel to the outlet at the bottom of the cyclone. The gas, however, reverses its flow direction and exits the cyclone from the top outlet at the center of the cyclone.

In a uniflow cyclone, the gas and solids enter the cyclone in a manner similar to that in a reverse-flow cyclone, but both the solids and the gas exit from the bottom of the cyclone. Specifically, the solids move downward along the wall of the barrel to the outlet along the wall of the cyclone, whereas the gas moves downward to the outlet at the center of the cyclone.

The reverse-flow cyclone is the type most commonly used because of its high efficiency. There are three types of reverse-flow cyclones based on the inlet configuration: tangential inlet, axial inlet, and volute inlet cyclones, as shown in Figure 3.19(a). For both the tangential inlet and the volute inlet

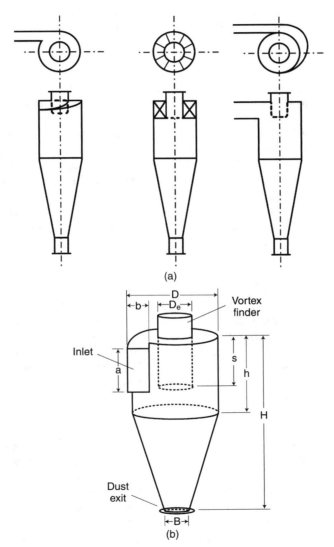

Figure 3.19. Cyclones: (a) inlet configuration of a cyclone; (b) schematic diagram of a cyclone.[78]

TABLE 3.5. Common Designs of Reverse-Flow Cyclone[78]

Design Type	Duty	a/D	b/D	De/D	S/D	h/D	H/D	B/D
Stairmand	HE[a]	0.5	0.2	0.5	0.5	1.5	4.0	0.375
Swift	HE	0.44	0.21	0.4	0.5	1.4	3.9	0.4
Lapple	GP[b]	0.5	0.25	0.5	0.625	2.0	4.0	0.25
Swift	GP	0.5	0.25	0.5	0.6	1.75	3.75	0.4
Stairmand	HT[c]	0.75	0.375	0.75	0.875	1.5	4.0	0.375
Swift	HT	0.8	0.35	0.75	0.85	1.7	3.7	0.4

[a]HE, high efficiency; [b]GP, general purpose; [c]HT, high throughput.

cyclones, the gas–solids mixture enters the cyclone tangentially. The volute inlet cyclone is applied to minimize the impact of the inlet solids on the outlet gas pipe for cases in which the diameter of the outlet gas pipe is large. For the axial inlet cyclone, the gas–solids mixture enters the cyclone axially, and rotary centrifugal motion is impelled by the axial swirl vanes. The tangential inlet and the volute inlet are the two common inlet configurations for cyclones applied in industry. Figure 3.19(b) shows the schematic diagram of a tangential inlet reverse-flow cyclone with typical design variables. These variables include the cyclone diameter, D, and seven dimension ratios—a/D, b/D, De/D, S/D, h/D, H/D, and B/D. Table 3.5 presents the values of these variables for designing high-efficiency, general-purpose, or high-throughput reverse-flow cyclones.[78] The detailed design procedure for a cyclone is available in the literature.[56,77]

Standpipe-Flow and Solid- Flow Control Valves
The solids flow in a standpipe is characterized by the downward flow of solids with the aid of gravitational force against a gas pressure gradient. Solids fed into the standpipe are often from hoppers, cyclones, or fluidized beds. A standpipe can be either vertical or inclined, and its outlet can be simply an orifice or can be connected to a valve or a fluidized bed. There can be aeration along the side of the standpipe.[56] Figure 3.20 illustrates an application of two types of standpipes: overflow and underflow standpipes. In the overflow standpipe shown in Figure 3.20(a), the flow of solids goes through a standpipe between two beds, that can be operated in a moving bed or in a fluidized bed mode. Solids enter the standpipe from the surface of the upper bed. To reach a high-pressure head, the outlet of the standpipe is immersed in the lower bed. No valve is required at the outlet of the pipe because the solids flux depends on the overflow rate in the fluidized bed. The pressure drop in the standpipe, ΔP_{sp}, equals the pressure drop in the column encompassing the pressure drop contributions from the lower fluidized bed, ΔP_{LB}, the grid or distributor, ΔP_{grid}, and the upper fluidized bed, ΔP_{UP}, as illustrated on the righthand side of Figure 3.20(a). H_{SP} is the bed height in the standpipe. The pressure distribution profiles in a standpipe for a given particle depend on the gas velocity. The pressure

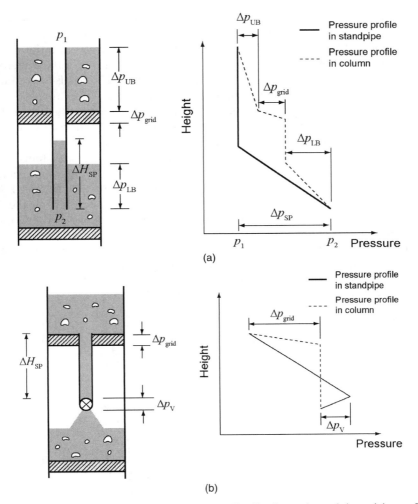

Figure 3.20. Schematic diagrams and pressure distributions of standpipes: (a) overflow standpipe; (b) underflow standpipe.[56]

distribution profiles reflect the difference in the solids concentration for the upper bed, the lower bed, and the standpipe.

In an underflow standpipe, the flow takes the solids from the bottom of an upper bed to a standpipe, and the solids are discharged to a freeboard region of a lower fluidized bed, as shown in Figure 3.20(b). The flow rate is controlled by a valve at the outlet of the standpipe. The standpipe usually is operated in a moving-bed mode and the pipe is designed high enough to provide a seal pressure for the solids flow. As shown on the righthand side of Figure 3.20(b), the pressure drop in the standpipe, including the pressure drop built by the particles in the pipe and the pressure loss from the valve, ΔP_V, equals the pressure drop in the column, which only includes the pressure drop in the grid or distributor, ΔP_{grid}. Small particles basically will be discharged into the underflow standpipe if the particle size distribution is not uniform in the system.

In a standpipe system, the solids flow rate is controlled by solids flow control valves, which can be mechanical, nonmechanical, or a combination. The mechanical valves, as the name implies, use mechanical means such as a slide valve or a butterfly valve to control the solids flow rate.[79] Particle clogging and attrition, valve erosion by solid particles, and maintaining the gas seal can be challenges for the operation of mechanical valves in solids flow systems. Thus, nonmechanical valves are attractive options. Figure 3.21 shows six arrangements of nonmechanical valves for solid flow control in CFB systems. For a nonmechanical valve, only aeration gas is needed, and by adjusting the gas aeration rate, the solids flow rate can be controlled to a dense or dilute environment in a flow system.[80]

The loop seal or reversed V-valve has a restriction on the extent of the solids circulation rate in the CFB system, but can have a steady operation. The L-valve has a large range of operating solids circulation rate, but is constrained by the geometry and design of the valve system. The J-valve has a low restric-

Loop seal, reverse
V-valve or fluoseal

L-valve

J-valve

H-valve

Inclined inlet with
mechanical valve
(slide valve, butterfly valve)

Inclined inlet with concentric
slide valve in the riser

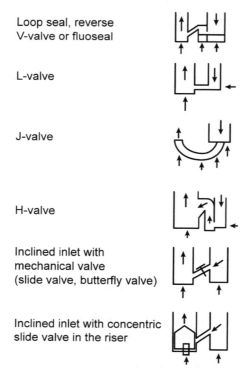

Figure 3.21. Nonmechanical valve applications for solids feeding in CFB risers.[79]

tion on the extent of the solids circulation rate. The H-valve is limited in the solids circulation rate, but can have a steady operation. For an inclined inlet with a mechanical slide valve or a butterfly valve, the solids flow rate is controlled by varying the valve opening. This arrangement, however, has a poor stability of operation. The inclined inlet with a concentric slide valve in the riser can achieve a large solids circulation rate and a small pressure drop. For a LFCFB, all types of inlet configurations in Figure 3.21 can be applied, because the pressure loss is not a critical issue.[79] For a HFCFB, the configurations of a J-valve and an inclined inlet with a concentric slide valve in the riser could be a feasible choice, because both yield a small pressure loss and can achieve a high solids circulation rate in the CFB system.

3.3.4   Gaseous Fuel CLC Systems and Operational Results

Promising results from the laboratory- and bench-scale experiments prompted efforts for testing the CLC process at larger scales. Prior operational experiences in industrial CFB processes provide a logical basis and a fundamental framework for their extension to chemical looping combustion applications. The parameters governing the design of the CFB are discussed in the previous sections. The focus in this section is on the available key results from the subpilot and above scale CLC systems being operated worldwide. To evaluate the performance of various processes, parameters defined in Section 3.3.1 including fuel conversion (Equations [3.3.8] and [3.3.9]) and solids conversion (Equation [3.3.4]) are used.

Chalmers University 10-kW$_{th}$ CLC System
Chalmers University of Technology in Sweden, as part of the European Union (EU) project GRACE (Grangemouth Advanced CO_2 Capture), designed and constructed a 10-kW$_{th}$ dual fluidized bed CLC system to validate the feasibility of the CLC Process and to evaluate the performance of their oxygen carriers (OCs) under looping operating conditions. The schematic diagram and a photograph of their system are shown in Figure 3.22.

Reactor Design and Standard Operation Procedure. In Figure 3.22, the interconnected fluidized bed CLC system consists of a reducer (fuel reactor, A in Figure 3.22 (a)), oxidizer (air reactor, B), riser (C), cyclone, and two loop seals. This system was designed to use gaseous fuels such as natural gas. The oxygen carrier oxidizes the fuel in the reducer, generating CO_2 and H_2O. The reduced oxygen carrier then enters the first loop seal where the carrier particles move to the oxidizer. The loop seal prevents the leakage of gaseous fuels as well as air to the oxidizer and to the reducer, respectively. The preheated air supply to the oxidizer regenerates the reduced oxygen carrier exothermally and conveys the oxygen carriers through the riser to the cyclone. The cyclone separates regenerated oxygen carrier from spent air. The carrier then is recycled to a second loop seal, thus preventing the spent air from being entrained

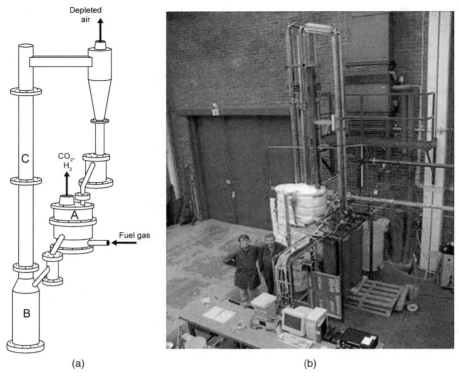

(a)   (b)

Figure 3.22. (a) Schematic diagram and (b) photograph of the $10\,kW_{th}$ dual fluidized bed CLC system constructed at Chalmers University.[81] With permission from Prof. Anders Lyngfelt of Chalmers University of Technology.

to the reducer. This complete loop results in a CO_2-rich stream from the reducer after steam condensation and hot spent airstream from the cyclone. The key design and operation parameters are given in Table 3.6.

The reducer (A), a bubbling fluidized bed, is where the fuel is converted using the oxygen from the oxygen carrier. The size of the reducer is determined from such information as the terminal settling velocity (U_t), minimum fluidization velocity (U_{mf}), solids flux, and gas volumetric flow rate. The cross-sectional area of the reducer increases with the reducer height with a maximum diameter of 0.25 m to accommodate the gas volume increase caused by the combustion of natural gas. The reactor height is 0.35 m with the freeboard height designed at about twice the dense bed height. The fluidization gas velocity in the reducer ranges from $5U_{mf}$ to $15U_{mf}$ and is always higher than $0.25U_t$. Although it is required in a commercial CLC system, a separator that separates particles from the gaseous product of the reducer is not installed in the present setup.

TABLE 3.6. Operating Parameters and Design Values for the Chalmers University 10-kW$_{th}$CLC System[81]

Parameter	Reducer/Fuel Reactor	Oxidizer/Air Reactor	Riser
Reactor Inner Diameter (m)	0.25	0.14	0.072
Reactor Height (Reactor Bed Height) (m)	0.35 (0.13)	0.53	1.85
Gas Velocity U/U_t	—	1.2–3	4–10
Gas Velocity U/U_{mf}	5–15	—	—
Particle Apparent Density (kg/m^3)		2,500–5,400	
Particle Mean Diameter (m)		$(100–200) \times 10^{-6}$	
Designed Operating Pressure (Pa)		100,000	
Designed Operating Temperature (°C)		950	
Designed Fuel		Methane	
Designed Air/Fuel Ratio		1.2–2.6	
Designed Fuel Processing Capacity/Fuel Power (kW$_{th}$)		10	

The regeneration and conveying of the reduced oxygen carrier particles obtained from the reducer are carried out in an oxidizer (B) and a riser (C) operated in the bubbling or turbulent and dilute or dense transport fluidization conditions, respectively. The oxidizer (B) is where oxidation of the particles using preheated air occurs. Given a large airflow requirement, a large oxidizer was chosen, 0.14 m in diameter, so that a desired fluidization regime can be established. The height of the oxidizer is influenced by the particle reactivity, reaction time, and the need for cooling. On the top of the oxidizer is a conical section that serves as a transition area from the oxidizer to the riser (C). The riser diameter (0.072 m) nearly halves that of the oxidizer to achieve a high air velocity for transporting oxygen carrier particles to the reducer. This design, combining the oxidizer and the riser, avoids a secondary air injection system.

The Chalmers University 10-kW$_{th}$ unit was operated continuously for multiple days using various oxygen carrier particles. In the start-up stage, the temperature of the oxidizer first is increased by the combustion of gaseous fuel. In the meantime, air is used to circulate the oxygen carrier particles in the system while distributing the heat. After achieving the required temperature in the looping system through solids circulation, the reducer is fluidized using inert N$_2$ gas, whereas airflow is continued in the oxidizer. After a stable temperature profile is observed, the gaseous fuel is sent into the reducer in place of the inert N$_2$ gas. Continuous operation for ~8–11 hours is carried out under these conditions to obtain operational data such as oxygen carrier reactivity, conversion, and attrition. In the shutdown stage, the fuel flow is stopped, and the reducer is fluidized using N$_2$ for a short period to purge the remaining fuel. Both reactors then are fluidized using air. This procedure ensures complete oxidation of oxygen carriers before the next start-up begins.

TABLE 3.7. Oxygen Carrier Properties Used in the 10-kW$_{th}$ Chalmers University CLC Unit

Parameter	Particle A	Particle B
Production Method	Freeze-Granulation	Spin-Flash Drying
Active Material	NiO	NiO
Support Material	$NiAl_2O_4$	$NiAl_2O_4$
Fraction of Active Material (mass%)	40	60
Size Range of Particles (µm)	90–212	45–250
Apparent Density of Particles (kg/m³)	3,800	4,400
Crush Strength (N)	2.5	5.3

Test Results. The results obtained from the different experiments performed in the 10-kW$_{th}$ system using different oxygen carriers and operating parameters are discussed in this section along with an estimate of the solids inventory required for commercial-scale operations.

Oxygen Carrier Used: The performance of a CLC system is highly dependant on the oxygen carrier performance. Table 3.7 indicates the properties of the two types of the oxygen carriers, Types A and B, used in the Chalmers University 10-kW unit.

Fuel and the Solids Conversion: Lyngfelt and Thunman[81] used Particle A from Table 3.7 to carry out a high conversion of fuel for 100 hours. The gaseous product stream from the reducer contains 0.5% CO, 1% H$_2$, and 0.1% methane, giving a heating-value-based fuel conversion efficiency of 99.5%. In these tests, N$_2$ was used to fluidize the loop seals, resulting in the dilution of CO$_2$. In the commercial system, steam would be used in the loop seals to avoid such dilution. Linderholm et al.[82] used Particle B in Table 3.7 for 160 hours of operation of the 10-kW$_{th}$ unit. The CO and methane concentrations obtained from the experiments in the reducer outlet were 0.3–1% and 0.15–0.6%, respectively. The CO$_2$ concentration under an N$_2$-free basis was ~99%. The heating value-based fuel conversion efficiency is ~99.1%. Lyngfelt and Thunman[81] also used iron-based particles as the oxygen carrier for an operating time of 17 hours and found that the CO and CH$_4$ concentrations were in the range of 2%–8%. Johansson et al.[83] investigated the oxidation extent of elutriated samples of Particle A from Table 3.7 and found that they were not completely oxidized. Postexperiment investigations on the oxygen carrier revealed that the reactivity of the particles was maintained.

Effect of Operating Conditions: Extensive testing of the 10-kW$_{th}$ unit indicates that the temperature and the solids circulation rate of the reducer significantly influence the outlet composition of the reducer. Lyngfelt and Thunman,[81] using Particle A from Table 3.7, observed that greater than 790°C, the CO concentration increased with temperature and was similar to those obtained from equilibrium considerations. Linderholm et al.,[82] using Particle B from Table 3.7, made a similar observation studying the variation in the CO concentration under changing temperatures. It was found that an increase in temperature led to a decrease in the CH_4 outlet fraction for a constant solids circulation rate and air-to-fuel ratio.

Linderholm et al.[82] studied the dependence of CH_4 conversion on the solids circulation rate. Results indicated that a lower circulation rate, which leads to lower average mass conversion (ω) of the oxygen carrier in the reducer, yielded a higher CH_4 conversion for a constant temperature. This result is from to the increased metallic nickel phase at lower circulation rates. This metallic Ni catalyzes the methane conversion, resulting in lower methane fractions in the reducer outlet.

Gas Leakage and Carbon Deposition: The leakage of gas from the oxidizer to the reducer or *vice versa* will result in dilution of the CO_2 yield or loss in carbon capture, respectively. The gas leakage to the oxidizer can be monitored by observing the concentration of CO_2 in the cyclone outlet. Lyngfelt and Thunman[81] and Linderholm et al.[82] did not detect significant CO_2 at the outlet. The absence of CO_2 from the oxidizer also indicates that carbon deposition is not significant because carryover of the deposited carbon from the reducer to the oxidizer would also result in CO_2 formation. A direct carbon deposition test was performed by Linderholm et al.[82] In this test, air was introduced to the reducer after CLC operation and N_2 purging. By measuring the amount of CO_2 produced after air injection, the amount of deposited carbon was estimated. Several grams of carbon were deposited after 8-h continuous operation, much lower than the amount of carbon converted, which corresponds to 9 g of carbon per minute.

Attrition and Agglomeration: Attrition and agglomeration directly affect the extent of particle replacement and particle flow during CLC operation. To reduce the particle purge rate, it is desirable for the oxygen carrier particles to maintain chemical reactivity and physical integrity. The attrition rate of the particles is an important parameter that affects flow performance and the economics of the CLC process. A low attrition rate of the particles is highly desirable. Minimal agglomeration also is desired so that the property of particle flow is stable in the fluidized bed CLC operation. The attrition rate was determined from the collection of particles smaller than 45 μm. Under this condition, Lyngfelt and Thunman[81] calculated the attrition rate for Particle A to be 0.0023% per hour. This figure corresponded to an approximated particle lifetime of 40,000 hours. Linderholm et al.[82] investigated the attrition for Particle B and found that it was 0.022% per hour. This figure corresponded to an approximated particle lifetime of 4,500 hours. Linderholm et al.[82] also performed substoichiometric runs that resulted in extensive agglomeration of the particles. The reason was associated with the excess conversion of the metal oxide to metal. The agglomerates were a couple of centimeters in size, causing the defluidization phenomenon in the bed.

Solid Particles Cost, Circulation Rate, and Inventory: Based on these considerations, Lyngfelt and Thunman[81] estimated the cost of the particles is \$5.2/kg (4€/kg), requiring a solids inventory in the system of 100–200 kg/MW$_{th}$. Linderholm et al.[82] estimated the required circulation rate to maintain adiabatic operation at a large scale to be 4 kg/s·MW$_{th}$, assuming a difference in mass conversion ($\Delta\omega$) of 0.2.

CSIC 10-kW$_{th}$ CLC System

A 10-kW$_{th}$ CLC system was constructed at Instituto de Carboquimica (CSIC) in Zaragoza, Spain, to determine the fuel conversion, mass balance, agglomeration, attrition, and reactivity of Cu-based oxygen carrier particles under long-term CLC operating conditions.[53] The schematic diagram and a photograph of the 10-kW$_{th}$ unit are shown in Figure 3.23.

Reactor Design and Standard Operation Procedures. Figure 3.23(a) is the schematic diagram illustrating major parts of the system. They include a 0.1-m ID bubbling fluidized bed as the reducer, a 0.16-m ID bubbling fluidized bed as the oxidizer, a riser conveying the oxygen carrier particles to the reducer, two loop seals, three cyclones recovering the oxygen carrier particles, a solids reservoir, and a solids valve above the reducer controlling the solids flow rate. An external oven mount over the reducer is used to provide necessary heat during start-up. Filters are used to recover the fine powders after the cyclone. In the reducer (A in Figure 3.23[a]), the fuel, methane, acts as the fluidizing agent with the oxygen carrier particles as the bed material. The bed height is 0.5 m, and the freeboard height is 1.5 m. Methane reacts with the Cu-based oxygen carriers to produce CO_2 and H_2O. After the oxygen carrier is reduced, the particles are transferred to the oxidizer through a loop seal. The oxidizer (B) is where air, as the fluidizing agent, oxidizes the reduced oxygen carrier from the reducer. The bed height for the oxidizer is 0.5 m, and the freeboard height is 1 m. The residence time of the oxygen carrier in the oxidizer is sufficient for complete oxidation of the particles. These completely regenerated oxygen carrier particles from the oxidizer are conveyed pneumatically to the solids reservoir through the riser. Nitrogen is used in the two loop seals.

The CSIC 10-kW$_{th}$ CLC unit was operated using different sizes of copper-based oxygen carriers. The continuous testing began with utilization of preheated air to fluidize the oxygen carriers both in the reducer and the oxidizer and to heat the oxygen carriers to 500°C. To increase the temperature to 800°C, the fluidizing agent in the reducer was changed to nitrogen, and external heating was used. To distribute the heat throughout the system, particle circulation was initiated. When oxygen was not detected in the outlet gas from the reducer, the fluidized gas in the reducer was switched from N_2 to 75% N_2 and 25% CH_4. To terminate the reaction, the reducer fluidizing gas was switched to N_2. The process for each type of oxygen carrier particles was operated

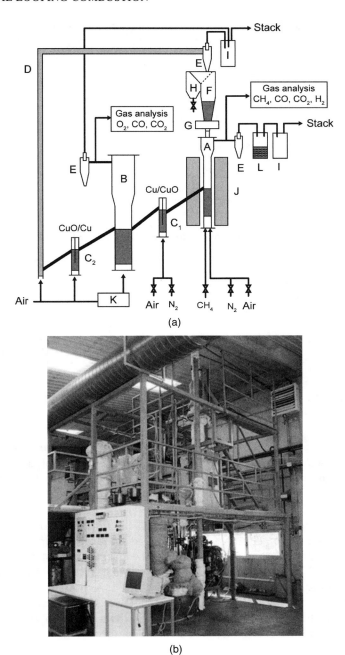

(a)

(b)

Figure 3.23. (a) Schematic diagram showing A = reducer, B = oxidizer, C = loop seals, D = riser, E = cyclone, F = solids reservoir, G = solids valve, H = diverting solids valve, I = filters, J = oven, K = air preheater, and L = water condenser; (b) photograph of the 10-kW$_{th}$ CLC system constructed at the Instituto de Carboquimica. With permission from Dr. Luis F. de Diego of CSIC.

TABLE 3.8. Operation Condition for the CSIC 10-kW$_{th}$CLC Unit[53]

Parameter	Reducer/Fuel Reactor	Oxidizer/Air Reactor	Riser
Bed Height (m)	0.5	0.5	
Pressure Differential (cm H$_2$O)	30–40	30–40	
Gas Velocity (U, m/s)	0.07–0.14	0.3–0.85	5–7
Temperature (°C)		700–800	
Solids Circulation Flux (G_s, kg/hr)		60–250	
Solids Inventory (kg)		21	
Designed Fuel Processing Capacity/Fuel Power (kW$_{th}$)		10	

TABLE 3.9. Oxygen Carrier Particle Properties Used in the CSIC 10kW$_{th}$CLC Unit[53]

Parameter	Particle A	Particle B
Diameter of Particles (d_p, μm)	100–300	200–500
Apparent Particle Density (kg/m^3)	1,500	1,560
Porosity	54.8	53.7
Active Material (CuO wt.%)	14	
Support	Commercial γ-Alumina	

continuously for 100 hours with 60 hours of fuel injection. More information on the operation conditions is given in Table 3.8.

Test Results. The 10-kW$_{th}$ CLC unit was reported to be operated smoothly using Cu-based oxygen carriers, with methane as the fuel under varying solids circulation fluxes. The key results from the testing of the unit are as follows.

Oxygen Carrier: In the CLC process at CSIC, the impregnation method using commercial γ-Alumina (Puralox NWa-155; Sasol Germany GmbH, Hamburg, Germany) support particles was used for the preparation of oxygen carrier particles with a CuO content of 14 wt.%.[53,84] Unlike other oxygen carriers, the Cu-based oxygen carrier undergoes exothermic reactions in both the reduction and oxidation stages.[85] This unique property of Cu-based oxygen carriers simplifies the heat integration in the system. Two types of oxygen carrier particles with different sizes were prepared. The properties of these particles are given in Table 3.9.

Fuel and Solids Conversion: The fuel conversion in the reducer is one of the most important parameters in the CLC Process. The fuel conversion is affected by the normalized molar flow rate ratio (φ, the reaction stoichiometry adjusted flow rate ratio between the oxygen carrier and the fuel).[86] For a CLC process using Cu-based oxygen carrier particles, ϕ is given by

$$\phi = \frac{F_{CuO}}{4F_{CH_4}} \tag{3.3.19}$$

where F_{CuO} and F_{CH_4} are the molar flow rates of Cu-based oxygen carrier particles and methane, respectively. The value of ϕ is varied by varying the solids circulation rate, and its influence on the fuel conversion is studied. With an increase in ϕ from 0.75 to 1.5, a notable increase in CH_4 conversion is observed. When ϕ reaches 1.4, the fuel conversion plateaus at nearly 100% when Particle B from Table 3.9 is used. The oxygen carrier samples were collected at different operating times and tested in a TGA for reactivity using a gas mixture of 15 vol.% CH_4, 20 vol.% H_2O, and 65 vol.% N_2 for reduction and pure air for oxidation. The test revealed a decrease in oxygen-carrying capacity in samples with an increasing operating time. It was found during the experiment that the elutriated fines were rich in CuO. Therefore, the loss in oxygen-carrying capacity was partially attributed to the more severe attrition of CuO compared with the support. It also was found that the oxygen-carrying capacity as well as the CuO content in the particles stabilized after 50 hours of operation. The stabilized CuO content was reported to be 9.5 wt.% for Particle A and 10 wt.% for Particle B.

Gas Leakage and Carbon Deposition: The gas leakage between the reducer and the oxidizer was monitored by analyzing the gas outlet stream composition from the individual reactors. No CO and/or CO_2 were reported in the analysis of the outlet stream from the oxidizer, indicating that there was no gas leakage from the reducer to the oxidizer. The result also indicates that the carbon deposition is insignificant.

Attrition and Agglomeration: For the CSIC CLC process using a Cu-based oxygen carrier, the attrition rate of the oxygen carriers were reported with a cutoff powder size of 40 μm. The attrition rate at the beginning of the operation was very high and decreased with time to a stable value of 0.04 wt.%/h after 50 hours for both particles A and B.[86] With the condition of the attrition rate described on oxygen carrier particles, the estimated lifetime of the particles would be ~2,400 hours. With this lifetime of the particles, the solids in the CLC unit need to be replaced 3.4 times per year, which was indicated to be suitable for industrial applications. Because of the large weight fraction of support, no signs of agglomeration were identified during the CSIC CLC operation using the Cu-based oxygen carrier.[53,84]

VUT 120-kW$_{th}$ CLC System
A dual circulating-fluidized-bed (DCFB) system was designed and constructed at the Vienna University of Technology (VUT) in Austria to study the CLC

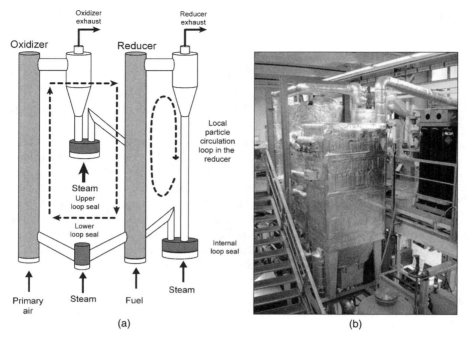

Figure 3.24. (a) Schematic diagram and (b) photograph of the dual circulating fluidized bed (DCFB) system constructed at the Vienna University of Technology. With permission from Dr. Tobias Pröll of the Vienna University of Technology.

process. The DCFB system has a designed fuel power of $120\,\mathrm{kW_{th}}$.[87,88] The schematic diagram and photograph of the DCFB system are given in Figure 3.24.

Reactor Design and Standard Operation Procedure. As can be seen from Figure 3.24, the DCFB system consists of mainly a reducer (fuel reactor), an oxidizer (air reactor), an upper loop seal, a lower loop seal, an internal loop seal, and two cyclones. The system forms two particle circulation loops: the global particle circulation loop, and the local particle circulation loop. The global particle circulation loop encompasses the reducer and the oxidizer connected with the oxidizer cyclone and the upper and lower loop seals. The local particles circulation loop encompasses the reducer, the reducer cyclone, and the internal loop seal. The oxidizer is operated at the dilute or dense transport regime, and the reducer is operated at near the turbulent regime. The cyclone and internal loop seal placed around the reducer allow for local particles circulation within the reducer, independent of the global particles circulation between the oxidizer and the reducer. The fluidization agents for the reducer and the oxidizer are gaseous fuel and air, respectively. Gaseous fuel is injected at the bottom of the reducer, and air is injected at both the bottom (primary

TABLE 3.10. Design and Operating Parameters of the DCFB System[55]

Parameter	Oxidizer/Air Reactor	Reducer/Fuel Reactor
Reactor Inner Diameter (m)	0.15	0.159
Reactor Height (m)	4.1	3
Height of Primary Gas Inlet (m)	0.025	0.06
Height of Secondary Gas Inlet (m)	1.325	n/a
Inlet Gas Flow (Nm^3/hr)	138	12
Outlet Gas Flow (Nm^3/hr)	113.9	35.9
Designed Operating Temperature (°C)	940	850
Archimedes Number	7.55	9.13
Superficial Gas Velocity (m/s)	7.32	2.08
Gas Velocity Ratio U/U_{mf}	1,280.4	315.4
Gas Velocity Ratio U/U_t	15.5	3.8
Designed Fuel	Natural Gas	
Designed Air/Fuel Ratio	1.2	
Designed Fuel Processing Capacity/Fuel Power (kW_{th})	120	

air inlet) and the middle portion (secondary air inlet) of the oxidizer. The staged injection of air enables effective control over the oxidizer particles entrainment rate, which is equal to the global solids circulation rate. At a fixed overall air injection rate, a larger primary air injection rate leads to a higher global solids circulation rate and *vice versa*. Because it is compatible with both the oxidizer and the reducer, steam is used as the fluidization gas in all the loop seals. The design and operating parameters of the DCFB system are given in Table 3.10.

The fundamental characteristics of the VUT DCFB system and the systems developed by Chalmers University and CSIC are similar, even though the fluidization regimes, either bubbling or turbulent, under which the reducer is operated, are different. Because the operating gas velocities for the bubbling and turbulent fluidized beds are different, it gives rise to different gas–solid contact efficiencies between these fluidized beds, and, hence, different reactant conversions and different cross-sectional areas required of a reactor for a given volumetric gas flow rate. Table 3.11 compares the specific operating parameters of the VUT reducer and those of the bubbling fluidized bed reducers in the CLC unit by Chalmers University.[87]

Test Results. The VUT DCFB system was constructed in 2007. Various tests have been performed since early 2008. Different types of fuels such as hydrogen, syngas, natural gas, and propane are converted in the DCFB with both ilmenite particles and nickel-oxide-based particles. In the following, the start-up and shutdown procedures for the VUT DCFB system are described, as well as the test results obtained using ilmenite-based oxygen carrier particles

TABLE 3.11. Comparison of the Operating Parameters of VUT Reducer and Chalmers University Reducer; Both Reactors are Normalized to 120-kW$_{th}$ Scale[87]

Parameter	VUT Turbulent Bed Reducer	Chalmers Bubbling Bed Reducer
U/U_{mf}	315	~10
Gas Residence Time (s)	1.3	~0.9
Solids Inventory (kg/MW$_{th}$)	250	~420
Diameter (m)	0.159	~0.850

and NiO-based oxygen carrier particles. Practical issues encountered during the DCFB operation such as gas leakage and carbon depositions also are discussed.

Start-up and Shutdown Procedures: Prior to the injection of the gaseous fuel, the system is loaded with oxygen carrier particles. The system then is pre-heated by a 35-kW natural gas burner attached to the reducer and a 15-kW air preheater installed on the oxidizer. During the preheating, the particles are fluidized with air and circulated throughout the system. The temperature of the system increases gradually. Once the safe ignition temperature of hydrogen is reached, the fluidizing gas in the reducer is switched to N_2, and H_2 is added gradually. When the system reaches ~730°C, which is the safe ignition temperature of propane, propane is injected into the oxidizer via a separate nozzle for direct preheating. The natural gas burner simultaneously is switched off. Once the temperature of the system exceeds 780°C, natural gas also can be introduced to the oxidizer as the preheating fuel. Upon reaching the operating temperature of the system, the fluidizing gas in the loop seals is switched to steam. Meanwhile, the gaseous fuel is introduced to the reducer, and the preheating fuel gradually is switched off. Because the system is highly exothermic and the solids circulation rate is high, it can be operated without external heat input, provided that the particles are active enough in converting the fuels. Cooling jackets are used to maintain a stable operating temperature. To shut down the system, the fluidizing gas in the reducer first is switched to N_2 to purge the fuel in the reducer. Once the purging step is completed, air is used to circulate the particles throughout the system during system cooling.

DCFB Opeartion Using Ilmenite Particles: To examine the feasibility of cheap, naturally occurring ore as the oxygen carrier for applications in the DCFB system, ilmenite ($FeTiO_3$) particles with a mean diameter of ~220μm are used to convert H_2, syngas, and natural gas. A fuel conversion of >96% is obtained when hydrogen is used as the fuel. The corresponding reducer operating temperature is 950°C with a total solids inventory of a fuel processing capacity (fuel power) of ~24 kW$_{th}$. Because of the relatively low reactivity of the ilmenite ore, an increase in fuel power leads to a decrease in the hydrogen fuel conversion. For example, the hydrogen conversion is lower than 87.5%, when the DCFB system is operated at $90 kW_{th}$.[89] When syngas with 1:1 H_2 to CO molar ratio is used, a similar trend is observed. When the reducer is operated at 960°C with a system solids inventory of 70 kg, the syngas conversion decreases from ~85% at $40 kW_{th}$ to ~72% at $90 kW_{th}$. Throughout the test, the hydrogen conversions are at least 10% higher than the carbon monoxide conversions[90] indicating that ilmenite is more reactive while using hydrogen-rich fuel.

Similar to the previous cases, the degree of the oxidation of natural gas is characterized by both the fuel conversion and the CO_2 yield. The effects of several operating parameters including fuel processing power, the system solids inventory, and catalyst addition on the natural gas conversion performance are tested. The reducer is operated at 950°C in all these tests. It again is found that an increase in the fuel processing capacity leads to a decreased methane conversion and CO_2 yield. The change in the system solids inventory from 70 kg to 85 kg, however, shows little effect on the natural gas conversion. An addition of olivine, which is active in converting hydrocarbons catalytically, led to a notable increase in the natural gas conversion. These results indicate that the reactivity of the ilmenite particles is the key factor that affects the process performance. Throughout the test, the methane conversion and the CO_2 yield are low with the highest being ~72% and ~67%, respectively. Such conversions are achieved at a low fuel processing capacity $(20 kW_{th})$ with the addition of olivine (ilmenite:olivine = 4.7:1 by weight).[90] Tests performed using the VUT DCFB system indicate that ilmenate, a cheap naturally occurring material, has the potential to be used in the CLC systems, but it is desirable to improve the fuel conversions further. Such improvements can be made by both optimization of the current CLC process system and by enhancement of the ilmenite particle reactivity. The results also suggest that further treatment of the reducer flue gas from the DCFB-based CLC system is necessary because of the presence of a significant amount of unconverted fuel.

DCFB Using NiO-Based Particles: NiO-based oxygen carrier particles also are tested in the VUT DCFB system. The oxygen carrier particles used in the tests are a 50:50 mixture of $NiAl_2O_4$-supported NiO particles and $MgAl_2O_4/NiAl_2O_4$-supported NiO particles.[55,91] The active NiO in the mixture of particles amounts to 40% of the particle weight. The mean particle size is 135 μm.[55] In most of the tests concerning NiO-based particles, the solids inventory is kept at 65 kg. Because the NiO-based particles have a much higher reactivity compared with ilmenite particles, the system can reach or exceed the designed fuel processing capacity of $120 kW_{th}$ without sacrificing fuel conversion. Experiments performed on the DCFB achieved up to $140-kW_{th}$ fuel process capacity using natural gas, and both the fuel conversion and the CO_2 yield were greater than 90%. When hydrogen and/or carbon monoxide are used as the fuel, the air/fuel ratio has the most significant impact on the fuel conversion. When the air/fuel ratio is lower than 1, an increase in the air/fuel ratio leads to a significant increase in the fuel conversion in a close to linear manner. A fuel conversion higher than 99% for both CO and H_2 can be achieved when the air/fuel ratio is higher than 1.1.[89] When natural gas is used as the fuel, important factors that affect the fuel conversion and CO_2 yield are reducer operating temperature and air/fuel ratio.[55,89,91]

Experiments carried out at the fuel processing capacity of $140\,kW_{th}$ and air/fuel ratio of 1.1 indicate that an increase in reducer operating temperature leads to an increase in both the fuel conversion and the CO_2 yield. Approximately 98% methane conversion and 94% CO_2 yield are achieved when the reducer is operated at 977°C. In comparison, the effects of the air/fuel ratio on the methane conversion and the CO_2 yield are more complex. When the air/fuel ratio is higher than 0.8, the changes in the fuel conversion and the CO_2 yield with respect to an increase in the air/fuel ratio show opposite trends. An increase in the air/fuel ratio decreases methane conversion, but enhances CO_2 yield. It is believed that a high air/fuel ratio increases the oxidation state of the nickel-based particles in the reducer. Because metallic nickel is more active in the catalytic conversion of methane, the increase in the oxidation state of the nickel particles adversely affects the methane conversion.

In another set of experiments, the effects of the fuel processing capacity on the methane conversion and the CO_2 yield are investigated. One interesting observation is that, when all other operating parameters are held equal, an increase in the fuel processing capacity leads to a slight increase in both the fuel conversion and the CO_2 yield. The fuel processing capacities investigated in the test are in the range of $55–140\,kW_{th}$. These results indicate that a further increase is possible in the fuel processing capacity of the DCFB system without compromising the fuel conversion and the CO_2 yield. This behavior may result from the improved fuel-particle contact in the reducer that results from a higher fuel injection rate. Another possible explanation is that an increase in the fuel processing capacity decreases the oxidation state of the nickel-based particles and, hence increases the catalytic activity of the particles for methane conversion.

A conversion of propane tested in the VUT DCFB system shows that propane is easier to convert compared with natural gas. This can be explained by the high chemical stability of methane compared with other hydrocarbons. Table 3.12 generalizes the operational parameters and results of the VUT DCFB system using both ilmenite particles and NiO-based particles. Based on the test results obtained at a fuel processing capacity of $140\,kW_{th}$, the solids inventory for a commercial DCFB system is estimated to be $240\,kg/MW_{th}$. The corresponding global solids circulation rate is $11.4\,kg/s \cdot MW_{th}$ or $41,000\,t/h$ for a $1,000$-MW_{th} CLC plant. Such a high solids circulation rate results from the operating mode of the DCFB system, which maximizes the gas and solids reaction rates at the expense of the extents of the solids conversion. An opposite case is the moving bed systems, which are discussed in Chapters 4 and 5. In these chemical looping systems, higher gas and solids conversions are

TABLE 3.12. Operational Results of the VUT DCFB Unit (Data with Air/Fuel Ratio < 1 are Not Shown)[55,89–91]

	Type of Particle						
	Ilmenite			NiO			
Parameter	H_2	Syngas	NG	H_2	CO	NG	Propane
Fuel Processing Capacity (kW_{th})	24–90	40–92	20–130	n/a	n/a	60–140	127
Reducer Temperature (°C)	894–966	878–976	839–969	850–900	850–900	800–950	900
Fuel Conversion (%)	87.5–96	72–85	28–72	80–99+	95–99	93–96	n/a
CO_2 Yield (%)	—	57–77	25–66	—	95–99	72–94	94
System Solids Inventory (kg)	70–90	70–85	70–90	55–75	65–75	65	65
Olivine Addition (kg)	0	0	0–15	0	0	0	0
Air/Fuel Ratio	1–1.2			1–1.2		1.1	

achieved with longer gas–solid contact time. During the VUT DCFB tests, the highest solids flux in the oxidizer is $90 kg/m^2 \cdot s$.

As discussed in Section 3.3.2, the large airflow rate and solids circulation rate in a commercial CLC oxidizer requires either an oxidizer with a large overall cross-sectional area or with a higher operating pressure and solids flux. Therefore, the design of the commercial DCFB system may face a different challenge when compared with that for the current test unit. Note that in all tests performed with natural gas as the fuel, at least 5 vol.% (dry basis) of the reducer exhaust gas is combustible fuel. Although a higher fuel conversion and a CO_2 yield may be achieved in a scale-up unit of the current DCFB system, further treatment to the reducer exhaust gas is likely to be necessary to avoid an energy penalty and to minimize environmental effects. Such treatments include further combustion of the reducer exhaust gas with oxygen.

Gas Leakage and Carbon Deposition: Gas leakage between the reducer and the oxidizer is undesirable because it reduces the effectiveness of the CO_2 capture capability and/or the energy conversion efficiency of the CLC system. Carbon deposition on particles can cause similar effects. Moreover, excessive carbon deposition on particles can deform and deactivate the particles. It is reported that no CO_2 has been detected from the oxidizer exhaust throughout the tests, indicating that neither gas leakage from the reducer to the oxidizer

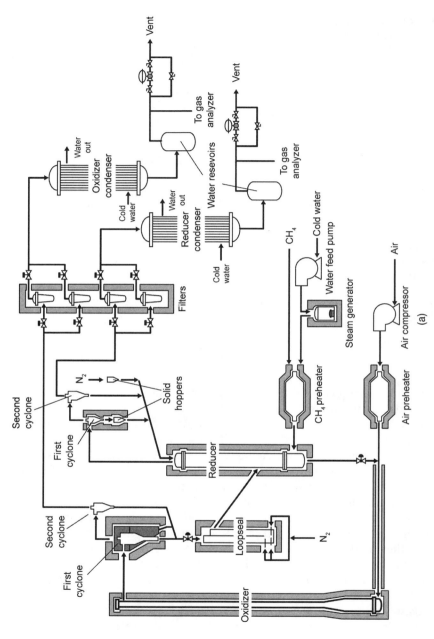

Figure 3.25. Chemical looping combustion system at the Korean Institute of Energy Research: (a) schematic diagram; (b) photograph of the first generation unit; and (c) photograph of the second generation unit. With permission from Dr. Ho-Jung Ryu of KIER.

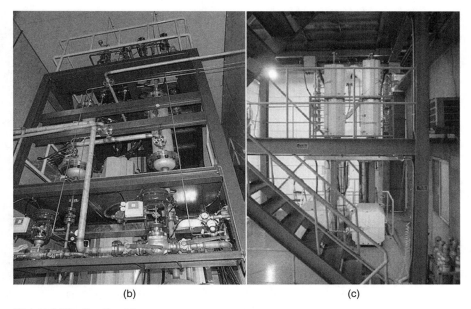

(b)   (c)

Figure 3.25. Continued

nor carbon deposition in the reducer is significant. The gas leakage from the oxidizer to the reducer is quantified by testing the N_2 concentration in the reducer exhaust. It is determined that 0.2–0.4 vol.% of the gas introduced to the oxidizer leaks into the reducer. Such a leakage rate is deemed acceptable.[55]

KIER 50-kW$_{th}$ CLC System

A 50-kW$_{th}$ chemical looping combustion system was constructed by the Korea Institute of Energy Research (KIER) (Daejeon, Korca) in 2002. Two types of oxygen carriers, NiO supported on bentonite and Co_xO_y supported on $CoAl_2O_4$, were tested in the system. The test results for the NiO/bentonite oxygen carriers indicate that the system is capable of converting 99.7% of CH_4 while releasing a low concentration of CO and H_2 at the outlet of the reducer. The system, however, yields a high attrition rate of the oxygen carrier particles. With the Co_xO_y/$CoAl_2O_4$ oxygen carriers, an average fuel conversion of 99.6% was achieved during 25 hours of continuous operation.[54] The emission of NO_x was negligible during the oxidation of particles with air in the riser. The schematic diagram and photograph of the 50-kW$_{th}$ chemical looping combustion unit are given in Figure 3.25(a, b). This unit was advanced to a second generation 50-kW$_{th}$ CLC unit constructed by KIER in 2006, as shown in Figure 3.25(c). No experimental results on the second-generation unit are published to date. The following section discusses only the first-generation unit.

TABLE 3.13. Reactor Configuration and Operating Conditions of the KIER 50-kW$_{th}$CLC Unit[54]

	Reactor Type	Gas Injection (m/s)	Diameter (m)	Height (m)	Temperature (°C)
Reducer	Bubbling Fluidized Bed	CH$_4$, 0.05	0.143	5	869
Oxidizer	High-Velocity Fluidized Bed	Air, 3.0	0.078	2.5	890
Loop Seal		N$_2$, 0.02	0.078	1.5	

Reactior Design and Standard Operation Procedure. The system consists of a bubbling fluidized bed reducer and a high-velocity fluidized bed oxidizer. Two-stage cyclones are installed on the reducer for oxygen carrier recovery, and loop seals are installed between the reactors to prevent gas mixing. Preheated methane or other gaseous fuel enters the bubbling fluidized bed and is distributed through a perforated distributor. The inner diameter of the reducer is 0.143 m and its height is 2.5 m. The flow rate of the CH$_4$ fuel is 0.048 m^3/min, and the reducer is operated at 869°C. The riser, with an inner diameter of 0.078 m, carries the particles back to the reducer using preheated air at a gas velocity of 3.0 m/s. The temperature in the oxidizer is maintained at 890°C. Table 3.13 summarizes the reactor configuration and operation conditions of the 50-kW$_{th}$ CLC.

Test Results. The 50-kW$_{th}$ unit was operated to test the continuous and longterm operability of the CLC system. Initially, NiO/bentonite particles were used. The unit was operated for only 3.5 hours because of the high attrition rate of the particles. The oxygen carrier then was replaced with Co$_x$O$_y$/CoAl$_2$O$_4$, which has better attrition resistance. The unit was operated successfully for 25 hours. The circulation rate of solid particles was ~8 kg/m^2 s. The oxygen carrier was prepared using the coprecipitation and impregnation method with 70% metal oxide content. The total solids inventory for the 50-kW$_{th}$ unit was 33 kg.

From the pressure drop measurement during the 25-h operation, the riser and the loop seal maintained a constant pressure drop, but the pressure in the reducer exhibited a small decrease, as the attrited particles were entrained through the cyclones. Therefore, the particle attrition rate needs to be improved for longterm operation. Table 3.14 shows the properties of the cobalt-based oxygen carrier used for the 50-kW$_{th}$ unit.

Fuel and Solids Conversion: The gas composition in the exhaust streams of the reducer indicates an average CH$_4$ conversion of 99.6% and a CO$_2$ yield of higher than 97% during the 25-h continuous test. The exhaust stream consisted of 0.04% CH$_4$, 3.24% CO, and 97% CO$_2$. The recyclability and conversion

TABLE 3.14. Particle Properties of the Oxygen Carrier Used in the KIER 50 kW$_{th}$CLC Unit[54]

Particle	$Co_xO_y/CoAl_2O_4$
Preparation Method	Coprecipitation/Impregnation
Metal Oxide Content	70%
Size Range of Particles	106–212 μm
Bulk Density	970 kg/m^3
Solids Circulation Rate	8 kg/m^2 s
Solids Inventory	33 kg

of the particles were determined by monitoring the O_2 consumption ratio, which is the ratio of the O_2 consumption rate detected by a gas analyzer to the theoretical O_2 consumption rate projected based on fuel conversion:

$$O_2 \text{ Consumption Ratio} = \frac{\text{Actual } O_2 \text{ Consumption}}{\text{Theoretical } O_2 \text{ Consumption}} \qquad (3.3.20)$$

The O_2 consumption ratio approached 1, implying that the recyclability of the particles was retained during the operation. The average NO_x concentrations at the riser outlet were measured to be 0.8 ppm, 6.0 ppm, and 5.8 ppm for N_2O, NO, and NO_2, respectively. These concentrations are significantly lower than those in conventional combustion processes because the chemical looping combustion was carried out at a relatively low temperature and under flameless conditions.

To prevent carbon deposition on the oxygen carrier, steam was injected along with CH_4. The purpose of the steam is to gasify the deposited carbon into CO and H_2.

$$H_2O + C \rightarrow CO + H_2 \qquad (3.3.21)$$

Because almost no H_2 was observed in the reducer, it was projected that no carbon deposition was taking place on the particles because of CH_4 decomposition and/or reverse Boudouard reaction. Moreover, because no CO or CO_2 is observed from the oxidizer outlet, the carbon deposition on the particles was concluded to be minimal.

Summary on Gaseous Fuel CLC Subpilot-Scale Testing
The subpilot-scale testing on gaseous fuel CLC systems to date suggests that the reactivity, recyclability and strength of the supported oxygen carrier particles is suitable for longterm CLC operations. The test results also indicate insignificant carbon deposition and minimal gas leakage between the CLC reactors. Thus, CLC system operation for the gaseous fuel conversion is feasible, at least for smaller-scale units. The CO_2 capture efficiencies achieved in

these tests, however, are still notably lower than 100% for larger-scale units. For example, a CO_2 capture efficiency of close to 99% is achieved in 10-kW$_{th}$ scale units, whereas the highest CO_2 capture efficiency achieved in the 50- and 120-kW$_{th}$ scale CLC units was 97%. The incomplete conversion of the fuel to CO_2 in the tests suggests that an additional oxygen polishing step may be necessary for fully converting the remaining gaseous fuel into CO_2. Because of the large solids and airflow rates required for operating the CLC system, design and operational issues such as gas–solid contact mode, solids handling, system temperature and pressure, and reaction kinetics may dictate the ultimate scale that can be applied for commercial CLC operation.

3.3.5 Solid Fuel CLC Systems and Operational Results

Solid fuels such as coal, petroleum coke, and biomass are notably cheaper than gaseous fuels such as natural gas and syngas. Therefore, it is desirable to convert solid fuels directly in the CLC systems, circumventing the need of using syngas or gaseous fuel feedstock and/or costly syngas-generation systems. The rate of the solid–solid reaction between the oxygen carrier particles and the solid fuel is slow; however, it can be augmented by employing such gasifying agents as steam and CO_2.[41] The enhanced solid–solid reaction scheme is discussed in Section 2.2.12. Beyond solid–solid reaction enhancement, the fates of impurities in solid fuels such as ash and other pollutants need to be addressed for solid fuel CLC systems. Because of the aforementioned challenges, direct solid fuel CLC is at a relatively early developmental stage. However, laboratory experiments using various types of solid fuels reacting with oxygen carrier particles in circulating-fluidized-bed systems revealed the possibility of converting solid fuels in a CLC scheme.[39,41–43] Two solid-fuel circulating-fluidized-bed CLC systems, both at 10-kW$_{th}$ scale, are discussed in this section.

Modified Chalmers University 10-kW$_{th}$ Solid-Fuel CLC Unit for Solid-Fuel Conversion
The solid fuel testing in the 10-kW$_{th}$ unit at Chalmers University was intended to explore issues associated with the operation of the solid-fuel CLC system. Two solid fuels of different characteristics, South African coal[92] and petroleum coke,[93] were tested under continuous operation for 22 hours and 11 hours, respectively.

Reactor Design and Operating Procedures. The reactor system used for solid fuel testing is similar to the Chalmers University 10-kW$_{th}$ unit used for testing gaseous fuel. The major modification of the unit involves the division of the reducer or fuel reactor into three chambers to handle two solids, namely, the fuel and the oxygen carrier particles. The modified fuel reactor design houses a low-velocity chamber for fuel volatilization and conversion, a high-velocity chamber for solids recirculation, and a carbon stripper chamber to recover the unconverted fuel from the oxygen carrier particles to be transferred to the oxidizer or air reactor. Auxiliary units added around the reducer were a coal feeder and an internal particle circulation system consisting of a small riser and cyclone. The schematic of the modified Chalmers University 10-kW_{th} solid-fuel CLC system is given in Figure 3.26.

Test Results. The entire reactor system is placed inside an oven. During the start-up stage, the reactor system is slowly heated to 950°C. Meanwhile, air is used to fluidize the particles in all the reactors. After the temperature of the reactors is stabilized, the fluidization gas is switched to nitrogen. Steam, which is a gasifying agent for solid fuels, then is injected into the fuel conversion chamber of the reducer as the fluidizing gas. Solid fuel then is injected at a constant rate. Stable operation of the reactor system is achieved after 30 to 45 minutes.

The objective of the solid fuel tests in a CLC unit was to examine the key issues such as fuel conversion, carbon capture efficiency, and particle attrition that affect the smooth operation of the unit.[92] Because of the difference in fuel types, the parameters assessing solid-fuel CLC systems are defined differently from those in gaseous-fuel CLC systems. Specifically, the gas conversion is defined as the ratio of CO_2 concentration to the combined concentration of all the carbonaceous gases exiting the reducer. It is similar to the CO_2 yield defined previously and is expressed as follows:

$$f_{CO_2} = \frac{(CO_2)_{\text{Reducer}}}{(CO_2)_{\text{Reducer}} + (CO)_{\text{Reducer}} + (CH_4)_{\text{Reducer}}} \qquad (3.3.22)$$

where $(X)_Y$ refers to the measured concentrations of the gas species X exiting reactor Y. The solid fuel conversion, η_{Fuel}, is defined as the ratio of solids converted in both the reactors to the inlet solid flow. The equation used to calculate this parameter is given by

$$\eta_{\text{Fuel}} = \frac{F_{C,\text{Oxidizer}} + F_{C,\text{Reducer}}}{F_{C,\text{Fuel}}} \qquad (3.3.23)$$

where $F_{C,\text{Fuel}}$ is the total flow rate of carbon in the feedstock, and $F_{C,\text{Reducer}}$ and $F_{C,\text{Oxidizer}}$ are the flow rates of gaseous carbon from the reducer and oxidizer, respectively. The expression represents the ratio of the total gaseous carbonaceous flow leaving the two reactors to the total carbon entering the system. Note that the solid conversion as defined represents the conversion of solid fuel to gaseous products, which can be CO, CO_2, and/or CH_4. Another parameter used by the researchers at Chalmers University is the carbon capture efficiency, which is defined as the ratio of gaseous carbon exiting the reducer to that exiting the reducer and oxidizer combined as given by

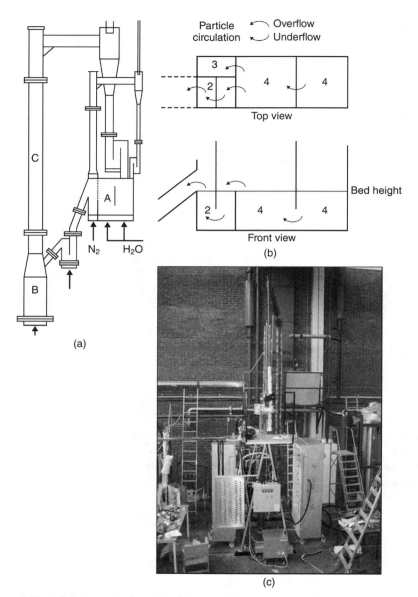

Figure 3.26. (a) Schematic diagram of the modified Chalmers University 10-kW$_{th}$ solid fuel CLC unit; (b) detailed schematics of the reducer (A); (c) photograph of the 10-kW$_{th}$ solid fuel CLC unit.[92] With permission from Professor Anders Lyngfelt at Chalmers University of Technology.

$$\eta_{CC} = \frac{F_{C,Reducer}}{F_{C,Reducer} + F_{C,Oxidizer}} \tag{3.3.24}$$

Equation (3.3.24) notably does not consider the unconverted solid carbon exiting the system. Furthermore, because the gaseous carbon from the reducer is a mixture of CO_2 and gaseous carbonaceous fuels such as CO and CH_4, the carbon capture only can reflect the CO_2 capture when all the unconverted carbonaceous fuels in the gaseous products from the reducer are converted to CO_2 through an additional oxygen polishing step. Such a polishing step is not present in the current setup. The carbon capture efficiency for the tests was reported in their study based on Equation (3.3.24).

Oxygen Caarier and Fuel: The efficiency of the solid fuel CLC depends on the oxygen carrier and the type of solid fuels used. Other factors that play an important role are the gasifying agent, the flow rate and the effectiveness of the carbon stripper. Along with the need to react immediately with the products from fuel gasification, the oxygen-carrier particles also have to tolerate the effect of pollutants and contaminants associated with the solid fuel. Thus, it is expected that more oxygen carriers will be lost per cycle compared with a gaseous fuel CLC. Leion et al.[41] found that the reactivity of the oxygen carriers derived from natural ilmenite mineral had a similar performance to synthesized iron oxide-based particles for various solid fuels. The finding led to the employment of particles made from natural ilmenite mineral for 10-kW_{th} unit operation to assess the process feasibility with these inexpensive oxygen carriers. The oxygen carrier properties are shown in Table 3.15.

The solid fuels that were tested in the 10-kW_{th} unit were South African coal and petroleum coke. South African coal has a high ash and volatiles content, whereas petroleum coke has high carbon and sulfur content. The fuel characteristics are given in Table 3.16. Continuous tests with multiple short periods of 1–4.5 hours were performed with South African coal and petroleum coke, which add up to 22 hours of operation for South African coal[92] and 11 hours of operation for petroleum coke.[93]

TABLE 3.15. Properties of the Oxygen Carrier Particles Used in the Modified Chalmers University 10-kW_{th} Solid Fuel CLC Unit[92]

Type of Iron Oxide-Based Particlae	Ilmenite (Concentrated Form)
Molar Ratio of Iron to Titanium	~1:1
Size Fraction	90–250 µm
Density	2,100 kg/m³
Surface Area	$0.11 \times 10^3 \, m^2/kg$
Total Solids Used	13 kg

TABLE 3.16. Properties and Nomenclature of the Solid Fuels Used in the Modified Chalmers University 10-kW$_{th}$ Solid-Fuel CLC Unit

Properties	South African Coal[92]	Petroleum Coke[93]
Proximate Analysis (wt.%)		
Moisture	—	8.00
Volatiles	21.6	—
Ash	15.9	0.46
Ultimate Analysis (wt.%)		
C	62.5	81.32
H	3.5	2.87
O	7.7	0.45
N	1.4	0.88
S	0.7	6.02
Heating Value (MJ/kg)	23.9	31.75
Solid Flow Rate Used (kg/hr)	0.500	0.655
Average Particle Diameter (µm)	170	—

Fuel Conversion: In the experiments that used South African coal,[92] the solid oxygen carrier inventory was 13 kg, and a steady coal flow rate of 500 g/h was used. This condition corresponds to a 3.32-kW$_{th}$ power rating considering a complete fuel conversion. The actual solid fuel conversion observed for this fuel ranged from 50% to 80%. The low conversion was attributed to a low efficiency of the cyclone used in the carbon stripper system. This system, including the carbon stripper cyclone, was modified before the test, using petroleum coke as the fuel. The operating conditions remained the same for petroleum coke except for an increased fuel flow rate of 655 g/h, which increased the power rating to 5.8 kW$_{th}$.[93] The actual solid fuel conversion observed for this test was ~70%. The low fuel conversion was attributed to the low reactivity of the fuel and the unsatisfactory performance of the modified carbon stripper cyclone.

Gas Conversion and CO$_2$ Capture: The gas conversion refers to the conversion of the gaseous products from solid fuel gasification into CO$_2$. It is given in terms of the CO$_2$ fraction defined in Equation (3.3.22). The experiments with South African coal resulted in a gas conversion range of 78%–81%.[92] The CO$_2$ capture efficiency, defined in Equation (3.3.24), is reported to be 82%–96%. The gas conversion observed using petroleum coke is ~75%, whereas the CO$_2$ capture efficiency varies between 60% and 75%.[93] The low CO$_2$ capture efficiency is attributed to the low reactivity of the fuel. Thus, to achieve a higher CO$_2$ capture efficiency, increased residence time and solids inventory are required.

Oxygen Carrier Particle Performance: The oxygen carrier particles did not exhibit a tendency to agglomerate for experiments using both solid fuels.

This observation is considered an initial indication of smooth operation with these oxygen carriers, but a short duration of testing renders it difficult to assess conclusively the performance of the particles. A preliminary lifetime study was performed for the oxygen carriers used in the petroleum coke fuel experiments. The loss of fines (particles $< 45 \mu m$) was in the range of 0.01–0.03 %/h, corresponding to an oxygen carrier lifetime of 3,000–9,500 hours.[93]

The results obtained from the solid fuel tests indicate that the reducer solids inventories used in the experiment, ~1,506 kg/MW$_{th}$ for South African Coal and ~862 kg/MW$_{th}$ for petroleum coke, are inadequate for an effective solid fuel conversion. Continued testing with solid fuels for extended, stable, continuous operation is required before the solid fuel CLC system can be concluded to be feasible.

Southeast University 10-kW$_{th}$ Solid-Fuel CLC Unit

The objective of the solid-fuel testing in a 10-kW$_{th}$ unit at Southeast University in Nanjing, China was to examine the spouted bed reducer design, and the effect of the reducer temperature and fuel impurities on the continuous operation of the CLC unit. Experiments using two different solid fuels, coal and biomass, were performed using several oxygen carriers for a cumulative operating time of 130 hours[94,95] and 30 hours,[96] respectively.

Reactor Design and Standard Operating Procedures. The reactor assembly consists of interconnected fluidized beds for performing the continuous operation of the CLC unit. The CLC unit includes the reducer, the oxidizer, and the cyclone. A schematic of the Southeast University 10-kW$_{th}$ solid fuel CLC unit is shown in Figure 3.27. The reducer is made up of a two-chamber rectangular spouted bed. The main chamber, that is, the reaction chamber, is where the fuel conversion and oxygen carrier reduction reaction take place. The minor chamber, that is, the inner seal, serves to facilitate the flow of reduced oxygen carriers to the oxidizer via an overflow pipe. The bottom of the conical section in the fuel reactor is where the recycled CO_2 or CO_2/steam and solid fuel are introduced to the reactor using the screw feeder. The oxidizer is a cylindrical high-velocity riser with a perforated plate at the bottom acting as the air distributor. The solids entering the oxidizer from the reducer, are carried to the top of the riser, and then are separated by the cyclone. The oxidized solid carriers return to the reducer to complete the loop.

The individual reactors have external heaters to maintain constant temperatures in the system. The temperature control assists in studying the influence of temperature on fuel reactor performance. The external heating is utilized to bring the reactors to and maintain the required temperature. Once a constant reactor temperature is achieved, the solid particles in the reactors are fluidized and the fuel injection then starts. The fuel inlet flow is maintained at a constant level, and stabilized operation is achieved within hours.

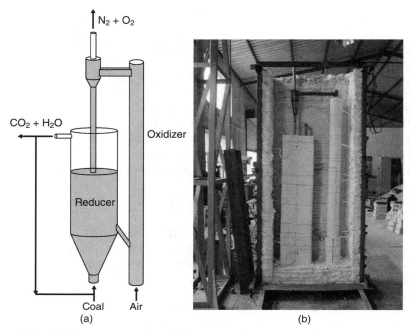

Figure 3.27. (a) Schematic diagram of the 10-kW$_{th}$ Southeast University solid fuel CLC unit; (b) photograph of the CLC unit. With permission from Prof. Laihong Shen of Southeast University in China.

Test Results. The applicability of a spouted bed for use in the 10-kW$_{th}$ unit is examined in this study. The applicability is evaluated in terms of two parameters: carbon capture efficiency and CO_2 capture efficiency. In this study, the carbon capture efficiency is defined as:

$$\eta_C = \left(1 - \frac{\text{carbon mass flow of fly ash}}{\text{carbon flow of coal}}\right) \times f_{CO_2} \qquad (3.3.25)$$

The term in brackets represents the total carbon flow in the system that has been converted, whereas the second term, f_{CO_2}, represents the gas conversion as expressed by Equation (3.3.22). Note that the carbon capture efficiency defined by Equation (3.3.25) assumes that no carbon enters the oxidizer. Thus, the equation does not account for the CO_2 fraction that could be generated in the oxidizer. The CO_2 capture efficiency is defined as the ratio of the CO_2 flow leaving the reducer to the total carbonaceous gas flow leaving the two reactors as expressed by:

$$\eta_{CO_2,C} = \frac{F_{CO_2,\text{Reducer}}}{F_{C,\text{Reducer}} + F_{C,\text{Oxidizer}}} \qquad (3.3.26)$$

Here, the numerator is the flow of carbon dioxide exiting the reducer and does not include other gaseous products such as CO and CH_4. The terms in the denominator represent the carbon flow exiting the fuel and air reactors. These two efficiency parameters, which are different from those given in Equations (3.3.22–3.3.24), are used to characterize the performance of the oxygen carriers and the process.

Oxygen Carrier and Fuel: Multiple experiments with varying solid fuels and oxygen carriers were conducted. Two types of nickel oxide-based oxygen carriers were tested with coal as the fuel. The high reactivity and thermal stability of nickel were deemed desirable to convert effectively fuels with higher fixed carbon content such as coal. A combined coal CLC operation of 30 hours and 100 hours were carried out using nickel-based oxygen carriers synthesized by impregnation (Type A)[95] and coprecipitation (Type B)[94] methods, respectively. In both tests, Shenhua bituminous coal was used.

An iron-based oxygen carrier (Type C in Table 3.17) was used to convert biomass whose fixed carbon content was relatively low. Iron-based oxygen carriers were selected because of their favorable thermodynamic properties and low cost. A total of 30 hours of continuous operation was conducted in the 10-kW_{th} unit.[96] In this test, iron oxide powders without any support material were used to convert pine sawdust. The properties of iron oxide and other oxygen carrier particles and the types of solid fuels used are given in Tables 3.17 and 3.18, respectively.

Carbon Capture Efficiency: The chemical looping conversion of coal using Type A particles given in Table 3.17 yields an increased carbon capture efficiency, when the reducer temperature increases, reaching a maximum efficiency of 92.8% at 970°C. Although the thermodynamic limit for the gas conversion of 95.2% in the reducer was attained, the elutriation of fine char particles in the fuel reactor reduced the carbon capture efficiency. A recirculation of

TABLE 3.17. Properties of the Oxygen Carrier Particles Used in the 10-kW_{th} Southeast University Unit

Properties	Type A[95]	Type B[94]	Type C[96]
Metal Oxide Used	NiO	NiO	Fe_2O_3
Support Used	$NiAl_2O_4$	Al_2O_3	n/a
Synthesis Method	Impregnation	Coprecipitation	n/a
Size Fraction (mm)	0.2–0.4	0.2–0.4	0.3–0.6
Density (kg/m^3)	2,350	2,350	2,460
Surface Area (m^2/kg)	8.694×10^4	5.081×10^4	5.465×10^2
Porosity (m^3/kg)	3×10^{-4}	1.801×10^{-4}	1.536×10^{-4}
Hours of Operation (hr)	30	100	30
Total Solids Used (kg)	11	11	12

TABLE 3.18. Nomenclature and Properties of the Solid Fuels Used in the 10-kW$_{th}$ Southeast University Unit

Properties	Coal[94,95]	Biomass[96]
Proximate Analysis (wt.%)		
Moisture	6.98	11.89
Fixed Carbon	53.85	14.77
Volatiles	33.59	75.78
Ash	5.58	1.56
Ultimate Analysis (wt.%)		
C	65.06	40.06
H	4.34	5.61
O	15.98	39.88
N	0.86	0.90
S	1.2	0.10
Lower Heating Value (MJ/kg)	24.80	14.47
Solid Flow Rate Used (kg/h)	1.2	3.0
Power Rating of Unit (kW$_{th}$)	8.27	12.06
Sauter Mean Diameter (mm)	0.38	1.5

recovered fine particles from the reducer or oxidizer could assist in the carbon capture efficiency. A similar behavior was observed for Type B particles when reacted with coal. Type C particles were used to convert biomass; a carbon capture efficiency varying from 53.7% to 65.1% was obtained.[96]

CO$_2$ Capture Efficiency: The CO$_2$ capture efficiency was observed to increase with an increase in the reducer temperature and reach a plateau at 80% and 960°C for Type A particles from Table 3.17 tested using coal.[95] The bypassing of gases and the residual char transported from the fuel reactor to the air reactor constrained the attainment of a higher efficiency. Improved inner seal design and improved particle reactivity are among the possible solutions in achieving higher CO$_2$ capture efficiencies. Similar trends also were observed for Type B particles tested using coal.

Oxygen Carrier Particle Performance: Type B particles were operated in the 10-kW$_{th}$ unit for 100 hours using coal as the fuel. Significant deterioration in oxygen carrier performance was observed during this operation. The deterioration in performance was reflected indirectly by the decrease in the CO$_2$ concentration at the fuel reactor outlet from 93.42% to 81.04%.[94] The studies indicated that the sintering of the oxygen carrier particles is the major reason for performance degradation. Most significant sintering effects were observed in the high-temperature exothermic oxidation of Type B particles in the oxidizer. The utilization of a higher steam:CO$_2$ ratio in the reducer gas inlet improved the performance of the particles. Sintering effects also were observed in Type C particles.

The overall solids inventories used in the experiments correspond to ~1,330.65 kg/MW$_{th}$ for coal and ~995 kg/MW$_{th}$ for biomass. The experimental results indicate that such solids inventories are not adequate for effective fuel conversion and CO_2 capture. Improved inner seal design and longer duration of stable continuous operations are desirable.

Summary on Solid-Fuel CLC and Ongoing Scale-Up Activities
Apart from the two aforementioned solid fuel CLC systems, a 1-MW$_{th}$ coal-fired CLC prototype is being developed under the ÉCLAIR project sponsored by European Commission's Research Fund for Coal and Steel program. Numerical modeling of reactor system hydrodynamics, cold flow model testing, and reaction kinetics studies are being carried out. According to ALSTOM, which is a leading partner of the project, the 1-MW$_{th}$ prototype will be constructed and operated in 2010.[97]

Because of the difficulty of direct conversion of solid fuels with an oxygen carrier, the existing solid fuel CLC systems have a lower fuel conversion and a lower CO_2 capture efficiency than do the gaseous fuel CLC systems. Continued research on oxygen carrier particle synthesis, solid fuel conversion enhancement, and design of an effective residue char stripping and reducer unit is required for the development of solid-fuel CLC systems.

3.4 Concluding Remarks

The combustion of gaseous or solid fossil fuels using air in traditional approaches requires an energy-intensive CO_2 separation step from the downstream flue gas steam, when carbon emission control is considered. The approach based on chemical looping provides a specific reaction path that naturally generates a concentrated stream of CO_2 and steam from the metal oxide reaction with fossil fuels while yielding reduced metal oxide. The subsequent carbon-free combustion of reduced-metal oxide with air produces heat and, hence, electricity. The combustion process based on the chemical looping approach also yields higher energy conversion efficiency than do the traditional approaches.

The projection of chemical looping combustion operation using Ni, Cu or Fe particles indicates that considering various operating factors, including oxygen carrier reactivity, airflow rate, and oxygen carrier flux, the required solid-handling system will be so large that it may practically constrain its application to such a large-scale system as 1,000 MW$_{th}$. However, it may be significantly less constrained under a chemical looping gasification route that produces hydrogen under pressures. In general, the chemical looping systems will require processing a high flow rate of solid particles and air in reactors and/or transport lines. It yields a high-density flow condition when the operating pressure is elevated. Thus, large chemical looping combustion systems commonly will possess high risers and high downcomers or low risers with a series of low downcomers that provide a large solids inventory capacity and a large pressure difference in the downcomers for high solids flows.

Various designs of the CLC systems at subpilot scale and beyond are introduced in this chapter. They are represented by the Chalmers University 10-kW$_{th}$ system, the CSIC 10-kW$_{th}$ system, the VUT 120-kW$_{th}$ system, the KIER 50-kW$_{th}$ system, the modified Chalmers University 10-kW$_{th}$ solid fuel system, and the Southeast University 10-kW$_{th}$ solid-fuel system. The operation of these processes revealed varied degrees of success in performance relative to operating variables. The variables of particular importance include particle reactivity and recyclability, particle attrition, fuel conversion, carbon capture efficiency, CO_2 capture efficiency, and carbon deposition rate. Although the subpilot test results exhibit optimism for the CLC Process, long-term testing, system reliability, and the economics of the CLC system remain the key factors that will dictate the process viability for its eventually commercial applications. A common issue pertaining to the current CLC systems, especially those with solid fuels as feedstock, is the incomplete conversion of the fuels. To enhance fuel conversion and CO_2 capture, an oxygen polishing step may be required. Continued research on the CLC using solid fuels, such as coal, petroleum coke, and biomass, is necessary to increase the cost attractiveness of the process.

References

1. Stultz, S. C., and J. B. Kitto, *Steam, Its Generation and Use*, 40th edition, Babcock & Wilcox (1992).

2. Energy Information Administration, "International Energy Outlook 2006," DOE, (2006).

3. Energy Information Administration, "Emission of Greenhouse Gases in the United States 2007," DOE/EIA – 05732007, (2007).

4. International Energy Administration, "World Energy Outlook 2006" (2006).

5. Intergovernmental Panel on Climate Change, "Special Report, Carbon Dioxide Capture and Storage: Summary for Policymakers" (2005).

6. Figueroa, J. D., T. Fout, S. Plasynski, H. McIlvried, and R. D. Srivastava, "Advances in CO_2 Capture Technology—the U.S. Department of Energy's Carbon Sequestration Program," *International Journal of Greenhouse Gas Control*, 2(1), 9–20 (2008).

7. Rao, A. B., and E. S. Rubin, "A Technical, Economic, and Environmental Assessment of Amine-Based CO_2 Capture Technology for Power Plant Greenhouse Gas Control," *Environ. Sci. Technol*, 36, 4467–4475 (2002).

8. Rochelle, G. T., "Technology Opportunities for Post-Combustion Capture by Amine Scrubbing," EPRI workshop (2009).

9. Iijima, M. I., K. Takashina, T. Tanaka, H. Hirata, and T. Yonekawa, *Carbon Dioxide Capture Technology for Coal Fired Boiler*, Proceedings of the 7th International Conference on Greenhouse Gas Technologies (2004).

10. Singh, D., and P. L. Douglas, "Techno-Economic Study of CO_2 Capture from an Exiting Coal-Fired Power Plant: MEA Scrubbing VS. O_2/CO_2 Recycle Combustion," *Energy Conversion and Management*, 44, 3073–3091 (2003).

11. IEA, "Potential for Improvement in Power Generation with Post-Combustion Capture of Carbon Dioxide. Interim Final Report Prepared for IEA GHG R&D Programme" (2004).

12. Bohm, M. C., "Capture-Ready Power Plants–Options, Technologies, and Economics," MS Thesis, Massachusetts Institute of Technology (2006).

13. Woods, M. C., P. J. Capicotto, J. L. Haslbeck, N. J. Kuehn, M. Matuszewski, L. L. Pinkerton, M. D. Rutkowski, R. L. Schoff, and V. Vaysman, "Cost and Performance Baseline for Fossil Energy Plants," DOE/NETL-2007/1281 (2007).

14. Ramezan, M., T. J. Skone, N. Y. Nasakala, and G. N. Liljedahl, "Carbon Dioxide Capture from Existing Coal-Fired Power Plants," DOE/NETL-401/110907 (2007).

15. Lu, Y., S. Chen, M. Rostam-Abadi, R. Varagani, F. Chatel-Pelage, and A. C. Bose, "Techno-Economic Study of the Oxy-Combustion Process for CO_2 Capture from Coal-Fired Power Plants," Paper Presented at the Annual International Pittsburgh Coal Conference (2005).

16. Rubin, E. S., C. Chen, and A. B. Rao, "Cost and Performance of Fossil Fuel Power Plants with CO_2 Capture and Storage," *Energy Policy*, 35(9), 4444–4454,(2007).

17. Hutchinson, H., "Old King Coal," *Mechanical Engineering*, 124(8), 41–45 (2002).

18. Massachusetts Institute of Technology, "The Future of Coal: Options for a Carbon-Constrained World" (2007).

19. Davison, J., "Performance and Costs of Power Plants with Capture and Storage of CO_2," *Energy*, 32, 1163–1176 (2007).

20. Yeh, J. R., K. Rygle, and K. Penniline, "Semi-Batch Absorption and Regeneration Studies for CO_2 Capture by Aqueous Ammonia," *Fuel Processing Technology*, 86, 1533–1546 (2005).

21. Kohl, A. L., and R. Nielsen, *Gas Purification*, 5th edition, Gulf Pubulisher (1997).

22. Higman, C., *Gasification*, 2nd edition, Gulf Professional (2008).

23. Lewandowski, D., and D. Gray, "Potential Market Penetration of IGCC in the East Central North American Reliability Council Region of the U.S.," Paper Presented at the Gasification Technologies Conference (2001).

24. Kanniche, M., and C. Bouallou, "CO_2 Capture Study in Advanced Integrated Gasification Combined Cycle," *Applied Thermal Engineering*, 27(16), 2693–2702 (2007).

25. Damen, K., M. van Troost, A. Faaij, and W. Turkenburg, "A Comparison of Electricity and Hydrogen Production Systems with CO_2 Capture and Storage: Part A: Review and Selection of Promising Conversion and Capture Technologies," *Progress in Energy and Combustion Science*, 32, 215–246 (2006).

26. National Energy Technology Laboratory, "Carbon Sequestration: CO_2 Capture" (2006).

27. Horn, F. S., M, "Control of Carbon Dioxide Emissions from a Power Plant (and Use in Enhanced Oil Recovery)," *Fuel*, 61(5), 415–422 (1982).

28. Carbon Capture Journal, "First Oxyfuel Pilot Plant Opens at Schwarze Pumpe in Germany," (6), 3–4 (2008).

29. Châtel-Pélage, F., R. Varagani, P. Pranda, N. Perrin, H. Farzan, S. J. Vecci, Y. Lu, S. Chen, M. Rostam-Abadi, and A. C. Bose, "Applications of Oxygen for Nox Control and CO_2 Capture in Coal-Fired Power Plants," Paper Presented at the Second International Conference on Clean Coal Technologies for our Future (2005).

30. Lewis, W. K., and E. R. Gilliland, "Production of Pure Carbon Dioxide," U.S. Patent 2,665,971 (1954).

31. Knoche, K. F., and H. J. Richter, "Improvement of the Reversibility of Combustion Processes," *Brennstoff-Waerme-Kraft*, 20(5), 205–210 (1968).

32. Ishida, M., and H. G. Jin, "A New Advanced Power-Generation System Using Chemical-Looping Combustion," *Energy*, 19(4), 415–422 (1994).

33. Ishida, M., D. Zheng, and T. Akehata, "Evaluation of a Chemical-Looping-Combustion Power-Generation System by Graphic Exergy Analysis," *Energy*, 12(2), 147–154 (1987).

34. Hossain, M. M., and H. I. de Lasa, "Chemical-looping combustion (CLC) for inherent CO_2 separations-a review," *Chemical Engineering Science*, 63(18), 4433–4451 (2008).

35. Ryu, H. J., and G.-T. Jin, "Criteria for Selection of Metal Component in Oxygen Carrier Particles for Chemical-Looping Combustor," *Hwahak Konghak*, 42(5), 588 (2004).

36. Chuang, S. Y., J. S. Dennis, A. N. Hayhurst, and S. A. Scott, "Development and Performance of Cu-Based Oxygen Carriers for Chemical-Looping Combustion," *Combustion and Flame*, 154(1/2), 109–121 (2008).

37. Mattisson, T., F. Garcia-Labiano, B. Kronberger, A. Lyngfelt, J. Adanez, and H. Hofbauer, "Chemical-Looping Combustion Using Syngas as Fuel," *International Journal of Greenhouse Gas Control*, 1(2), 158–169 (2007).

38. Cao, Y., B. Casenas, and W. P. Pan, "Investigation of Chemical Looping Combustion by Solid Fuels:2. Redox Reaction Kinetics and Product Characterization with Coal, Biomass, and Solid Waste as Solid Fuels and CuO as an Oxygen Carrier," *Energy & Fuels*, 20(5), 1845–1854 (2006).

39. Cao, Y., and W. P. Pan, "Investigation of Chemical Looping Combustion by Solid Fuels: 1. Process Analysis," *Energy & Fuels*, 20(5), 1836–1844 (2006).

40. Thomas, T., L.-S. Fan, P. Gupta, and L. G. Velazquez-Vargas, "Combustion Looping Using Composite Oxygen Carriers," U.S. Provisional Patent Series No.11/010,648 (2004); U.S. Patent No. 7,767,191 (2010).

41. Leion, H., T. Mattisson, and A. Lyngfelt, "Solid Fuels in Chemical-Looping Combustion," *International Journal of Greenhouse Gas Control*, 2(2), 180–193 (2008).

42. Rubel, A., K. L. Liu, J. Neathery, and D. Taulbee, "Oxygen Carriers for Chemical Looping Combustion of Solid Fuels," *Fuel*, 88(5), 876–884 (2009).

43. Zhao, H., L. Liu, B. Wang, D. Xu, L. Jiang, and C. Zheng, "Sol-Gel-Derived NiO/$NiAl_2O_4$ Oxygen Carriers for Chemical-Looping Combustion by Coal Char," *Energy & Fuels*, 22(2), 898–905 (2008).

44. Mattisson, T., A. Jardnas, and A. Lyngfelt, "Reactivity of Some Metal Oxides Supported on Alumina with Alternating Methane and Oxygen-Application for Chemical-Looping Combustion," *Energy & Fuels*, 17(3), 643–651 (2003).

45. Mattisson, T., M. Johansson, and A. Lyngfelt, "The Use of NiO as an Oxygen Carrier in Chemical-Looping Combustion," *Fuel*, 85(5–6), 736–747 (2006).

46. Lyngfelt, A., B. Leckner, and T. Mattisson, "A Fluidized-Bed Combustion Process with Inherent CO_2 Separation; Application of Chemical-Looping Combustion," *Chemical Engineering Science*, 56(10), 3101–3113 (2001).

47. Mattisson, T., M. Johansson, and A. Lyngfelt, "Multicycle Reduction and Oxidation of Different Types of Iron Oxide Particles–Application to Chemical-Looping Combustion," *Energy & Fuels*, 18(3), 628–637 (2004).

48. Johansson, M., T. Mattisson, and A. Lyngfelt, "Investigation of Fe_2O_3 with $MgAl_2O_4$ for Chemical-Looping Combustion," *Industrial & Engineering Chemistry Research*, 43(22), 6978–6987 (2004).

49. Cho, P., T. Mattisson, and A. Lyngfelt, "Comparison of Iron-, Nickel-, Copper- and Manganese-Based Oxygen Carriers for Chemical-Looping Combustion," *Fuel*, 83(9), 1215–1225 (2004).

50. Leion, H., T. Mattisson, and A. Lyngfelt, "The Use of Petroleum Coke as Fuel in Chemical-Looping Combustion," *Fuel*, 86(12–13), 1947–1958 (2007).

51. Johansson, E., T. Mattisson, A. Lyngfelt, and H. Thunman, "A 300 W Laboratory Reactor System for Chemical-Looping Combustion with Particle Circulation," *Fuel*, 85(10–11), 1428–1438 (2006).

52. Abad, A., T. Mattisson, A. Lyngfelt, and M. Johansson, "The Use of Iron Oxide as Oxygen Carrier in a Chemical-Looping Reactor," *Fuel*, 86(7–8), 1021–1035 (2007).

53. de Diego, L. F., F. Garcia-Labiano, P. Gayan, J. Celaya, J. M. Palacios, and J. Adanez, "Operation of a 10 KWth Chemical-Looping Combustor During 200 H with a CuO- Al_2O_3 Oxygen Carrier," *Fuel*, 86(7–8), 1036–1045 (2007).

54. Ryu, H. J., Y. Seo, and G.-T. Jin, *Development of Chemical-Looping Combustion Technology: Long-Term Operation of a 50 KWth Chemical-Looping Combustor with Ni- and Co- Based Oxygen Carrier Particles*, Proceedings of the Regional Symposium on Chemical Engineering (2005).

55. Pröll, T., P. Kolbitsch, J. Bolhàr-Nordenkampf, and H. Hofbauer, "A Novel Dual Circulating Fluidized Bed (DCFB) System for Chemical Looping Process," *AIChE Journal*, 55(12) 3255–3266 (2009).

56. Fan, L.-S., and C. Zhu, "Principles of Gas-Solids Flows," Cambridge University Press (1998).

57. Yerushalmi, J., and N. T. Cankurt, "Further Studies of the Regimes of Fluidization," *Powder Technology*, 24(2), 187–205 (1979).

58. Bi, H. T., and L.-S. Fan, "Existence of Turbulent Regime in Gas-Solid Fluidization," *AIChE Journal*, 38(2), 297–301 (1992).

59. Grace, J. R., H. T. Bi, and M. Golriz, "Circulating Fluidized Beds," *Handbook of Fluidization and Fluid-Particle Systems*, edited by W.-C. Yang, Marcel Dekker (2003).

60. Lewis, W. K., and E. R. Gilliland, "Conversion of Hydrocarbons with Suspended Catalyst," U.S. Patent 2,498,088 (1950).

61. Squires, A. M., "The Story of Fluid Catalytic Cracking: The First Circulating Fluidized Bed," *Circulating Fluidized Bed Technology*, edited by P. Basu, Pergamon Press (1986).

62. Bi, H. T., and J. R. Grace, "Flow Regime Diagrams for Gas-Solid Fluidization and Upward Transport," *International Journal of Multiphase Flow*, 21(6), 1229–1236 (1995).

63. Bi, H. T., J. R. Grace, and J. X. Zhu, "Regime Transitions Affecting Gas-Solids Suspensions and Fluidized-Beds," *Chemical Engineering Research & Design*, 73(A2), 154–161 (1995).

64. Kunii, D., and O. Levenspiel, *Fluidization Engineering*, 2nd edition, Butterworth-Heinemann (1991).

65. Bi, H. T., and J. X. Zhu, "Static Instability Analysis of Circulating Fluidized-Beds and Concept of High-Density Risers," *AIChE Journal*, 39(8), 1272–1280 (1993).

66. Werther, J., "Fluid Mechanics of Large-Scale CFB Units," Paper Presented at the 4th International Conference on Circulating Fluidized Beds (1993).

67. Grace, J. R., A. S. Issangya, D. R. Bai, H. T. Bi, and J. X. Zhu, "Situating the High-Density Circulating Fluidized Bed," *AIChE Journal*, 45(10), 2108–2116 (1999).

68. Knowlton, T., J. Matsen, D. King, and D. Geldart, "Benchmark Modeling Test (Challenge Problem 1)," *Fluidization VIII*, Tours (1995).

69. Parssinen, J. H., and J. X. Zhu, "Particle Velocity and Flow Development in a Long and High-Flux Circulating Fluidized Bed Riser," *Chemical Engineering Science*, 56(18), 5295–5303 (2001).

70. Lu, Y., "Study of a 2-D CFB Unit," BE Thesis, Tsinghua University (1988).

71. Yousfi, Y., and G. Gau, "Aerodynamique De L'ecoulement Vertical De Suspension, Cocentrees Gaz Solids : I. Regimes D'ecoulement Et Stabilite Aerodynamique," *Chemical Engineering Science*, 29, 1939–1194 (1974).

72. Drahos, J., J. Cermak, R. Guardani, and K. Schugerl, "Characterization of Flow Regime Transitions in a Circulating Fluidized-Bed," *Powder Technology*, 56(1), 41–48 (1988).

73. Hirama, T., H. Takeuchi, and T. Chiba, "Regime Classification of Macroscopic Gas Solid Flow in a Circulating Fluidized-Bed Riser," *Powder Technology*, 70(3), 215–222 (1992).

74. Mori, S., K. Kato, E. Kobayashi, and D. Liu, *Effect of Apparatus Design on Hydrodynamics of Circulating Fluidized Bed*, Proceedings of the AIChE Meeting (1991).

75. Kirbas G., S. W. Kim, H. T. Bi, C. J. Lim, and J. R. Grace, "Radial Distribution of Local Concentration Weighted Particle Velocities in High Density Circulating Fluidized Beds," Paper Presented at the 12th International Conference on Fluidization–New Horizons in Fluidization Engineering (2007).

76. Kulah G., X. Song, H. T. Bi, C. J. Lim, and J .R. Grace, "A Novel System for Measuring Solids Dispersion in Circulating Fluidized Beds," Paper Presented at the 9th International Conference on Circulating Fluidized Beds (2008).

77. Knowlton, T. M., "Cyclone Separators," *Handbook of Fluidization and Fluid-Particle Systems*, edited by W.-C. Yang, Marcel Dekker (2003).

78. Jumah, R. Y., and A. S. Mujumdar, "Dryer Emission Control Systems," *Handbook of Industrial Drying*, edited by A. S. Mujumdar (2006).

79. Cheng, Y., F. Wei, G. Q. Yang, and Y. Jin, "Inlet and Outlet Effects on Flow Patterns in Gas-Solid Risers," *Powder Technology*, 98(2), 151–156 (1998).

80. Knowlton, T. M., "Solids Transfer in Fluidized Systems," *Gas Fluidization Technology*, edited by D. Geldart, Wiley (1986).

81. Lyngfelt, A., and H. Thunman, "Construction and 100 h of Operational Experience of a 10-KW Chemical Looping Combustor," *CO₂ Capture and Storage Project (CCP) for Carbon Dioxide Storage in Deep Geologic Formations for Climate Change Mitigation*, edited by D. Thomas, Elseveir Science (2005).

82. Linderholm, C., A. Abad, T. Mattisson, and A. Lynyfelt, "160 H of Chemical-Looping Combustion in a 10 KW Reactor System with a NiO-Based Oxygen Carrier," *International Journal of Greenhouse Gas Control*, 2(4), 520–530 (2008).

83. Johansson, M., T. Mattisson, and A. Lyngfelt, "Use of NiO/NiAl₂O₄ Particles in a 10 KW Chemical-Looping Combustor," *Industrial & Engineering Chemistry Research*, 45(17), 5911–5919 (2006).

84. de Diego, L. F., P. Gayan, F. Garcia-Labiano, J. Celaya, M. Abad, and J. Adanez, "Impregnated CuO/Al₂O₃ Oxygen Carriers for Chemical-Looping Combustion: Avoiding Fluidized Bed Agglomeration," *Energy & Fuels*, 19(5), 1850–1856 (2005).

85. Abad, A., J. Adanez, F. Garcia-Labiano, L. F. de Diego, P. Gayan, and J. Celaya, "Mapping of the Range of Operational Conditions for Cu⁻, Fe⁻, and Ni-Based Oxygen Carriers in Chemical-Looping Combustion," *Chemical Engineering Science*, 62(1–2), 533–549 (2007).

86. Adanez, J., P. Gayan, J. Celaya, L. F. de Diego, F. Garcia-Labiano, and A. Abad, "Chemical Looping Combustion in a 10 KWth Prototype Using a CuO/Al₂O₃ Oxygen Carrier: Effect of Operating Conditions on Methane Combustion," *Industrial & Engineering Chemistry Research*, 45(17), 6075–6080 (2006).

87. Kolbitsch, P., J. Bolhàr-Nordenkampf, T. Pröll, and H. Hofbauer, "Design of a Chemical Looping Combustor Using a Dual Circulating Fluidized Bed (DCFB) Reactor System," Paper Presented at the 9th International Conference on Circulating Fluidized Beds (2008).

88. Pröll, T., K. Rupanovits, P. Kolbitsch, J. Bolhàr-Nordenkampf, and H. Hofbauer, "Cold Flow Model Study on a Dual Circulating Fluidized Bed (DCFB) System for Chemical Looping Processes," Paper Presented at the 9th International Conference on Circulating Fluidized Beds (2008).

89. Kolbitsch, P., T. Pröll, J. Bolhàr-Nordenkampf, and H. Hofbauer, "Operating Experience with Chemical Looping Combustion in a 120 KW Dual Circulating Fluidized Bed (DCFB) Unit," Paper Presented at the 9th International Conference on Greenhouse Gas Control Technologies (2008).

90. Pröll, T., T. Mattisson, K. Mayer, J. Bolhàr-Nordenkampf, P. Kolbitsch, A. Lyngfelt, and H. Hofbauer, "Natural Minerals as Oxygen Carriers for Chemical Looping Combustion in a Dual Circulating Fluidized Bed System," Paper Presented at the 9th International Conference on Greenhouse Gas Control Technologies (2008).

91. Bolhàr-Nordenkampf, J., P. Kolbitsch, T. Pröll, and H. Hofbauer, "Performance of a NiO-Based Oxygen Carrier for Chemical Looping Combustion and Reforming in a 120 KW Unit," Paper Presented at the 9th International Conference on Greenhouse Gas Control Technologies (2008).

92. Berguerand, N., and A. Lyngfelt, "Design and Operation of a 10 KWth Chemical-Looping Combustor for Solid Fuels–Testing with South African Coal," *Fuel*, 87(12), 2713–2726 (2008).

93. Berquerand, N., and A. Lyngfelt, "The Use of Petroleum Coke as Fuel in a 10 KWth Chemical-Looping Combustor," *International Journal of Greenhouse Gas Control*, 2(2), 169–179 (2008).

94. Shen, L., J. Wu, Z. Gao, and J. Xiao, "Reactivity Deterioration of NiO/Al$_2$O$_3$ Oxygen Carrier for Chemical Looping Combustion of Coal in a 10 KWth Reactor," *Combustion and Flame*, 156(7), 1377–1385 (2009).

95. Shen, L., J. Wu, and J. Xiao, "Experiments on Chemical Looping Combustion of Coal with a NiO Based Oxygen Carrier," *Combustion and Flame*, 156(3), 721–728 (2009).

96. Shen, L., J. Wu, J. Xiao, Q. Song, and R. Xiao, "Chemical-Looping Combustion of Biomass in a 10 KWth Reactor with Iron Oxide as an Oxygen Carrier," *Energy & Fuels*, 23, 2498–2505 (2009).

97. Beal, C., L. Maghdissian and A. Brautsch, "Development Status of Metal Oxides Chemical Looping Process for Coal-Fired Power Plants," Proceedings of the 34th International Technical Conference on Clean Coal & Fuel Systems (2009).

CHAPTER 4

CHEMICAL LOOPING GASIFICATION USING GASEOUS FUELS

F. Li, L. Zeng, S. Ramkumar, D. Sridhar, M. Iyer, and L.-S. Fan

4.1 Introduction

Several coal gasification plants are in operation worldwide for electricity, H_2, and liquid fuels production. They include the 313-MW (250-MW to grid) Tampa Electric's Polk Integrated Gasification Combined Cycle (IGCC) Plant in Polk County, FL, the 292-MW (262 MW to grid) Wabash River Gasification Repowering IGCC Project in Wabash River, Indiana, and the 253-MW Nuon Buggenum IGCC Power Plant in Buggenum, the Netherlands. There are also several proposed coal gasification plants such as Duke Energy's 630-MW Edwardsport IGCC plant in Knox County, IN, and Excelsior Energy Inc.'s 602-MW Mesaba IGCC plant in the Iron Range of Northeast Minnesota. Despite extensive ongoing activities, coal-gasification-related technologies still represent only a very small fraction of the total world fossil energy generation. This is in part a result of the intensive capital requirement of the gasification systems. Without carbon capture, the IGCC plant's capital-intensive nature is illustrated by the proposed Mesaba plant, which is expected to be 30% more expensive than a coal-fired power plant with a similar electricity generation capacity.[1] Under a carbon-constrained scenario, the capital investment for IGCC plants can be cheaper than coal-fired power plants.[2] Through process intensification and enhanced energy conversion efficiency, the total capital and operating costs can be reduced further.

Chemical Looping Systems for Fossil Energy Conversions, by Liang-Shih Fan
Copyright © 2010 John Wiley & Sons, Inc.

In this chapter, the traditional coal gasification processes for the production of electricity, hydrogen, and liquid fuels are presented in Section 4.2. Special attention to the IGCC Process for electricity generation is given in Section 4.2.1. Section 4.3 describes the historical developments of the iron-based chemical looping processes using gaseous fuels, including the Lane Process, the Messerschmitt Process, the U.S. Bureau of Mines Process, and the Institute of Gas Technology Process. The current development of the Syngas Chemical Looping (SCL) Process using gaseous fuels, including coal-derived syngas and C_1–C_4 hydrocarbons, is introduced in Section 4.3.4. The design, analysis and optimization of the SCL Process are discussed in Section 4.4 where results of both the thermodynamic analyses and experimental studies of the SCL reactor systems along with their simulation using the ASPEN PLUS (Aspen Technology, Inc., Houston, TX) are given. In Section 4.5, the ASPEN PLUS simulation, results, and analyses for the traditional gasification processes and the SCL process are described. An example of the SCL application to a coal-to-liquid (CTL) configuration then is presented in Section 4.6. The Calcium Looping Process using gaseous fuels is elaborated in Section 4.7 where the reaction characteristics of the process are described. The Calcium Looping Process applications to the CTL configuration with sulfur and CO_2 capture are given in Section 4.7.4. Comparisons of the chemical looping processes using gaseous fuels with the traditional coal gasification processes also are given in this chapter with consideration of process configurations, CO_2 capture, energy conversion efficiencies, and capital and operating cost requirements.

4.2 Traditional Coal Gasification Processes

Coal gasification can convert coal to syngas. The syngas can be processed further to various products such as electricity, hydrogen, liquid fuels, and chemicals in a general process scheme as shown in Figure 4.1 (previously given as Figure 1.10). It is shown in the figure that, depending on the end products desired, gasification process schemes can vary as elaborated in this section.

4.2.1 Electricity Production—Integrated Gasification Combined Cycle (IGCC)

The IGCC Process can be described generically in terms of three operating blocks according to their functions. These blocks, as shown in Figure 4.2, are the coal preparation and gasification block, the gas conditioning and cleanup block, and the power generation block. For illustration purposes, the functions for each block presented are based on an IGCC system resembling those of the Nuon Power Buggenum gasification plant with an entrained flow, dry feed, and oxygen blown slagging gasifier resembling that of the Shell Coal Gasification Process (SCGP).

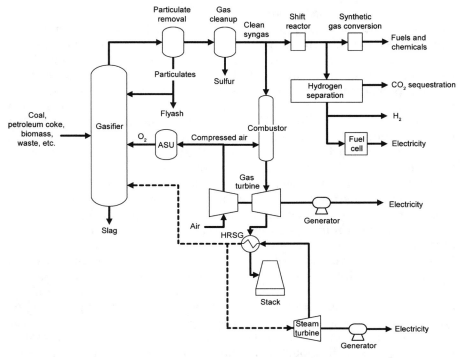

Figure 4.1. Schematic diagram of coal gasification processes.[3]

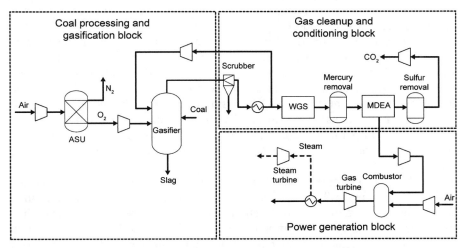

Figure 4.2. Flow diagram of an IGCC power plant.[4]

Coal Processing and Gasification Block

In the coal processing and gasification block, coal first is pulverized and dried. The dry coal powder then is fed to the lock hoppers for pressurization. When the pressure is increased to the gasifier's conditions, the dry coal powder is conveyed pneumatically with N_2 or CO_2 to the gasifier. Oxygen generated from the air separation unit (ASU) and pressurized coal is injected horizontally via four feeders in the central section of the gasifier.[5] The oxygen injected to the gasifier has a typical purity of 95%–99.5% (by volume). Steam also can be introduced into the gasifier to enhance the carbon conversion.

The Shell gasifier is a slagging gasifier that typically operates at temperatures between 1,400°C and 1,600°C and pressures between 30 and 40 atm.[5,6] Under the high operating temperature of the gasifier, most of the coal ash melts and flows toward the bottom of the gasifier where it is removed as slag. A membrane wall cooled by water is used in the gasifier; therefore, steam is generated simultaneously. The vaporization of water lowers the temperature of the membrane wall, and thus the lifetime of the gasifier wall is prolonged. In the gasifier, pulverized coal and oxygen move upward concurrently and partially are oxidized by oxygen, forming raw syngas that contains various contaminants including particulates, H_2S, NH_3, HCl, mercury, and other heavy metals. Before exiting from the top of the gasifier, the high-temperature raw syngas is quenched with a recycled low-temperature syngas stream. The gas quenching step quickly lowers the temperature of raw syngas to 800°C–900°C. By doing so, the small amount of ash entrained in the raw syngas is hardened quickly, minimizing the stickiness problem during fly ash solidification. After being quenched, the raw syngas passes through a syngas cooler, which is essentially a heat exchanger or boiler that generates steam (50–180 atm) from the high-temperature syngas. After the syngas cooler, the temperature of the raw syngas is decreased to ~300°C.

The raw syngas obtained at the outlet of the syngas cooler contains CO, H_2, CO_2, H_2O, and contaminants. The exact composition of the raw syngas may vary depending on the characteristics of the coal used and the operating conditions of the gasifier. When bituminous coal is used, the typical CO:H_2 ratio in the raw syngas is at 2:1 with ~1% sulfur compounds.[6] Note that the pollutants in the raw syngas stream are different from those in the tail gas from a coal-fired power plant because of the reducing environment in the gasifier. The cold gas efficiency (LHV, lower heating value) of the Shell gasifier is higher than 80%.[6]

The ASU is an important unit. The cryogenic ASU operation is both energy- and capital-intensive. The cryogenic ASU is essentially a distillation column that separates oxygen from nitrogen based on the differences in their boiling points. In its operation, the ambient air first is introduced to a filtration unit for removing particulates. It then is compressed using a multistage compressor with intercoolers from 5 to 7 atm or up to 13 atm depending on whether low-pressure products or high-pressure products are desired. After drying and cleaning, the purified airstream is split into several substreams. The substreams either are expanded for cooling or are refrigerated to reach the cryogenic temperature. Finally, a distillation column is used to separate air into oxygen and nitrogen.[7] Before leaving the ASU, the low-temperature, separated gaseous streams are recycled to chill the inlet air. When used in the gasification process, most of the oxygen obtained from the ASU is compressed to the pressure of the gasifier. A small portion of the oxygen also can be used for sulfur recovery. The pressurized nitrogen product can be used for coal conveyance. It also can be injected into the gas turbine to decrease the combustion temperature and to increase the mass flux through the gas expander for increased power generation.

Gas Cleaning and Conditioning Block

Because pollutants are present in the raw syngas, cleaning and conditioning steps are needed. There are two different scenarios for the gas cleaning and conditioning block: the traditional scenario and the carbon constrained scenario.

Traditional Scenario. In the traditional scenario, no carbon capture is required, and gas cleanup is targeted at pollutants such as mercury, sulfur, and nitrogen compounds. Under such a scenario, the raw syngas from syngas cooler passes through a dry solids removal (DSR) unit to remove the particulates. The DSR unit either can be a cyclone or a candle filter. The particulate-free syngas exiting from the DSR then will pass through a wet scrubber in which water is used to remove the remaining particulates and water-soluble contaminants such as HCl and NH_3. The syngas exiting from the wet scrubber then is sent to a hydrolysis unit operating at $200°C–400°C$.[8] The function of the hydrolysis reactor is to convert the trace quantities of organic COS to inorganic H_2S. This is because of the low solubility of COS in the solvents used in the Selexol, Rectisol, or methyl-diethanolamine (MDEA) units for sulfur scrubbing downstream. The reaction in the hydrolysis unit is:

$$COS + H_2O \rightarrow H_2S + CO_2 \qquad (4.2.1)$$

After the hydrolysis unit, the syngas stream is cooled down to ~40°C and is introduced to an activated carbon bed for removing the mercury and other trace quantities of heavy metals. The activated carbon bed is impregnated with sulfur, that reacts with elemental mercury:

$$Hg + S \rightarrow HgS \qquad (4.2.2)$$

The sulfur in the mercury-free syngas stream is removed in an acid gas removal (AGR) process such as the amine scrubbing, Rectisol, and Selexol processes. All these processes are based on either chemical or physical absorption and involve a similar concept that employs a lean solvent to capture acid gas pollutants (H_2S and/or CO_2) at a low temperature in the solvent absorption step and releases them at an elevated temperature in the solvent desorption step. These processes consume a significant amount of parasitic energy because of the extensive cooling and reheating requirements of the solvents.

Among the AGR processes, the amine scrubbing processes are based on chemical absorption. Solvents such as monoethanolamine (MEA) and MDEA are used in these processes. These solvents exhibit good absorption capacity when the acid gas partial pressure is low; however, when the acid gas partial pressure is high, the solvents can react with the acid gas to form heat-stable salts that plague the system. Moreover, the absorption capacity of amine-based solvents plateaus when the partial pressure of the acid gas is high. Besides, the desorption step requires a large amount of heat to break the chemical bonds between H_2S and the solvents.

In contrast to the amine solvents, the solvents in the methanol-based Rectisol Process and the glycol-based Selexol Process are used based on physical absorption. Although the absorption capacity of these physical solvents is relatively low at low acid gas partial pressures, it increases linearly with the acid gas partial pressure and is not subject to the absorption limit observed in chemical absorption. Moreover, no by-product is formed in these processes. The physical absorption usually is carried out at low temperatures. For instance, the typical absorption temperature in the Rectisol Process is ~−60°C and that in the Selexol Process is ~5°C. Therefore, refrigeration is required for both the Rectisol and Selexol processes. Because of a lower solvent temperature in the absorption step, the energy consumption for refrigeration in the Rectisol Process is higher than the Selexol Process, where only mild refrigeration is needed.[9] Although the physical absorption-based Rectisol and Selexol processes often are used in gasification systems where the acid gas partial pressures are high, the option of using the chemical absorption-based MDEA process in the gasification plant also has been explored because of its good reaction selectivity and low refrigeration requirements. It was reported that MDEA can deliver performance comparable with the physical absorption processes when the pressure of the gas stream is not very high.[10] Once the sulfur is removed, the clean syngas can be sent to the power generation block for combustion. Because CO is still in the syngas stream, carbon in the coal is not captured in this scenario and will be emitted from the combustor exhaust in the form of CO_2.

Carbon-Constrained Scenario. In the carbon-constrained scenario, carbon is captured to mitigate CO_2 emissions. As discussed in Chapter 3, there are three options for carbon capture, precombustion, postcombustion, and oxycombustion. In the IGCC process, carbon capture before combustion is a more economical option. Similar to the traditional scenario, the particulates in the raw syngas first are removed. The particulate-free syngas, at ~250°C, is introduced to a water–gas shift (WGS) reactor with sulfur tolerant catalyst, instead of an hydrolysis unit. Steam also is introduced to adjust the steam:CO ratio to greater than a 2 to avoid carbon deposition and to achieve a higher CO conversion.[11] In the WGS reactor, CO in the syngas stream is converted to H_2 and CO_2 as expressed by:

$$CO + H_2O(g) \rightarrow CO_2 + H_2 \qquad \Delta H = -41.1 \, kJ/mol \, 298 \, K \qquad (4.2.3)$$

Another option that avoids the addition of steam to the WGS reactor is to use water quenching, instead of gas quenching, to cool the high-temperature syngas. In this case, a large amount of water is used to bring down the temperature of the raw syngas from the gasifier. The raw syngas then will be saturated with steam and is ready for the WGS reaction. The main advantage of water quenching is the low capital and operating cost, but it negatively can affect the thermal efficiency of the process. More discussion on the WGS reaction is given in Section 4.2.2.

The gas stream exiting from the WGS reactor consists mainly of H_2 and CO_2 along with a small amount of unconverted CO and pollutants such as H_2S and mercury. An activated carbon bed similar to that in the traditional scenario is used to remove mercury. The acid gas treatment procedure, however, is different because of the need for CO_2 removal. In this case, the acid gas removal process is used to remove both H_2S and CO_2. When the Selexol process is used, the removal of sulfur and CO_2 is carried out in two interconnected absorbers, the H_2S absorber and the CO_2 absorber. As shown in Figure 4.3, the lean, refrigerated solvent flows through the CO_2 absorber and subsequently through the H_2S absorber countercurrently to the CO_2- and H_2S-containing syngas flows. Such a countercurrent flow pattern ensures that the solvent entering the H_2S absorber is saturated with CO_2. As a result, no further CO_2 absorption will occur in the H_2S absorber, avoiding an undesirable temperature increase in the H_2S absorber, that can affect the solvent H_2S absorption capacity and selectivity. The Selexol Process can strip H_2S to below 1 ppm and CO_2 to ~1%. The CO_2- and sulfur-rich solvent is regenerated in two separate systems to obtain two streams: one with a high concentration of CO_2 and the other with a high concentration of H_2S. The H_2S-rich stream is sent to the Claus Process to produce sulfur as a by-product.[12] The CO_2-rich stream is compressed and sent to a pipeline for transportation and sequestration.

Power Generation Block
The power generation block contains a combined cycle system for electricity generation. Here, the term "combined cycle" refers to the two or more thermodynamic power cycles used in the power generation block, that is, the Brayton cycle in the gas turbine (GT) and the Rankine cycle in the steam turbine (ST).[6] This is contrary to the coal-fired power plant where only the steam turbine is used. This block usually contains a GT, heat recovery steam generator (HRSG), and ST, as shown in Figure 4.2. Although the actual

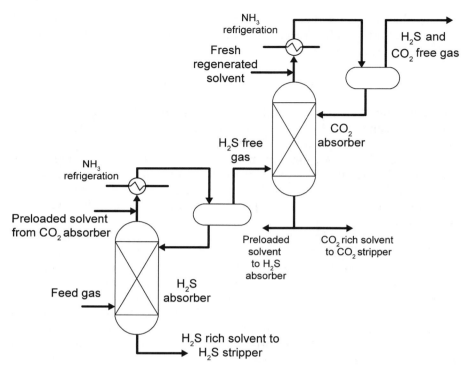

Figure 4.3. Schematics of the Selexol operation for CO_2 and H_2S removal.

arrangements of the power generation block can be complicated, the principle is simple. The fuel gas, either syngas in the traditional scenario or H_2-rich gas in the carbon-constrained scenario, is combusted in a gas turbine with part of the thermal energy extracted. To lower the firing temperature, to reduce NO_x emissions, and to increase the mass flow of the gas turbine, a stream of nitrogen from the ASU is mixed with the fuel gas. The gas turbine can be either independent or integrated with the ASU based on a specific process configuration.

The high-temperature gas exhaust from the gas turbine is used in the HRSG to generate steam to drive the steam turbines. Other heat recovered from the gasifier, syngas cooler, and WGS also can be used in the HRSG. Moreover, some supplemental fuels can be burned directly in the HRSG to enhance the generation of steam. All heat will be distributed to evaporate and superheat steam. The steam usually is ranked in three pressure levels, high pressure (HP, ~120 atm), intermediate pressure (IP, ~30 atm), and low pressure (LP, ~2 atm). Finally, the steam turbine converts the thermal energy contained in steam to electricity. The steam turbine exhaust will be condensed and pumped to the next steam generation cycle.

4.2.2 H₂ Production

Figure 4.4 shows the schematics of two coal gasification processes for H_2 production. As is shown, the syngas is generated, conditioned, and converted in a manner similar to the carbon-constrained IGCC process. The major difference between these two processes lies in the usage of the hydrogen-rich stream after acid gas is removed from the shifted syngas. In the hydrogen production case, the hydrogen-rich stream exiting from the acid gas removal unit is purified further in a pressure swing adsorption (PSA) unit. The high-purity hydrogen is used as an end product, while the tail gas from the PSA is combusted in a combined cycle system to generate electricity to meet the parasitic energy consumption. This section focuses on two key operations that are of particular relevance to the coal-to-hydrogen process, the water–gas shift reaction, and the pressure swing adsorption operation.

Water–Gas Shift Reaction
The WGS reaction, shown in Reaction (4.2.3), converts CO to H_2, and thus the hydrogen composition in the syngas is enhanced. The reaction is

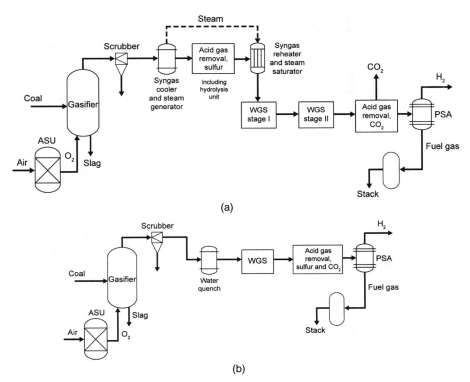

Figure 4.4. Coal gasification process for hydrogen production: (a) clean syngas WGS scheme and (b) raw syngas WGS scheme.

exothermic. Therefore, the equilibrium of this reaction is favored at lower temperatures. However, the rate of the water–gas shift reaction is rather slow at low temperatures. Therefore, a high-temperature WGS reactor usually is used to convert bulk quantities of CO to H_2. Depending on the sulfur tolerance of the catalyst, one of the two types of the WGS system, that is, a clean (sweet) syngas shift or a raw syngas shift, can be used.

Clean Syngas Shift. Coal-derived syngas usually contains a significant amount of sulfur compounds such as H_2S and COS. The WGS catalysts under the clean syngas shift type of operation are typically sensitive to sulfur. Therefore, sulfur cleaning in the syngas stream using such cleaning processes as MDEA, Rectisol, and Selexol are required prior to the shift reaction taking place. Under the type of operation shown in Figure 4.4(a), a syngas cooler is used to reduce the raw syngas temperature while generating steam. With COS in the raw syngas stream hydrolyzed, the raw syngas then will be cooled down before entering the desulfurization unit. The sulfur-free syngas then is reheated and saturated with steam before entering the shift reactors. Part of the heat for syngas reheating and steam generation is produced from the WGS reactors.

To enhance H_2 generation, the WGS reaction often is carried out in two stages. The first stage is the high-temperature shift (300°C–500°C) using iron catalysts. This stage converts the bulk of the CO to hydrogen at an elevated temperature in the presence of steam. The second stage is the low-temperature shift (210°C–270°C) that uses copper-based catalysts to enhance further the CO conversion at a low temperature.[13,14] Although the two-stage WGS scheme provides enhanced CO conversion and high selectivity toward H_2 production, it has several disadvantages such as:

1. Copper-based catalysts are extremely intolerant to sulfur (<0.1 ppm), which imposes stringent desulfurization requirements for the upstream acid gas removal system.[15]
2. Within the high-temperature range, a high steam:CO ratio is required to enhance the CO conversion and the consequent hydrogen production. At 550°C, the steam:CO ratio can be as high as 50 in a single-stage operation or 7.5 for a dual-stage process to obtain 99.5% pure H_2.[16] The large steam requirement leads to increased energy consumption for steam generation.

To perform hydrogen purification in PSA units and acid gas control in the process, the CO_2 acid gas needs to be removed after the WGS reactors and before the PSA units, whereas the sulfur acid gas needs to be removed before the WGS reactor, as shown in Figure 4.4(a). Thus, removing two different acid gases from two separate gaseous streams in a clean syngas shift process may pose challenges to the operation of the acid gas control system in a synergistic manner.

Raw (Sour) Syngas Shift. The WGS catalysts used under the raw (sour) syngas shift type of operation can tolerate high levels of sulfur. In fact, the catalyst needs to be sulfided to remain active. The raw syngas WGS catalyst usually consists of group VIB primary metals such as Mo and/or W promoted by group VIII metals such as Fe, Co, and/or Ni.[17] The operating temperatures of such catalysts range from 250°C to 500°C.[18] One WGS unit often is used in this case. Therefore, the CO concentration in the shifted gas can be significantly higher than that from the clean syngas shift. It is worth noting, however, that most of the COS can be converted to H_2S in the raw syngas shift reactor; thus, a dedicated COS hydrolysis step is avoided. Under this type of operation, shown in Figure 4.4(b), the raw syngas from the gasifier usually is quenched with water to generate steam-saturated syngas. Such a syngas is shifted in the raw-syngas WGS reactor. The two acid gas removal units (i.e., desulfurization and CO_2 removal units), thus, can be placed contiguously in the process to yield a synergistic, integrated acid gas treatment system.

Studies conducted by Becker et al.[19] on the gasification-based ammonia synthesis process showed that the two types of the WGS system presented have comparable performance in terms of both energy conversion efficiency and capital investment. Therefore, the choice of the WGS type is a process optimization issue that requires the consideration of such variables as the type of the feedstock, the configuration of key units, and the hydrogen product purity requirements.

Pressure Swing Adsorption (PSA)

Although the WGS reactor and acid gas removal units can produce a concentrated hydrogen stream, a PSA unit often is used to purify further the hydrogen product. Figure 4.5 shows the schematic of a PSA unit. The PSA unit consists of multiple fixed-bed adsorbers filled with adsorbents such as silica gel, molecular sieves, or activated carbon. The adsorbents are selective toward gases with high polarity and low volatility. Therefore, when a pressurized gas mixture is introduced into an adsorber, the individual components of the gaseous mixture will be adsorbed sequentially based on their inherent chemical and physical properties. H_2, because of its high volatility and low polarity, is less prone to being adsorbed. Therefore, a high-purity H_2 stream can be obtained at the outlet after the impurities are captured by the adsorbents. A H_2 product with a purity of 99.999% and greater can be produced from the PSA.[20] After extracting pure hydrogen, the impurities adsorbed on the adsorbents are purged, forming a tail gas stream and regenerating the adsorbents. The tail gas, which contains CO, CO_2, and H_2, can be used for heat and power generation.

As shown in Figure 4.5, each individual adsorber is operated in a discrete, cyclic manner. The operation cycle is called pressure swing cycle, which typically consists of five consecutive steps.

Step 1: Adsorption: The pressurized, H_2-rich gas from the acid gas removal unit is introduced to the adsorber inlet (feed side), which usually is located at

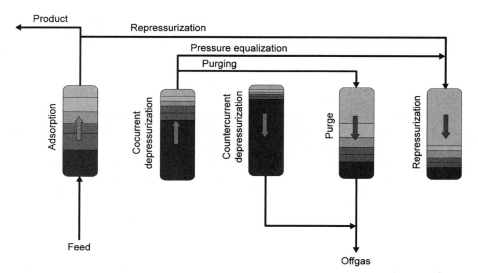

Figure 4.5. Pressure swing adsorption (PSA) unit flow diagram.

the bottom of the fixed-bed adsorber. The impurities in the H_2 stream are adsorbed on the fresh adsorbents, producing a high-purity hydrogen stream at the adsorber outlet (product side) located at the top of the fixed bed. During the adsorption step, the adsorbents become saturated with the impurities in a sequential manner. The adsorbents near the feed side of the adsorber will become saturated first, followed by those located closer to the product side. To prevent the leakage of impurities, feed gas is stopped before the adsorbents get consumed fully. This is followed by the cocurrent depressurization step (step 2).

Step 2: Cocurrent Depressurization: Hydrogen in the void space of the adsorbents is withdrawn from the product side with the feed side closed. As a result, the adsorber is depressurized. During this step, the impurities adsorbed on the adsorbents migrate toward the product side by desorption from the saturated adsorbents near the feed side and by readsorption on the unsaturated sorbents near the product side. The cocurrent depressurization step is stopped before the impurities start to exit the adsorber outlet. The hydrogen obtained in this step is used for the purging step (step 4) and for the repressurization step (step 5).

Step 3: Countercurrent Depressrization: Steps 3 and 4 regenerate the spent adsorbents resulting from steps 1 and 2. In the countercurrent depressurization step, the impurities adsorbed by the adsorbents are removed from the feed side of the adsorbent bed by depressurization. This step partially regenerates the adsorbents.

Step 4: Purge: The purified hydrogen obtained from step 2 is introduced to the product side of the adsorber. Simultaneously, effluent gas is removed from the feed side to maintain a constant adsorber pressure. In this step, the remaining impurities adsorbed by the adsorbents are purged. The spent adsorbents thus are regenerated. As mentioned previously, the tail gas obtained from the feed side in this step and the previous step can be used for heat and power generation.

Step 5: Repressurization: After purging, additional purified hydrogen is injected from the product side of the adsorber to repressurize it to the operating pressure of step 1. Upon completion of this step, the adsorber is ready for the next pressure swing cycle operation.

To maintain a constant product flow, the pressure swing adsorption unit usually consists of four or more (can be more than 10) adsorbers. An automatic rotating valve system is used to alternate gases among the adsorbers. The typical feed gas pressure for PSA ranges between 10 and 40 atm, whereas the tail gas pressure is usually between 1 and 10 atm. A modern PSA unit can recover 80%–92% of the hydrogen in the hydrogen-rich gas stream, turning it into a purified hydrogen product. The specific product yield is dependent on both the desired hydrogen product purity and the feed gas composition. Although the PSA can produce H_2 with a very high purity, the pressurization and depressurization steps can be energy-consuming. Other energy-intensive steps in the coal-to-hydrogen process include steam generation for the WGS reactions, acid gas removal, and CO_2 compression if CO_2 capture and sequestration (CCS) is required. The efficiency for hydrogen production in a gasification-WGS process is reported to be at ~60% (HHV) or even less if carbon capture is considered.[3]

4.2.3 Liquid Fuel Production

Synthetic liquid fuel also can be produced from coal-derived syngas. Such a process also is called indirect coal liquefaction as opposed to direct coal liquefaction in which coal is liquefied by direct hydrogen addition in the liquid-phase media. The indirect coal liquefaction process can be divided into four different blocks: a coal processing and gasification block, a gas conditioning and purification block, a liquid fuel production block, and a power generation block. Figure 4.6 shows the flow diagram of a generic indirect CTL Process. The coal processing and gasification block of the process given in the figure can be identical to that of the IGCC Process. The difference rests in the gas conditioning and liquid fuel production blocks. This is mainly a result of the difference between the syngas composition produced from an entrained flow gasifier ($CO:H_2$ ratio of 1.5–2) and the optimal composition (~0.5 for the cobalt-based catalyst) for liquid fuel synthesis in a Fischer–Tropsch reactor. In the following description, the Shell SCGP gasifier, which is discussed in Section 4.2.1, is used as an example.

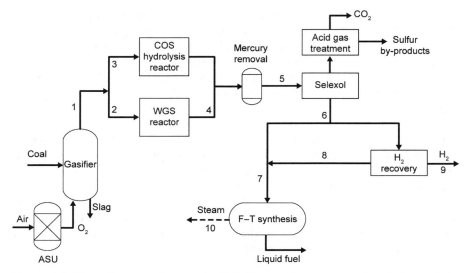

Figure 4.6. Flow diagram of a generic CTL Process (see text for descriptions of the numbered streams).

Gas Conditioning and Cleanup Block

To adjust the $CO:H_2$ ratio in the syngas to 0.5, stream 1 in Figure 4.6, which is the raw syngas exiting from the quench scrubbers, is split into two streams: stream 2 and stream 3. Stream 2, with ~50% of the untreated syngas, is introduced to a sour WGS reactor, which results in a gas stream with high concentrations of H_2 and CO_2 (stream 4). Additionally, most of the COS is converted to H_2S. Stream 3, which contains the other 50% of stream 1, passes through a hydrolysis unit that converts the trace quantities of COS to H_2S. After hydrolysis and the WGS reactors, the resulting streams are combined and pass through an activated carbon bed for mercury removal. Stream 5, the output of the mercury removal unit, then is introduced to a solvent-based acid gas removal unit to remove the sulfur and CO_2 in the stream. Sulfur removal is necessary to avoid the poisoning of Fischer–Tropsch catalysts, which are highly sensitive to sulfur. For example, cobalt-based catalysts require the H_2S concentration to be below 30 ppb. To ensure the desired low sulfur content (~30 ppb) in the Fischer–Tropsch feed stream, a zinc oxide guard bed usually is installed after the acid gas removal unit to strip the remaining sulfur and protect the Fischer–Tropsch reactor from spikes in H_2S concentrations. The syngas, stream 6, coming from the acid gas removal unit, has an optimized $CO:H_2$ ratio of 0.5 and is free of pollutants.

Stream 6 is split into two streams. Most of the syngas enters the Fischer–Tropsch reactor for liquid fuel synthesis. The remaining syngas is sent to a PSA unit to generate purified hydrogen. The low-pressure tail gas from the PSA will be recompressed and introduced into the Fischer–Tropsch reactor.

Liquid Fuel Production Block
Slurry bubble column reactors are used for the production of liquid fuels through Fischer–Tropsch reactions. There are three phases in the reactor: a solid phase (mainly catalyst), a liquid phase (mainly hydrocarbon products), and a gas phase (mainly syngas feed). The gas–liquid–solid mass transfer and reactions that take place in the reactor are complex.[21,22] The reactor is operated at 220°C and 24 atm. The Fischer–Tropsch reaction can be represented by the following reaction:

$$CO + 2H_2 \rightarrow -(CH_2)- + H_2O \qquad \Delta H = -165\,kJ/mol. \qquad (4.2.4)$$

As is shown, the reactions are highly exothermic, and thus, a large amount of water is used for cooling. As a result, low-temperature, medium-pressure steam (220°C, 20 atm) is generated (stream 10). This steam is sent to a heat recovery steam generator (HRSG) for further heating to generate electricity.

The product from the Fischer–Tropsch reactor can be represented by a hydrocarbon mixture ranging from CH_4 to long chain hydrocarbons (wax). Depending on the activity of the catalyst, the gas residence time, the temperature, and the pressure conditions in the reactor, the per-pass-conversion of syngas can be as high as 80%.[23] The Fischer–Tropsch products are sent to refining units where the gaseous products, that is, the unconverted syngas and light (C_1–C_4) hydrocarbons, are separated from the liquid phase and the solid phase. Part of the gaseous product is recycled to the Fischer–Tropsch reactor to increase the overall conversion. The remaining product is combusted in the power generation block. The liquid phase and the solid phase of the Fischer–Tropsch product are treated and separated based on boiling point. The final products of the process include a wide variety of fuels ranging from naphtha (C_5–C_{12}) to kerosene (C_{12}–C_{15}) and to diesel (C_{10}–C_{22}).

Power Generation Block
A combined cycle system often is included in the CTL process to provide the energy needed to operate the ASU, AGR, and gas compression units. The combined cycle system in the CTL process is similar to that for the H_2 production process and is significantly smaller than that for the IGCC process.

The coal gasification processes are versatile; however, the processes involve elaborate unit operation, which consumes a significant amount of energy. Simplification of the coal gasification processes through process intensification and, hence, the reduction of the energy intensity is desirable for modern coal gasification processes. The chemical looping approach offers a viable process intensification example for coal gasification applications. In the following sections, chemical looping processes that use coal gasification products as feedstock are discussed. The steam-iron process, which is the predecessor of the iron-based chemical looping process, first is discussed. It is followed by the description of the two chemical looping technologies, that is, the Syngas Chemical Looping (SCL) Process and the Calcium Looping Process. The advantages and challenges for these technologies are elaborated.

4.3 Iron-Based Chemical Looping Processes Using Gaseous Fuels

Initiated in the late 19th century,[24] the steam-iron process (SIP) is one of the earliest chemical looping processes that produce hydrogen from gaseous fuels. The SIP technology was used for commercial hydrogen production from coal derived producer/water–gas until the 1930s. The development of more cost-effective hydrogen production techniques from the then cheap and abundant hydrocarbon fuels, such as oil and natural gas, phased out the steam-iron process. Since then, the process has seen intermittent development throughout the past century. Concentrated research and development efforts were seen whenever oil and natural gas prices surged.[25] Described below in the chronological order are four SIP processes, that is, the Lane Process and the Messerschmitt Process from the 1900s, the U.S. Bureau of Mines Process from the 1950s, and the Institute of Gasification Technology Process from the 1970s. The Syngas Chemical Looping Process, which more recently has been developed based on similar concepts to those of the steam-iron processes, is also discussed.

The steam-iron process involves the cyclic treatment of iron oxide-based looping medium with syngas and steam. The redox (reduction-oxidation) reactions involved in the steam-iron process are given as follows:

Reduction Stage

$$Fe_3O_4 + H_2 \rightarrow 3FeO + H_2O \qquad (4.3.1)$$

$$Fe_3O_4 + CO \rightarrow 3FeO + CO_2 \qquad (4.3.2)$$

$$FeO + H_2 \rightarrow Fe + H_2O \qquad (4.3.3)$$

$$FeO + CO \rightarrow Fe + CO_2 \qquad (4.3.4)$$

Oxidation Stage

$$Fe + H_2O \rightarrow FeO + H_2 \qquad (4.3.5)$$

$$3FeO + H_2O \rightarrow Fe_3O_4 + H_2 \qquad (4.3.6)$$

These reactions also are known as steam-iron reactions. Throughout the years, the development of the steam-iron process has progressed with improvements in syngas generation methods, looping reactor design, and looping particle synthesis.

4.3.1 Lane Process and Messerschmitt Process

The idea of using steam-iron reactions to generate hydrogen was conceived in the late 19th century.[24] In the early 20th century, aerial navigation using hydrogen balloons emerged and fueled the development of the steam-iron process by expanding the hydrogen gas market.[26] Two of the most notable steam-iron processes developed in this period were those developed by Howard Lane[27] and by Anton Messerschimitt.[28] As shown in Figure 4.7, both processes use fixed-bed reactors to produce hydrogen from syngas derived from coke.

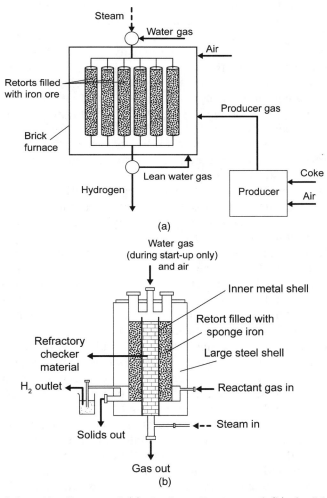

Figure 4.7. Schematic diagram of (a) the Lane Process and (b) the Messerschmitt Process.

The Lane Process, which is shown in Figure 4.7(a) (previously Figure 1.20(a)), has multiple retorts placed within a brick furnace.[27] The retorts are filled with coarse iron ore particles. The reactant gases are introduced to or removed from the retorts by cock valves. Coke is the fuel in this process. In the operation, part of the coke is fed to a producer reactor where it reacts with air to form producer gas, which consists of a significant amount of N_2, CO, and CO_2 along with a small amount of H_2 and H_2O. The producer gas subsequently is combusted to generate steam and to heat the retorts in the brick furnace. The rest of the coke reacts with steam to produce water gas, which consists mainly of H_2 and CO. The water gas is used as the reducing gas in the steam-iron process. The Lane Process has two stages of operations, reduction and oxidation. A brief purging step often is adopted to increase the purity of the hydrogen. During the reduction step, water gas is sent through the top of the retorts to react with the iron ore. The unconverted water gas is discharged from the bottom of the retorts to the brick furnace where it is combusted along with the producer gas to maintain the retort operating temperature at ~750°C. After the iron ore is reduced, the water gas is stopped, and the retort is purged with steam for approximately 35 seconds. In the purging step, the residual water gas is washed out and sent to the brick furnace for combustion. After purging, steam oxidation takes place. In this step, more steam enters from the top of the reactor to react with the reduced iron ore, generating the hydrogen product that exits through the bottom valve.

The Messerschmitt Process, shown in Figure 4.7(b) (previously Figure 1.20(b)), adopts a single reactor design. The Messerschmitt system contains a large steel shell lined with refractory material. Within this steel shell is an inner metal cylinder with a checker refractory. The refractory in the inner cylinder increases the heat capacity of the reactor. The top portion of the inner metal cylinder is connected to the large steel shell. The iron ore particles are loaded between the large steel shell and the inner metal cylinder.[28] The Messerschmitt Process also uses coke as the fuel. Unlike the Lane Process, all the coke is converted to water gas. Part of the water gas is used to heat the reactor, whereas the rest is used as the reactant gas.

The concept of the Messerschmitt Process is identical to that of the Lane Process. At the start-up stage, water gas is introduced from the top of the inner cylinder along with air. As a result, the refractory material in the inner cylinder is heated up. Heat is stored in the refractory materials and then conducted to the iron ores. When the iron ore bed reaches 750°C, the injection of water gas is stopped. The water gas is directed to the reactant gas inlet and enters the iron ore bed. The iron ore is reduced gradually in this step. The lean water gas that exits from the iron ore bed will enter the inner cylinder from the top where it is combusted with air to heat the refractory material. By doing so, the reactor temperature is maintained. After reduction is completed, air is stopped, and steam is introduced to the steam inlet located at the bottom of the inner cylinder. While moving up through the hot refractory material in the inner cylinder, the steam is superheated. The superheated steam then will enter the large steel shell where the reduced ore is located; the steam iron reaction takes place to generate hydrogen. When hydrogen with high purity is desired, a 30-second steam purging step is added between the reduction and oxidation steps.

The hydrogen product from both processes contains entrained iron ore particulates. Moreover, because coke is used as the fuel, sulfur compounds are introduced along with the water gas.[24] It was found by early researchers that the sulfur reduces the lifetime of the iron ore and also pollutes the H_2 stream. Furthermore, carbon deposition is catalyzed by the reduced iron oxides. To purify the H_2 product, a water scrubber was used to remove the particulates, as well as a small portion of CO_2 and H_2S. This was followed by using purifier boxes containing slaked lime to remove the CO_2 and H_2S impurities. The gas-cleaning steps purify the hydrogen stream from ~95% to 99%.[29] The reactions in the purifier boxes are:

$$Ca(OH)_2 + H_2S \rightarrow CaS + 2H_2O \qquad (4.3.7)$$

$$Ca(OH)_2 + CO_2 \rightarrow CaCO_3 + H_2O \qquad (4.3.8)$$

To improve the lifetime of the iron ore particles, the processes incorporated burn-off periods after certain hours of operation. In this step, the entire bed is burned in air to remove the accumulated iron carbide. It was believed by the early researchers that the iron carbide accumulates as a result of the relatively faster iron carbide formation rate [Reaction (4.3.9)] when compared with its removal with steam [Reaction (4.3.10)].[24,29]

$$Fe + 6CO \rightarrow FeC_3 + 3CO_2 \qquad (4.3.9)$$

$$3FeC_3 + 13H_2O \rightarrow Fe_3O_4 + 9CO + 13H_2 \qquad (4.3.10)$$

Table 4.1 compares the performance of the Lane and Messerschmitt processes. It is clear from the table that the chemical efficiencies of the two processes are similar. The higher water gas consumption for the Messerschmitt Process is from the use of water gas for both heating and particle reduction. The Messerschmitt Process' simpler design using a single retort results in a reduced

TABLE 4.1. Comparison of Lane and Messerschmitt Processes[29]

	Lane Process	Messerschmitt Process
Mode of Operation	Fixed Bed	Fixed Bed
Temperature	750°C	750°C
Water Gas Consumption ft^3/ft^3 H_2	2.5	3.5
Hydrogen Yield ft^3/t of Coke	6,500–7,000	6,500–7,000
H_2 purity	95–99%	95–99%

number of high-temperature gas–solid contact joints and, hence, easier maintenance. Moreover, the use of water gas for both heating and reduction contributes to the reduction in capital cost. The reduced capital and maintenance cost made the Messerschmitt Process a popular choice for hydrogen production during the early part of the 20th century.[24,29]

Despite being used for commercial hydrogen production for years, both the Lane and Messerschmitt processes faced difficulty in achieving high solids conversion owing to the low performance of the coarse iron ore particles[30] and the low hydrogen and CO content in the water gas.[24] The low solids conversion leads to a poor fuel gas conversion and steam to hydrogen conversion. Moreover, the lack of reliable gas-switching valves in the early years presented additional challenges to the successful operation of the process.[30] All these disadvantages increased the hydrogen production cost. Other developed and more cost-effective hydrogen production technologies such as natural gas-based reforming processes eventually replaced the steam-iron process.

4.3.2 U.S. Bureau of Mines Pressurized Fluidized Bed Steam-Iron Process

The mid-20th century saw an increase in hydrogen demand for oil refining. In the 1950s, the U.S. Bureau of Mines initiated work to demonstrate the feasibility of a high-pressure, fluidized bed steam-iron system to produce hydrogen continuously from gaseous fuel. It was expected that the fluidized bed reactor design would increase the solids conversion while eliminating the unreliable switchover valves required in the fixed bed steam-iron process.[31,32] A high-pressure fluidized bed reactor setup, as shown in Figure 4.8 was constructed to demonstrate both iron ore reduction and steam oxidation operations.

The reactor was 20 ft (6.1 m) high with a 2-in (5.1 cm) inner diameter. The effective bed height usually was maintained at ~16 ft (4.9 m), with the typical solids residence time of ~2 hours for the reduction operation and 30 minutes for the oxidation operation.[30] During a typical test, the reactor was heated electrically to 900°C and pressurized to 300 psi (20.4 atm). The solids were introduced at the top of the reactor and discharged from the bottom, whereas the gaseous reactant was introduced at the bottom and removed from the top. The reducing gas used in the experiments were simulated producer gases with compositions shown in Table 4.2. Magnetite powder pretreated by reducing gases and steam was used as the looping medium to avoid agglomeration. A mass spectrometer was used to analyze the gas composition, whereas the solids analysis was carried out via thermogravimetric analysis (TGA) coupled with the gas analysis.

Various tests were carried out in the reactor with the operating time varying from 5 to 50 hours. The effects of numerous conditions on the reactor performance evaluated by the gas and the solids conversions and H_2 purity were studied. The operating conditions studied include temperature, pressure, bed height, gas-to-solids ratio, and composition of gases and solids. Up to 70% of

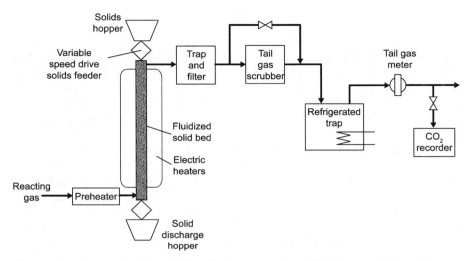

Figure 4.8. Schematic of the fluidized bed setup of the U.S. Bureau of Mines Process for studying both iron ore reduction and steam oxidation.

TABLE 4.2. Composition of Simulated and Commercial Producer Gas[30]

Fuel	Gas Composition (vol.%)				
	H_2	CO	CH_4	CO_2	N_2
Simulated Producer Gas	10–27	9–30	0–7	5–6	49–54
Anthracite Coal-Derived Producer Gas	16.6	27.1	0.9	5.0	50.4
Bituminous Coal-Derived Producer Gas	14.5	25.0	3.1	4.7	52.7
Coke-Derived Producer Gas	9.3	31.0	0.7	3.6	44.3

the reducing gas conversion was achieved during the reduction, with most of the Fe_3O_4 reduced to FeO. During the oxidation, a steam conversion of ~69% was achieved using the reduced solids. Hydrogen stream with a purity of 91%–98% (dry basis) was produced.

Although primarily operated as a fluidized bed, the system design also allowed the reactor to be operated under a countercurrent, free-falling bed mode. The gas and solids conversions for the reactor operating under such a mode also were studied. Although favored in thermodynamics, the free-falling bed operation led to a gas conversion lower than 44%, and the solids conversion was lower than 17% because of the short solids residence time in the reactor.[30]

The performance of the fluidized bed steam-iron process was encouraging, but the hydrogen production cost using this process was not attractive enough when compared with that of the natural gas-based hydrogen production processes. As a result, the demonstration was discontinued.

4.3.3 Institute of Gas Technology Process

A period of shortage in natural gas supply in the 1970s stimulated the develop-ment of processes to produce substitute natural gas (SNG) from coal. The HYGAS Process converts coal and hydrogen into methane by the methana-tion reaction. To improve the yield of the HYGAS Process, the Institute of Gas Technology (IGT) used the high-pressure steam-iron process to produce hydrogen.[33] Figure 4.9 shows the flow diagram of IGT's steam-iron process integrated with the HYGAS Process.

As shown in Figure 4.9, there are three major units in the IGT steam-iron process: the producer, the reducer, and the oxidizer. The producer gasifies the unconverted char from the HYGAS hydrogasifier with steam and air to generate a reducing gas (producer gas) stream. This reducing gas is sent to the reducer to reduce the iron ore. The reduced iron ore subsequently reacts with steam in the oxidizer to generate hydrogen. The hydrogen stream from the oxidizer then is sent to the hydrogasifier of the HYGAS Process for methane production.

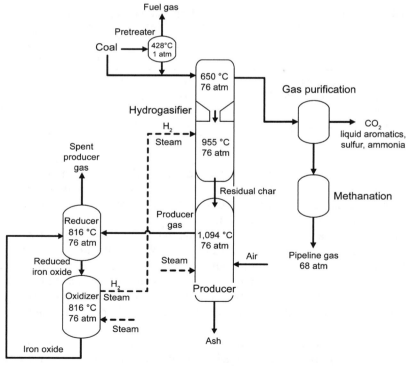

Figure 4.9. The HYGAS Process coupled with the high-pressure steam-iron process developed by IGT.[33]

The producer for the steam-iron process is a fluidized-bed reactor integrated with a staged fluidized-bed hydrogasifier for the HYGAS Process. Air, rather than oxygen, is used as the oxidant for the char gasification in the steam-iron process producer. For the nonslagging, nonagglomerating type of the steam-iron process producer, the designed operating temperature for the producer is 1,100°C with a char preheat zone to vaporize water in the slurry feed and to preheat the char from 315°C to 930°C. The producer gas stream has a high $(CO + H_2)/(CO_2 + H_2O)$ molar ratio (~4), making it effective in reducing the iron oxide. The producer operates at 76 atm, which is similar to the operating pressure of the hydrogasifier. The high operating pressure favors the formation of methane in the hydrogasifier.

The gas from the producer is introduced to the reducer to react with Fe_3O_4, forming a mixture of FeO and Fe_3O_4. The reducer is a two-stage fluidized-bed reactor that can be operated at up to 870°C. The FeO and Fe_3O_4 mixture that comes out of the reducer then is sent to the oxidizer to react with steam, generating hydrogen and Fe_3O_4. The oxidizer is also a two-stage fluidized-bed reactor with a maximum operating temperature of 870°C. Two-stage, countercurrent fluidized-bed reactors were used for both reducer and oxidizer to enhance the gas and solids conversions. Another feature of the IGT steam-iron process is that instead of iron ore, a modified iron oxide particle with 4% silica, 10% magnesia, and 86% hematite ore was used. The modified particle has higher particle reactivity and strength when compared with untreated iron ore. During the process demonstration, H_2 with a maximum purity of 95% (on a dry, N_2-free basis) was produced continuously for a period of 130 hours.[33] The solids conversion was between 5% and 15%. The performance of the IGT process again could not compete with the natural gas-based hydrogen production techniques, which also had been evolving throughout the years.

4.3.4 Syngas Chemical Looping (SCL) Process

The Syngas Chemical Looping (SCL) Process coproduces hydrogen and electricity from syngas. The SCL Process is based on the cyclic reduction and oxidation of specially tailored metal oxide composite particles.[34,35] This process produces a pure hydrogen stream and a concentrated carbon dioxide stream in two separate reactors. Therefore, an additional CO_2 separation step is avoided. The SCL Process consists of five major components: an ASU, a gasifier, a gas cleanup system, a reducer, and an oxidizer. Figure 4.10 (previously Figure 1.23) shows a simplified block diagram of the SCL Process.

In the SCL Process, a high-purity oxygen stream (oxygen concentration > 95%) from an ASU is sent to the gasifier. The gasifier uses oxygen from the ASU and steam to oxidize coal partially, forming high-temperature raw syngas. After gas quenching and particulate removal, a hot gas cleanup unit (HGCU) is used to remove most of the sulfur in the raw syngas. As can be seen from Figure 4.10, the SCL Process uses existing syngas generation and

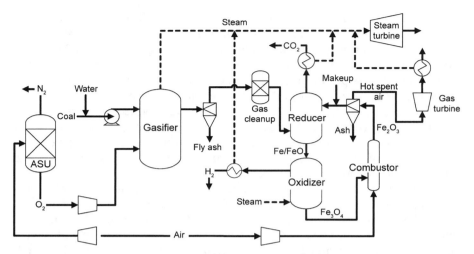

Figure 4.10. Simplified schematic of the Syngas Chemical Looping Process for hydrogen production from coal using slurry feed gasifier.

cleanup systems. The difference between the SCL Process and the conventional coal-to-hydrogen process lies in the manner in which H_2 is generated. The SCL Process can carry out various functions performed in the traditional gasification process, such as water–gas shift, CO_2 separation, and pressure swing adsorption for H_2 purification (purity > 99.95%) using two key looping reactors (i.e., the reducer and the oxidizer). Thus, the overall energy conversion scheme in the SCL Process is significantly simplified over the traditional process. There are fundamentally three different operating modes for looping unit operation: fluidized bed, moving bed, and fixed bed. The fluidized bed setup can be operated in circulating (dense or dilute pneumatic transport), bubbling, or turbulent fluidization regimes,[36] all of which are discussed extensively in Chapter 3 in connection with the chemical looping applications to combustion systems. The following discussion emphasizes the moving bed mode of operation, while recognizing that looping particle recycling often involves dense or dilute pneumatic transport, and in essence, the looping system under discussion is mostly a moving-bed/fluidized-bed system. The syngas is converted in a looping system to hydrogen and electricity in three steps that involve a reducer, an oxidizer, and a combustor.

Reducer
The purified syngas from the gas cleanup units is introduced to the reducer, which is a moving bed of iron oxide composite particles operated at 750°C–900°C and 30 atm. In this reactor, the syngas is converted completely to carbon dioxide and steam, whereas the iron oxide composite particles are reduced to a mixture of Fe and FeO [Reactions (4.3.11–4.3.14)].

$$Fe_2O_3 + CO \rightarrow 2FeO + CO_2 \qquad (4.3.11)$$

$$FeO + CO \rightarrow Fe + CO_2 \qquad (4.3.12)$$

$$Fe_2O_3 + H_2 \rightarrow 2FeO + H_2O \qquad (4.3.13)$$

$$FeO + H_2 \rightarrow Fe + H_2O \qquad (4.3.14)$$

The overall reaction in the reducer can be either slightly endothermic or slightly exothermic depending on the syngas composition, reaction temperature, and particle reduction rate. The mild endothermic to mild exothermic nature of the reducer simplifies the heat integration scheme of the reducer reactor because heat can be carried readily in or out of the reactor by the chemical looping particles.

In the reducer, the syngas is oxidized nearly completely by the Fe_2O_3 composite particles. Thus, the exhaust gas from the reducer contains mainly CO_2 and steam. Steam can be condensed out by extracting the heat from the high-temperature exhaust gas, resulting in a concentrated, high-pressure CO_2 stream that can be transported for sequestration.

Oxidizer
After being reduced in the reducer, a portion of the particles are introduced to the oxidizer. In the oxidizer, the reduced particles react with steam to produce a gas stream that contains solely H_2 and unconverted steam. Once the steam is condensed out from the gas stream, a H_2 stream of very high purity (>99.95%) can be obtained. The reactions involved in the reducer reactor include:

$$Fe + H_2O(g) \rightarrow FeO + H_2(g) \qquad (4.3.15)$$

$$3FeO + H_2O(g) \rightarrow Fe_3O_4 + H_2(g) \qquad (4.3.16)$$

The steam used in the oxidizer is produced from the syngas cooling units and combustor. The oxidizer reactor operates at 30 atm and 500°C–750°C. By introducing the low-temperature steam, the oxidizer is adjusted to be heat neutral. The heat released from the oxidation of the particles to Fe_3O_4 is used in the same reactor to heat the feed water/steam. The lower operation temperature of the oxidizer favors the steam-to-hydrogen conversion. The oxidizer also can be operated to generate syngas by introducing CO_2 produced from the reducer along with steam. By changing the ratio between the CO_2 and the steam, the $CO:H_2$ ratio in the syngas produced can be altered. The reactions between CO_2 and the reduced composite particles are:

$$FeO + CO_2(g) \rightarrow FeO + CO(g) \qquad (4.3.17)$$

$$3FeO + CO_2(g) \rightarrow Fe_3O_4 + CO(g) \qquad (4.3.18)$$

The CO and H_2 mixture can be used for chemical and liquid fuel synthesis. For example, a syngas product with a $H_2:CO$ ratio of 2:1 can be converted directly to liquid fuels in a Fischer–Tropsch reactor.

In the SCL Process, hydrogen and/or CO are produced using the chemical looping gasification concept in which syngas is converted indirectly to hydrogen with the assistance of iron oxide particles. This is fundamentally different from the traditional coal-to-hydrogen process in which the syngas is shifted to make hydrogen and followed by a CO_2 removal operation. By using the Syngas Chemical Looping Process, CO_2 is produced from a reactor different from where hydrogen is produced, hence, eliminating the energy-consuming CO_2 separation step. Because a significant portion of the carbon management cost is associated with the separation and compression of CO_2, the SCL Process offers a major advantage over the traditional coal-to-hydrogen process.

Combustor

Fe_3O_4 formed in the reducer is regenerated to Fe_2O_3 in a combustor. The combustor is a riser that conveys particles to the reducer with pressurized air. When the residence time in the riser is inadequate to allow the oxygen carrier to be oxidized completely, a dense bed with a larger diameter is placed below the riser. The dense bed is operated in the turbulent fluidization regime to provide sufficient reaction time to oxidize the oxygen carrier. The combustor also serves as a heat generator because a significant amount of heat is produced during the combustion of Fe_3O_4 to Fe_2O_3:

$$4Fe_3O_4 + O_2 \rightarrow 6Fe_2O_3 \qquad (4.3.19)$$

The high-pressure, high-temperature gas produced from the combustor can be used for electricity generation to compensate for parasitic energy consumption. In yet another configuration, part or all of the reduced particles from the reducer reactor can be sent directly to the combustor without reacting with steam in the reducer. By doing so, more heat would be available for electricity generation at the expense of decreased hydrogen production.

The step of combusting the reduced particles characterizes chemical looping combustion, which differs from chemical looping gasification in which the particles are regenerated using steam, instead of oxygen. Hence, the usage of both chemical looping gasification and chemical looping combustion concepts in the SCL system allows flexibility in adjusting the product yield between H_2 and electricity. Moreover, only one type of particles, engineered to have high recyclability and reactivity, is circulating among the three units, rendering the process highly efficient. The overall efficiency for the SCL process is estimated to be equivalent to ~67% (HHV, higher heating value) for hydrogen generation. A process concept similar to that of the SCL Process is being explored by ENI S.p.A (Rome, Italy).[37] In the ENI process, countercurrent, free-falling bed reactors are used for both the reducer and the oxidizer.[38] The reduction of Fe_2O_3 to a mixture of FeO and Fe_3O_4 was demonstrated in lab-scale reactor testing for the ENI process.

4.4   Design, Analysis and Optimization of the Syngas Chemical Looping (SCL) Process

All processes described in Section 4.3 convert gaseous fuels to hydrogen and/ or electricity based on similar reaction schemes. The difference in these processes and, hence, their performance lies in the properties and performance of the particles, the system configuration, and the reactor design. Among these processes, the SCL Process significantly differs from other steam-iron processes in that the SCL Process can be operated at substantially higher energy conversion efficiency. A key factor for such a difference concerns the improvement of the oxygen-carrying particles. The details of the preparation methods and physical and reactive properties of the particles are discussed in Chapter 2. Other factors are the gas–solid contact modes of the looping reactors and the process integration. In the following sections, these factors along with the experimental results of the SCL operation are discussed and analyzed. The variations of the SCL Process covered are the most effective for gaseous fuel conversion.

4.4.1   *Thermodynamic Analyses of SCL Reactor Behavior*

The thermodynamic properties of iron, as well as other metals, are introduced in Chapter 2 in the context of the primary metal/metal oxide selection. Here, the thermodynamic properties of iron are examined in relation to its reactions with gaseous reactants and reactor operations. Figure 4.11 illustrates a simplified iron oxide conversion scheme in the SCL Process. As is shown, the SCL process produces hydrogen and/or heat through the reduction and oxidation of iron (oxide) particles in a cyclic manner. The maximum gas and solids conversions, which are determined by thermodynamics, are closely related to the performance of the chemical looping process. Higher gas and solids conversions will lead to a lower particle circulation rate, higher product purity, lower parasitic energy requirements, and hence, improved energy conversion efficiency.

Thermodynamic Properties of Iron Oxides
As noted in Chapter 2, iron has four oxidation states: metallic iron, wüstite, magnetite, and hematite. In chemical looping processes, iron can swing between any two of the four oxidation states or the mixtures thereof. For example, in the IGT steam-iron process, iron swings between Fe_3O_4 and a mixture of FeO and Fe_3O_4. Thermodynamic diagrams show that the two oxidation states between which iron swings determine the extent of the gas and solids conversions.

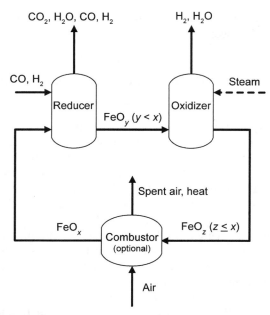

Figure 4.11. Schematic flow diagram of the Syngas Chemical Looping Processes.

Figure 4.12 shows the equilibrium gas compositions of both the iron–carbon–oxygen system and the iron–hydrogen–oxygen system at different temperatures.[39,40] It is noted that FeO is used here to represent wüstite, whose exact formula varies with temperature. From the phase diagrams, the equilibrium gas concentrations for different oxidation states of iron may vary significantly at any given temperature. The equilibrium gas compositions in both iron–carbon–oxygen and iron–hydrogen–oxygen systems at 850°C are shown in Table 4.3.

The implications of the equilibrium gas concentrations are twofold. From the gas conversion viewpoint, the fact that 99.9955% CO_2 equilibrates with Fe_2O_3/Fe_3O_4 suggests that at 850°C, the presence of excess Fe_2O_3 will lead to a 99.9955% conversion of CO to CO_2 for an Fe–C–O system. From the solids conversion viewpoint, provided that the CO concentration is higher than 45 ppm, all or part of the Fe_2O_3 in the system will be reduced to a lower oxidation state. To compare, the CO concentration must be higher than 62% to reduce FeO to Fe. Similar cases can be observed for an Fe–H–O system. Therefore, iron at higher oxidation states is more effective in oxidizing H_2/CO to H_2O/CO_2, whereas iron at lower oxidation states is more favorable for the conversion of H_2O/CO_2 to H_2/CO.

In the iron-based chemical looping processes generalized in Figure 4.11, the iron/ iron oxide reduced in the reducer is used in the oxidizer to convert steam into hydrogen. From its thermodynamic properties, iron under high oxidation states is a desirable feedstock for the reducer to achieve high syngas conver-

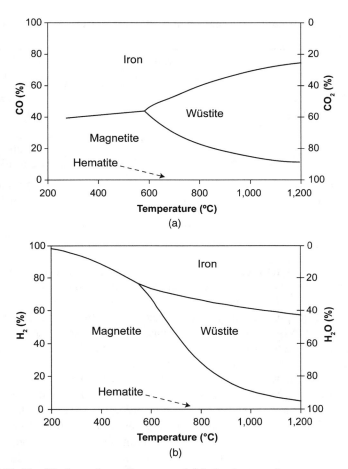

Figure 4.12. Equilibrium phase diagrams of (a) the iron–carbon–oxygen system and (b) the iron–hydrogen–oxygen system. Note that the lines separating the Fe_2O_3 phase from the Fe_3O_4 phase are too close to the x-axis to be shown.

sions. Meanwhile, the iron produced from the reducer should have low oxidation states to maximize the steam to hydrogen conversion in the oxidizer. Therefore, a reducer that maximizes solids and gas conversions ($x - y$ in Figure 4.11) has the potential to maximize the overall energy conversion efficiency of the process. As illustrated in the next section, the reactor design and the gas–solid contacting pattern play an important role in maximizing the gas and solids conversions in chemical looping processes.

Reactor Design and Gas–Solid Contacting Patterns
A key challenge to the SCL process lies in the design of the reducer and the oxidizer. Unlike the oxidation reaction in the combustor, which is intrinsically

TABLE 4.3. Equilibrium Gas Compositions with Different Oxidization States of Iron at 850°C

Iron Phase	Equilibrium Concentration of CO_2	Equilibrium Concentration of CO
Fe_2O_3/Fe_3O_4 Mixture	99.9955%	45 ppmv
Fe_3O_4/FeO Mixture	80.3%	19.7%
FeO/Fe Mixture	38.0%	62.0%

Iron Phase	Equilibrium Concentration of H_2O	Equilibrium Concentration of H_2
Fe_2O_3/Fe_3O_4 Mixture	99.995%	50 ppmv
Fe_3O_4/FeO Mixture	78.0%	22.0%
FeO/Fe Mixture	34.8%	65.2%

fast and is thermodynamically favored, the reactions in the reducer and the oxidizer are limited by the thermodynamic equilibrium and are relatively slow. It is, therefore, desirable to use an optimal reducer and oxidizer design that is reliable and less capital-intensive while maximizing the solid and gas conversions via a thermodynamically favored gas–solid contacting scheme. In this section, different designs for the reducer are analyzed. The oxidizer design analysis can be carried out in a similar manner.

Three reactor operating modes, fixed bed, moving bed, and fluidized bed, are investigated. A fixed-bed design similar to that employed in Lane's steam-iron process eliminates particle movement; however, such a design involves constant switching of reducing and oxidizing gases in an intermittent manner. The requirement for gas switching under high temperatures and high pressures makes it challenging for the design of the valve system. In addition, the reduced iron oxide particles tend to catalyze the reverse Boudard reaction, giving rise to carbon deposition on the particles.[35] Therefore, excessive carbon deposition may occur at the inlet of the fixed bed during the reducer operation. This carbon deposition reduces the purity of the hydrogen product. Moreover, it can affect the reactivity of the particles. Another challenge to a fixed-bed design is effective heat removal, which is required during the combustion step. In contrast, these problems can be minimized or avoided when a fluidized bed or a moving bed design is employed because of the continuous movement of both gas and solids. Therefore, a fluidized-bed or a moving-bed design is preferred for both the reducer and the oxidizer.

To compare the maximum gas and solids conversions using either a fluidized-bed design or a moving-bed design, a thermodynamic analysis was performed on a fluidized-bed reducer and a moving-bed reducer. It should be noted that thermodynamic analysis predicts the gas and solids conversions when a thermodynamic equilibrium among the reactants and the products is reached. The equilibrium can be reached under the conditions when the reaction is sufficiently fast and/or the gas–solid contact time in the reactor is sufficiently long. The reactions conducted in the chemical looping processes meet these criteria; therefore, the thermodynamic analyses can project the performance of the looping reactors with reasonable accuracy.

For illustration purposes, pure H_2 and pure Fe_2O_3 are used as the feedstock for the reducer, which is operated at 850°C. The ratio between the solid and gas molar flow rate is set to be s. The conversions of H_2 and Fe_2O_3 are denoted as x and y, respectively. The conversion of Fe_2O_3 is defined as the percentage of oxygen depleted from pure Fe_2O_3 as given by:

$$y = \frac{n_O/n_{Fe} - \hat{n}_O/\hat{n}_{Fe}}{n_O/n_{Fe}} \times 100\% \qquad (4.4.1)$$

where n_O/n_{Fe} corresponds to the molar ratio between the oxygen atom and the iron atom in Fe_2O_3, and $\hat{n}_O/\hat{n}_{Fe}$ corresponds to the molar ratio between the oxygen atom and the iron atom in the reduced solid product, that is, FeO_x ($0 < x < 1.5$). For instance, the reduction of Fe_2O_3 to Fe_3O_4 corresponds to a solids conversion of $(3/2 – 4/3)/(3/2) \times 100\% = 11.11\%$, the reduction to FeO corresponds to a conversion of 33.33%, and the reduction to Fe corresponds to 100% solids conversion. It is noted that the Fe_2O_3 conversion presented here corresponds to the oxygen capacity defined in Chapter 2. When H_2 is the only reducing gas, the possible reactions in the reducer include

$$H_2 + 3Fe_2O_3 \rightarrow H_2O + 2Fe_3O_4 \quad K_1 = P_{H_2O}/P_{H_2} = 1.92 \times 10^4 \quad \text{at 850°C}$$
$$(4.4.2)$$

$$H_2 + Fe_3O_4 \leftrightarrow H_2O + 3FeO \quad K_2 = P_{H_2O}/P_{H_2} = 3.5454 \quad \text{at 850°C}$$
$$(4.4.3)$$

$$H_2 + FeO \leftrightarrow H_2O + Fe \quad K_3 = P_{H_2O}/P_{H_2} = 0.5344 \quad \text{at 850°C}$$
$$(4.4.4)$$

where K_1–K_3 are the equilibrium constants that can be derived readily from Figure 4.12.

Fluidized-Bed Reducer
In a fluidized-bed reactor such as a dense-phase fluidized bed, significant mixing for the gas and the solid in the reactor occurs. Thus, in a fluidized-bed reducer (Figure 4.13[a]), the fresh syngas feedstock will be diluted by the gaseous product, which is rich in H_2O and CO_2. From Figure 4.12, the dilution of syngas by CO_2 and H_2O decreases the reducing capability of the syngas. Similarly, the mixing of solids in a fluidized bed results in a discharge of the low conversion solids from the fluidized bed. Therefore, the gas and solids

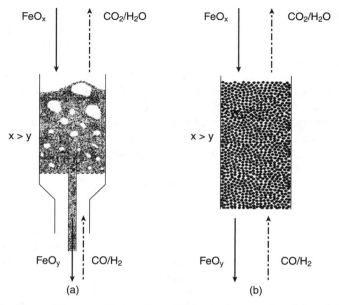

Figure 4.13. Gas–solid contacting pattern of the reducer using (a) a fluidized-bed design and (b) a moving-bed design.

conversions in the fluidized-bed reducer are constrained. A similar constraint applies when a fluidized bed reactor is used as the oxidizer.

The effect of mixing is illustrated by examining a sample fluidized-bed reactor. In the following analysis, it is assumed that both the solid and the gas are well mixed. The oxygen mass balance on the reactor can be given as

$$x = 3sy \tag{4.4.5}$$

This equation indicates that the oxygen depleted from the solid is transferred to the gas through the formation of steam or CO_2. Meanwhile, thermodynamic equilibrium gives

$$K_n = x/(1-x) \tag{4.4.6}$$

where n is 1, 2, or 3 depending on the phase of iron in the reducer. According to Figure 4.12, when excess Fe_2O_3 is present ($0 \leq y < 11.11\%$), the equilibrium constant will follow K_1; when the Fe_3O_4 and FeO mixture is present ($11.11\% \leq y < 33.33\%$), the equilibrium gas composition is determined by K_2; and when FeO and Fe coexist ($33.33\% \leq y < 100\%$), the equilibrium constant will follow K_3.

Equations (4.4.5) and (4.4.6) can be solved together to arrive at a relationship between the gas and the solids conversions (x, y) and the ratio of the solid-to-gas flow rates. It can be shown that x is a step function with respect to s as

$$x = \begin{cases} K_1/(K_1+1) & s > 3K_1/(K_1+1) \\ s/3 & 3K_1/(K_1+1) \geq s > 3K_2/(K_2+1) \\ K_2/(K_2+1) & 3K_2/(K_2+1) \geq s > K_2/(K_2+1) \\ s & K_2/(K_2+1) \geq s > K_3/(K_3+1) \\ K_3/(K_3+1) & K_3/(K_3+1) \geq s > K_3/3(K_3+1) \\ 3s & s \leq K_3/3(K_3+1) \end{cases} \qquad (4.4.7)$$

where y can be obtained from x using Equation (4.4.7):

$$y = x/3s \qquad (4.4.8)$$

For a fluidized bed reducer operated at 850°C, the gas and solids conversions can be obtained by substituting the values of $K_1 - K_3$ in Equations (4.4.2–4.4.4) into Equations (4.4.7) and (4.4.8):

$$x = \begin{cases} 1 & s > 3.0 \\ s/3 & 3.0 \geq s > 2.34 \\ 0.78 & 2.34 \geq s > 0.78 \\ s & 0.78 \geq s > 0.348 \\ 0.348 & 0.348 \geq s > 0.116 \\ 3s & s \leq 0.116 \end{cases} \qquad y = \begin{cases} 0.3333/s & s > 3.0 \\ 0.1111 & 3.0 \geq s > 2.34 \\ 0.26/s & 2.34 \geq s > 0.78 \\ 0.3333 & 0.78 \geq s > 0.348 \\ 0.116/s & 0.348 \geq s > 0.116 \\ 1 & s \leq 0.116 \end{cases}$$

Figure 4.14(a) shows the relationship between the gas and the solids conversions and the ratio between the solid and the gas molar flow rates. Figure 4.14(a′) shows the relationship between the gas and the solids conversions. Because for a fluidized bed reducer the gas and solids conversions at any steady-state operation conditions will fall on the curves shown in Figure 4.14(a) and Figure 4.14(a′), these curves are called the operating curves.[41] As is shown, the conversions of gas and solids are inversely correlated; that is, a higher solids conversion corresponds to a lower gas conversion and *vice versa*. In actual reactor operation, a full conversion of fuel gas is crucial. Figure 4.14(a′) shows a gas conversion of 100% corresponding to a solids conversion of less than 11.11%.

Aside from the analytical approach, the operating curve for the fluidized bed can be derived directly from the thermodynamic phase diagram. The rationale being that, because the solids and gases are well mixed and are at the thermodynamic equilibrium, the gas and solid concentrations should be at the equilibrium concentrations described by the thermodynamic phase diagram. Thus, the operating curve should coincide with the gas–solid equilibrium curve.

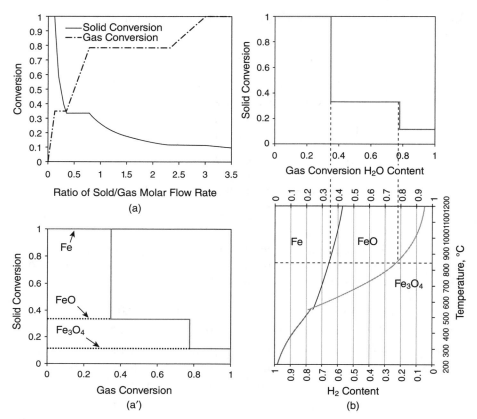

Figure 4.14. Operating curves for a fluidized-bed reactor: (a) solids and gas conversion vs. solid/gas molar flow rate ratio (a') relationship between solids conversion and gas conversion; and (b) derivation of operating curve from thermodynamic phase diagram.

The translation of the equilibrium phase diagram into the fluidized-bed operating curve is illustrated in Figure 4.14(b). The vertical dashed line describes the relationship between the phases of iron and the equilibrium gas concentration at a certain temperature (850°C in the figure). For a fluidized bed reducer, the equilibrium steam concentration is identical to the hydrogen conversion x in the operating curve. The phases of iron that equilibrate with such a gas concentration determine the solids conversion y defined by Equation (4.4.1). For instance, a solids conversion of 100% corresponds to pure iron, whereas a solids conversion of 33.33% corresponds to pure FeO. It is noted that the operating curve in the ideal case, which is also the equilibrium line, divides the graph into two regions. Because the gas and solids conversions always will be lower than or equal to the values at equilibrium in practical reactor operations, the gas and solids conversions will approach the equilibrium line from the lefthand side as the gas–solid contact time increases. Thermodynamics dictates that these conversions will not cross the equilibrium line.

Countercurrent Moving-Bed Reducer
Contrary to the fluidized-bed reactor, a moving bed reactor has minimal axial mixing of the gas and solid phases. When a moving-bed reactor with a countercurrent gas–solid contacting pattern is used as the reducer, as shown in Figure 4.13(b), a fresh syngas feed with high H_2 and CO concentrations will react with iron at lower oxidation states. Meanwhile, the partially converted syngas with low H_2 and CO concentrations will meet iron at higher oxidation states. Based on the thermodynamic diagram shown in Figure 4.12, such a contact pattern will maximize both solid and gas conversions. A similar case also can be expected when a moving-bed reactor is used as the oxidizer. For simplicity, with the mixing for both the solid and the gas phases neglected in the moving-bed reactor analysis, both the gas and solid compositions will vary with the axial position of the moving bed reactor. From Figure 4.13, the mass balance of oxygen in an infinitesimal layer between z and $z + \Delta z$ of the moving bed reactor at a steady state can be written as

$$3s(y_{z+\Delta z} - y_z) = (x_z - x_{z+\Delta z}) \tag{4.4.9}$$

Equation (4.4.9) can be expressed by

$$dx/dy = -3s \tag{4.4.10}$$

Thus, the relationship between the solids conversion y and the gas conversion x under a certain solid/gas molar flow rate ratio s is a straight line with a slope of $-3s$. Such a line represents the operating line. The operating line only is restricted by thermodynamic equilibrium. Specifically, at any point of the operating line, the ratio of the concentration between steam and hydrogen should not be higher than the equilibrium constant K:

$$x/(1-x) \le K_1 \quad \text{for } 0 \le y < 0.1111$$
$$x/(1-x) \le K_2 \quad \text{for } 0.1111 \le y < 0.3333$$
$$x/(1-x) \le K_3 \quad \text{for } 0.3333 \le y < 1$$

These equations can be expressed by

$$x \le K_1/(1+K_1) \quad \text{for } 0 \le y < 0.1111$$
$$x \le K_2/(1+K_2) \quad \text{for } 0.1111 \le y < 0.3333$$
$$x \le K_3/(1+K_3) \quad \text{for } 0.3333 \le y < 1$$

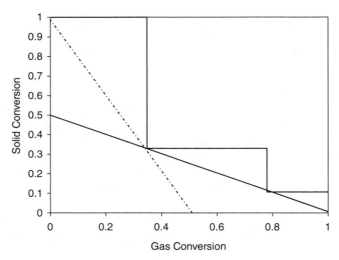

Figure 4.15. Operating lines in a countercurrent moving-bed reactor at a temperature of 850°C.

These restrictions are in conformity with the earlier discussion, which concluded that a practical reducer operating line should be located at the lefthand side of the equilibrium line and should not cross it. Thus, the feasible operating lines can be determined based on the mass balance and the thermodynamic phase diagram. Two possible operating lines are shown in Figure 4.15.

It can be shown that each point on the operating line corresponds to the gas/solids conversions at a certain axial position of the moving-bed reactor. Therefore, the gas and solids conversions can be obtained from the intercept of the operating line with the x and y axes. The intercept of the operating line with the y axis corresponds to the solids conversion at the solids outlet located at the bottom of the reactor where the highest possible solids conversion is achieved. Similarly, the intercept of the operating line with the x axis is the gas conversion at the gas outlet located at the top of the reactor. For example, the solid line in Figure 4.15 corresponds to a solid:gas molar flow rate ratio of 2:3. Under this operating condition, 100% of H_2 will be converted, and the solids will be reduced by ~50%. Because the operating line is restricted to the lefthand side of the equilibrium curve, it is not possible to achieve 100% conversion simultaneously for the gas and for the solids. In fact, the solid line corresponds to the maximum achievable solids conversion when H_2 is converted fully. Similarly, to achieve a full solids conversion, at least 92% excess H_2 more than the stoichiometric requirement needs to be introduced to the reactor, yielding a maximum gas conversion of ~52%.

With Figure 4.15, the optimal gas and solid flow rates for the SCL reducer can be determined. For example, a full conversion of syngas is essential for the reducer because incomplete syngas conversion will lead to a reduced energy conversion efficiency of the process. Therefore, the optimum operating line is the solid line in Figure 4.15, which corresponds to a solids conversion of ~50%. It also can be shown that multiple-stage interconnected fluidized bed reactors with a countercurrent gas–solid contact pattern can achieve a conversion similar to that of the moving bed. In contrast, a single-stage fluidized bed reducer can achieve merely an 11.11% solids conversion under the same operating conditions. When used as an oxidizer, a moving bed also has been shown to achieve higher conversions. To generalize, significantly improved gas and solids conversions can be achieved when a countercurrent moving bed is used as the reducer or as the oxidizer.

4.4.2 ASPEN PLUS Simulation of SCL Reactor Systems

Although the operating lines, as given in Figures 4.14 and 4.15, are useful, they are rather difficult to construct and apply to the analysis of the gas and solids conversions. The difficulty is caused by the multicomponent gas mixture involved, which is compounded further by the temperature variations in the axial direction of the reactor. Therefore, an alternative method is desired. One viable method is to use computer simulation based on software such as the advanced system for process engineering software, or ASPEN PLUS. The comprehensive physical and thermodynamic property databanks built into the ASPEN PLUS software render it suitable for reactor and process simulations. With appropriate modeling parameters, the ASPEN PLUS software can simulate simultaneously the flows of mass, heat and work in individual process units and the overall process.

Selection of ASPEN PLUS Modeling Parameters

Before employing the simulation model to analyze a fluidized bed or a moving bed, a set of common parameters needs to be determined. This section describes the necessary procedures for setting up these parameters. A built-in module in ASPEN, RGIBBS, is used to determine the equilibrium condition among the reactants and the various possible products. Other parameters that need to be selected include physical and thermodynamic property databanks and property methods, stream classes, chemical components, and calculation algorithms. The selection of appropriate parameters is essential for accurate simulation results. Tables 4.4 to 4.6 list the key parameters selected for the simulations. It is noted that modifications to the physical property data and physical property methods for the solids are often necessary to obtain consistent results from the literature and from the ASPEN PLUS simulation.

The Inorganic databank in the ASPEN PLUS software, which determines the physical and thermodynamic properties of the solids at various conditions, uses the Barin equation [Equation (4.4.11)] and its CPSXP $(a–h)$ coefficients[42] to obtain the Gibbs energy (G), enthalpy (H), entropy (S), and heat capacity (C_p) for the solids.

TABLE 4.4. Parameters for the ASPEN PLUS Model

Name of the Parameter	Parameter Setting
Reactor Module	RGIBBS
Physical and Thermodynamic Databanks	Combust, Inorganic, Solids, and Pure
Stream Class	MIXCISLD
Chemical Components	Listed in Table 4.5
Property Method (for Gas and Liquid)	PR-BM[a]
Calculation Algorithm	Sequential Modular (SM)

[a]Property methods for solids are discussed separately, and some correlative parameters are presented in Table 4.6.

TABLE 4.5. Components List in Reducer Simulation[a]

Component ID	Stream Type	Component Name
CO	Conv	Carbon Monoxide
CO_2	Conv	Carbon Dioxide
H_2	Conv	Hydrogen
H_2O	Conv	Water
Fe_2O_3	Solid	Hematite
Fe_3O_4	Solid	Magnetite
FeO	Solid	Wüstite
$Fe_{0.947}O$	Solid	Wüstite
Fe	Solid	Iron
C	Solid	Carbon, Graphite
Fe_3C	Solid	Triiron carbide
Hg	Conv	Mercury
HgS	Conv	Mercury Sulfide Red
S	Conv	Sulfur
H_2S	Conv	Hydrogen Sulfide
FeS	Solid	Iron monosulfide
$Fe_{0.877}S$	Solid	Pyrrhotite

[a]Species such as FeS_2 and HgO do not exist in the interested temperature range; therefore, they are not included.

$$G = a + bT + cT \ln T + dT^2 + eT^3 + fT^4 + gT^{-1} + hT^{-2}$$

$$H = a - cT - dT^2 - 2eT^3 - 3fT^4 + 2gT^{-1} + 3hT^{-2}$$

$$S = -b - c(1 + \ln T) - 2dT - 3eT^2 - 4fT^3 + gT^{-2} + 2hT^{-3}$$
(4.4.11)

$$C_p = -c - 2dT - 6dT^2 - 12fT^3 - 2gT^{-2} - 6hT^{-3}$$

Table 4.6 lists the CPSXP coefficients for iron and its oxides in the inorganic databank in ASPEN PLUS. The coefficients presented in this table were verified using HSC Chemistry 5.1 (Chemistry Software Ltd, Surrey, UK) in

TABLE 4.6. Parameters in the Inorganic Databank in ASPEN PLUS

Components	Fe_2O_3	Fe_3O_4	Fe	$Fe_{0.947}O$
Temperature Units	°C	°C	°C	°C
Property Units	J/kmol	J/kmol	J/kmol	J/kmol
T1	25.00	576.85	25.00	25.00
T2	686.85	1596.85	626.85	1376.85
a*	−9.28E+08	−9.71E+8	3.78E+07	−2.82E+8
a'**	−9.28E+08	−9.57E+8	3.78E+07	−2.82E+8
b*	1.98E+06	5.28E+5	−6.54E+05	4.02E+5
b'**	1.98E+06	5.36E+05	−6.54E+05	4.03E+05
c*	−2.58E+05	−5.02E+04	1.09E+05	−4.88E+04
c'**	1.98E+06	−5.09E+04	−6.54E+05	−4.86E+04
d	165.49	−35.97	−214.13	−4.18
e	−0.067	−6.02E−5	0.085	0.00
f	1.17E−05	6.13E−9	−1.95E−05	0.00
g	7.66E+09	−4.3E+10	−4.01E+09	1.40E+8
h	−3.76E+11	5.47E+9	1.98E+11	0.00

*The original values in the ASPEN databank before adjustment.
**The prime notation (') denotes the adjusted values for the parameters as shown in the shades.

addition to the literature sources.[43–46] In preparing for the table, the reference states in the literature sources were adjusted to match the specified ASPEN PLUS values, that is, 25°C and 1 atm. Minor differences, amounting to ~1% of the original values in the ASPEN databank, were found in coefficients a, b, and c for Fe_3O_4 and FeO ($Fe_{0.947}O$). Even though the differences are small, the simulation results can be affected significantly, particularly with respect to the phase transition conditions of various iron states.

Fluidized-Bed Reactor Model Setup and Simulations
A fluidized bed reactor can be approximated by a continuous stirred-tank reactor (CSTR), and thus a simple RGIBBS module can be used to simulate a fluidized bed operated under equilibrium conditions. With the modeling parameters selected, the fluidized bed model can be set up by connecting reactants and product streams to the RGIBBS module and by inserting the operating conditions and inlet compositions into the ASPEN flowsheet. The thermodynamic simulation is performed on the fluidized bed reactor to corroborate the theoretical analysis results obtained in Section 4.4.1. To obtain data comparable with those shown in Figure 4.14, a sensitivity analysis is performed to determine the relationship between the gas and solids conversions in a fluidized bed by varying the molar flow rate ratio between Fe_2O_3 and H_2. The operating temperature (850°C) for the simulation is identical to that in the theoretical analysis. The ASPEN simulation results are given in Figure 4.16. As is shown, the simulation results are almost identical to those obtained from the theoretical analyses in Figure 4.14(a).

Moving-Bed Reactor Model Setup and Simulations

Model Setup. The simulation of a countercurrent moving bed reactor is more complex than a fluidized bed simulation. Because no ASPEN PLUS simulator module is available for the simulation of a countercurrent moving bed reactor operated under equilibrium, a series of interconnected CSTR reactors based on the RGIBBS module is used to simulate the moving bed reactor. The model configuration is shown in Figure 4.17. As is shown, the solid entering stage k is the solid product discharged from stage $k + 1$, whereas the gas entering stage k is the gaseous product of stage $k - 1$. It can be shown that such a model configuration satisfies the mass balance and thermodynamic restrictions imposed on a countercurrent moving bed reactor. With a large number of RGIBBS blocks, the countercurrent moving bed reactor can be approximated.

The simulation of a moving bed system using an infinite number of RGIBBS blocks is not feasible. The simulation results, however, show asymptotic behav-

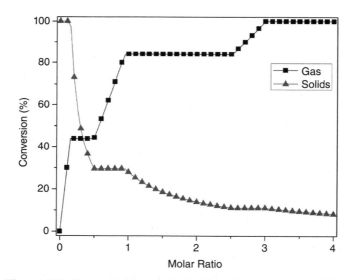

Figure 4.16. Operating curve of a fluidized bed reducer at 850°C.

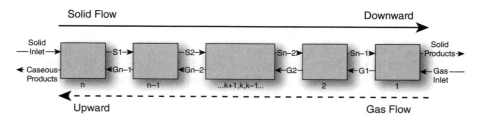

Figure 4.17. ASPEN simulation of an arrangement of a multistage fluidized bed system for the simulation of a moving bed.

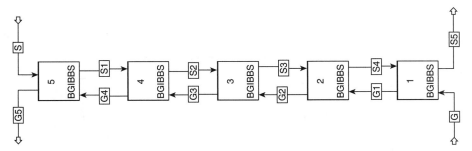

Figure 4.18. ASPEN simulation of a five-stage RGIBBS model for moving-bed simulations.

ior with an increasing number of RGIBBS blocks. Based on numerous case studies, a five-stage model configuration can simulate the countercurrent moving bed with good accuracy. This is verified by comparing results obtained from a five-stage model with those obtained from a six-stage model. The comparisons indicate that the two models are identical in all cases. Therefore, the five-stage RGIBBS model shown in Figure 4.18 is used to simulate the countercurrent moving bed reactor.

Case 1: Moving Bed with 100% H_2 Conversion and Maximized Solids Conversion. In this case, pure H_2 is the reducing gas. The goal for the case is to validate the (solid) operating line shown in Figure 4.15. The Fe_2O_3 to H_2 molar flow rate ratio, s, is set to be 2:3. The reactor is operated at 850°C. These conditions are identical to those denoted on the solid operating line in Figure 4.15, which shows a maximum solids conversion with a near complete conversion of gas. Figure 4.19 presents a cumulative gas/solids conversion along the five reaction stages of the reactor. The results indicate that the solids conversion is 49.98%, and the hydrogen conversion is 99.95%, which are consistent with the outcomes from the theoretical analyses in Section 4.4.1. The simulation results also indicate that the gas and solids conversions are independent of the operating pressure of the reactor across a range of 1–30 atm.

Case 2: Moving Bed with 100% Syngas Conversion. The goal for the Syngas Chemical Looping Process is to convert 100% of the gaseous fuel to hydrogen and/or electricity. The effect of the iron oxide flow rate on the conversion of syngas with a ratio of H_2:CO of 1:2 in a reducer operated at 30 atm and 900°C is investigated. Figure 4.20 shows the syngas and solids conversions under different solid–gas molar flow rate ratios with and without Fe_3C formation. As is shown, without Fe_3C formation, the solid-to-gas molar flow rate ratio should be more than 0.66 to convert all the syngas. This ratio corresponds to a maximum solids conversion of ~50.0%. Figure 4.21 shows the gas and solids conversions under such reaction conditions.

When Fe_3C formation is considered, the minimum solid/gas ratio, s, could increase to 1.1, and the maximum solids conversion is 30%. Due to the slow

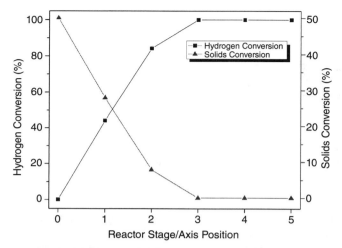

Figure 4.19. ASPEN simulation results for gas and solids conversions in a countercurrent moving-bed reactor as described in Case 1. Operating conditions: temperature, 850°C; pressure, 1–30 atm; reducing gas, H_2.

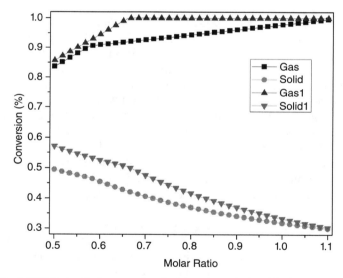

Figure 4.20. ASPEN simulation results showing the relationship between the gas and solids conversions and solid-to-gas molar flow rate ratio with (Gas and Solid) and without (Gas1 and Solid1) considering Fe_3C formation in a countercurrent moving-bed reactor as described in Case 2. Operating conditions are: temperature, 900°C; pressure, 30 atm; syngas composition, CO 66.6%, H_2 33.3%.

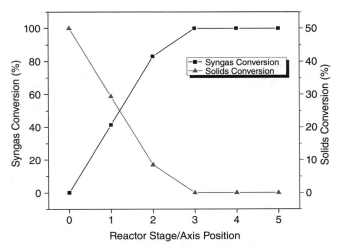

Figure 4.21. ASPEN simulation results showing gas and solid conversion profiles in a countercurrent moving bed reactor. Operating conditions are: temperature, 900°C; pressure, 30 atm; solid-to-gas molar flow rate ratio, 0.66; syngas composition, CO 66.7%, H_2 33.3%.

reaction kinetics, however, the formation of Fe_3C seldom is observed, especially in the presence of steam.[47,48] The simulation results are corroborated by experimental results, which are given in Section 4.4.3.

Further Applications
Based on the multistage ASPEN model, additional simulation conditions, results, and analyses are presented in the following sections.

Effect of Temperature. From the equilibrium consideration, a higher temperature favors the endothermic reaction, and a lower temperature favors the exothermic reaction. The reaction between Fe_2O_3 and CO is exothermic, whereas the reaction between Fe_2O_3 and H_2 is endothermic. Figure 4.22 shows the effect of temperature on the syngas conversion in a countercurrent moving bed reactor. The reactor is operated at 30 atm with a stoichiometric amount of gaseous and solid reactants. This figure indicates that the equilibrium conversions of CO and H_2 show opposite trends. Except for a significant increase at ~580°C, the overall syngas conversion slowly decreases when the temperature increases. The steep increase in reactivity at ~580°C is caused by the emergence of the wüstite phase, which does not exist below 550°C. Because the decrease in overall syngas conversion is relatively slow at temperatures above 600°C, the optimum operating temperature range for the reducer is determined to be 700°C–900°C as a result of the fast reaction kinetics at higher temperatures.

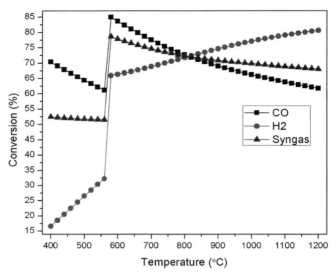

Figure 4.22. ASPEN simulation results showing the effect of temperature on the conversion of syngas, CO, and H_2 in a countercurrent moving-bed reactor. Operating conditions are: pressure, 30 atm; solid-to-gas molar flow rate ratio, 0.33; syngas composition, CO 66.7%, H_2 33.3%.

Fate of Sulfur and Mercury. Pollutant control is essential for any coal-conversion processes. H_2S, COS, and mercury are important pollutants that are present in coal-derived syngas. The multistage model can assist in determining the fates of these pollutants. An ASPEN simulation is conducted to examine the relationship between the sulfur content in the syngas and the formation of iron–sulfur compounds. The potential compounds considered include S, COS, SO_2, H_2S, FeS, and $Fe_{0.877}S$. At 900°C, 30 atm, and with an s of 0.66 (2:3), simulation results indicate that $Fe_{0.877}S$ is the only sulfur compound that may form in the solid stream. As shown in Figure 4.23, no iron–sulfur compound will be formed unless the H_2S level in syngas is higher than 600 ppm. Similarly, unless the COS level exceeds 650 ppm, COS will exit from the moving bed reducer without reacting with the Fe_2O_3 particles. The practical implication of these simulation results is significant; with the absence of sulfur attachment to the solids, the hydrogen product stream from the oxidizer will be sulfur-free. Thus, the sulfur control strategy for the SCL process is simplified. Specifically, a HGCU can be installed upstream of the reducer for the bulk of the sulfur removal. Because the available high-temperature sorbents can reduce the sulfur level in raw syngas to below 50 ppm with ease,[49] the remaining sulfur in the syngas will exit from the reducer along with CO_2 and H_2O. After condensing the steam, the sulfur-containing CO_2 will be ready for geological sequestration.[50] This process arrangement avoids the energy-intensive solvent-based

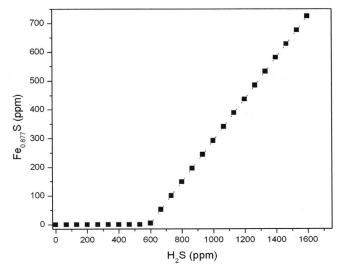

Figure 4.23. ASPEN simulation results for the relationship between the $Fe_{0.877}S$ formation and the syngas H_2S level in a countercurrent moving bed reactor. Operating conditions are: temperature, 900°C; pressure, 30 atm; solid-to-gas molar flow rate ratio, 0.66.

H_2S stripping processes. Cooling and reheating of the syngas also can be avoided, creating a more efficient sulfur control scheme.

Elemental mercury, the major form of mercury that is present in the raw syngas derived from coal, usually is captured using an activated carbon bed at low temperatures in the conventional coal gasification processes. From the ASPEN simulation, mercury will not react with any substances that are present in the reducer. Thus, all the mercury in the syngas stream will exit from the reducer flue gas and will not be present in the hydrogen stream from the oxidizer. The mercury-containing flue gas from the reducer can be treated with activated carbon before sequestration. The mercury separation technique in the SCL process is much more efficient than that in the traditional process, which involves cooling and reheating of the syngas.

4.4.3 Syngas Chemical Looping (SCL) Process Testing

As noted, a countercurrent moving bed with solid particles flowing downward and gas flowing upward represents an effective gas–solid contact mode for chemical looping gasification operations. The concept of the countercurrent moving bed also can be realized using a series of fluidized beds with countercurrent flows of the gas and solids.[51] For illustration, the moving bed configuration is used to characterize the looping reactor operation.

A bench moving-bed reactor setup in Figure 4.24 shows that the reactor assembly consists of a heated reaction zone, solids holding funnels, a screw

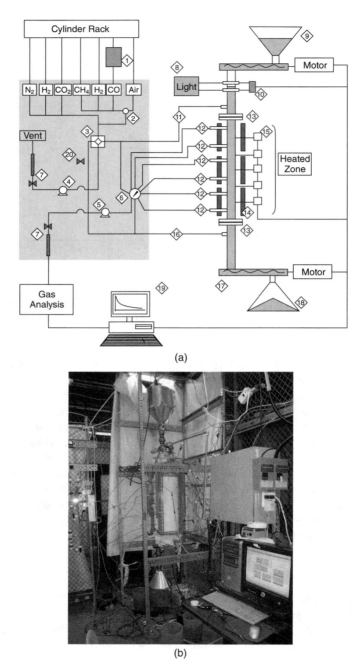

(a)

(b)

Figure 4.24. Bench-scale demonstration unit for SCL process. (a) Schematic flow diagram of the unit. (b) Picture of the unit. Reactor parts shown in (a): (1) CO pretreatment; (2) three-way safety valve; (3) cocurrent/countercurrent flow selector valve; (4) reactor gas-out miniature vacuum pump; (5) gas sample miniature vacuum pump;

feeder, a solids flow controller, an optical bed height control system, solids sampling ports, gas flow panels, gas sampling ports, the gas delivery and handling system, the gas analysis system, and a computer control system.[52] The material used for construction is primarily stainless steel type SS-304, which provides good inert characteristics for high-temperature reactions. The bench scale system shown in Figure 4.24 can be operated as either the reducer or the oxidizer in a semicontinuous mode. In a typical test, the oxygen carrier particle, either fully oxidized or reduced, is loaded into the top funnel. The particle then is moved steadily downward through the heated zone where it reacts with the reactant gas.

An integrated 25-kW$_{th}$ subpilot-scale Syngas Chemical Looping system is shown in Figure 4.25. The reactor system is capable of performing the operations of the reducer, the oxidizer, and the combustor in a simultaneous and continuous manner. Moreover, the arrangement of the reactor system is identical to that described in Section 4.3.4. As is shown in Figure 4.25, the reducer [heating section A; 3 in. (7.6 cm) ID, 36 in. (91.4 cm) in height] and the oxidizer [heating section B; 3 in. (7.6 cm) ID, 36 in. (91.4 cm) in height] are scaled-up versions of the bench unit. A rotary solids feeder is used to transport the reduced oxygen carrier discharged from the reducer to the oxidizer inlet. The partially regenerated oxygen carrier from the reducer then will be introduced to the combustor and pneumatically conveyed to the reducer inlet, thereby completing the redox cycle. A 250-kW$_{th}$ pilot-scale Syngas Chemical Looping system is being designed for operation under the sponsorship of ARPA-E (Advanced Research Projects Agency—Energy) of the U.S. Department of Energy.

The SCL Process involves conversion and generation of gases of hazardous nature such as CO, H$_2$, and methane at elevated temperatures and pressures. For instance, for hydrogen in air, the explosion limits are 18.3%–59%, and the flammability limits are 4%–74%. Therefore, safety considerations play a vital role in the design and operation of the SCL Process systems. This is especially true during the testing stages of the SCL Process, which involve a frequent start-up and shutdown test procedure. Specifically, the safety features adopted in the bench and sub-pilot system include: automatic reactor flushing and a preheating sequence during the startup stage for the safe introduction of combustible gas; built-in temperature sensors in the reducer and the oxidizer to monitor the reactor temperature to avoid combustible gas explosion; pressure transducers coupled with a pressure relief valve to prevent pressure

---

Figure 4.24. Continued (6) sample port selector valve; (7) needle valve and bubble flow meter; (8) light source; (9) top solid holding bin; (10) light photocell; (11) reactor gas out port/line; (12) gas and solid sample ports; (13) flanges; (14) heating coils; (15) thermocouples; (16) reactor gas in port/line; (17) bottom screw conveyor; (18) bottoms solid collection container; (19) computer control and data logging; (20) emergency shutoff valve.

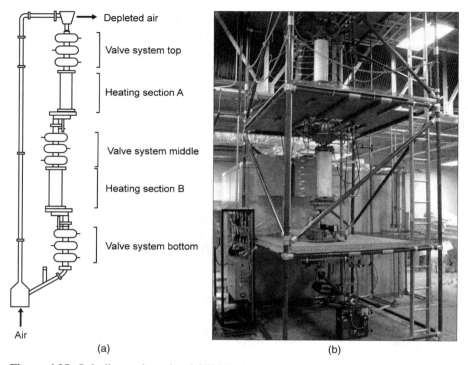

Figure 4.25. Subpilot-scale unit of $25\,kW_{th}$ for the SCL Process: (a) schematic flow diagram; (b) demonstration unit.

buildup; waste gas burners to prevent leakage of combustible gas in case of reactor malfunction; a ventilation fan to prevent the buildup of flue gas; and a fully automatic one-click shutdown sequence that simultaneously turns off reactant gases, switches on the N_2 flush gas, and decreases the reactor temperature. Numerous hazardous/combustible gas leak detectors also are placed in the vicinity of the reactor to alert the operator. These leak detectors are linked to the central controlling console and will trigger the automatic shutdown sequence upon preset threshold levels.

Reducer Performance Using Syngas as Feed
The reducer behavior in a countercurrent moving bed operation with syngas as the feed is discussed in this section. The operation based on the bench reactor, given in Figure 4.24, involves Fe_2O_3 composite particles that contain 60% Fe_2O_3 balanced with inert support. The composite particles are introduced from the top of the reactor at a flow rate of 12.87 g/min. The reactor is of 2 in. (5.1 cm) ID and 30 in. (76.2 cm) in height. The particles used for testing are cylindrical pellets with 5 mm in diameter and 2 mm in thickness. A gas

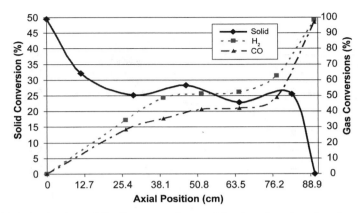

Figure 4.26. Reduction of Fe_2O_3 composite particles in a bench unit using syngas as the reducing gas.

mixture of 61.1% CO, 35.1% H_2, and 3.8% CO_2 (mol%) is sent to the bottom of the reactor at a rate of 1,686 mL/min (STP, Standard Temperature and Pressure). The gas composition is chosen to be similar to the syngas generated from a Shell SCGP gasifier. The average temperature along the heated zone of the reactor is 814°C, and the vessel is operated at atmospheric pressure. A steady solid/gas flow is maintained throughout the experiments. The steady-state gas and solids conversion profiles are shown in Figure 4.26.

As is shown in Figure 4.26, the syngas is converted steadily as it flows upward to react with partially reduced solids. Note that the syngas is converted by >99.9%. A nearly pure CO_2 stream is observed after the condensation of the steam at the gas outlet. The Fe_2O_3 composite particles also are converted steadily as they move downward. The conversion of Fe_2O_3 at the bottom of the demonstration unit is 49.5%. The carbon content in the particles, which affects the purity of the hydrogen produced in the reducer, is characterized to be 0.061 wt.%. The aforementioned gas and solids conversions are achieved with a solid residence time of ~57 minutes. The required solid residence time to reach such conversions can be reduced significantly under an elevated operating pressure (~30 atm). The results of the reactant conversion presented in Figure 4.26 are very close to the optimum gas and solids conversions predicted by the thermodynamic analyses described in Sections 4.4.1 and 4.4.2. Note that the gas and solids conversions in the moving bed reducer are significantly higher than those achieved in the IGT steam-iron process with a syngas conversion of 60% and solid reduction of 5%–15%.[33]

Reducer Performance Using Methane and Other Hydrocarbons as Feed
Iron oxide is active in cracking coal volatiles to methane,[53] the most stable hydrocarbon up to 1,030°C.[54] Therefore, if the reducer can fully convert

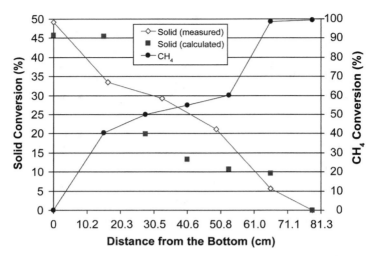

Figure 4.27. Reduction of Fe_2O_3 composite particles in a bench unit using methane as the reducing gas.

methane to CO_2 and steam using the Fe_2O_3 composite particle, then other hydrocarbons also can be converted fully. Close to 100% methane conversion is achieved in bench reactor testing.

The experimental conditions for methane conversion are similar to the conditions for syngas conversion. The Fe_2O_3 composite particle flow rate is 11.08 g/min, and the methane flow rate is 360 mL/min. H_2 also is introduced to the reactor at a flow rate of 133 mL/min to enhance methane conversion. The average reactor temperature is 930°C. Figure 4.27 shows the gas and solids conversion profiles under steady-state operation. It is seen that >99.8% of the methane is converted to CO_2 and H_2O, and the Fe_2O_3 particles are reduced by 49%. The ability of the SCL Process to convert methane effectively into CO_2 indicates its capability of converting a variety of other hydrocarbon fuels.

Oxidizer Performance
Particles reduced by syngas are circulated to the oxidizer where they react with steam to generate hydrogen. Specifically, the particles are introduced from the top of the oxidizer (which is of the same dimensions as the reducer discussed in the Reducer Performance Using Syngas as Feed section) at a rate of 13 g/min. Steam was injected at the bottom of the oxidizer at a rate of 1.355 g/min. The results show that more than 99.3% of the reduced particles are converted into Fe_3O_4 in the small preheating zone located above the reaction zone. Because the preheating zone was maintained at ~580°C, the results indicate that the reaction between steam and reduced Fe_2O_3 composite particles is fast even under a relatively low temperature.

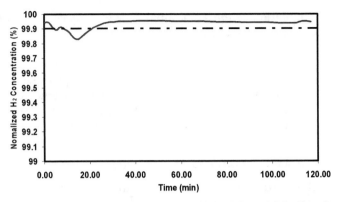

Figure 4.28. Hydrogen production using reduced Fe_2O_3 particles in a bench unit.

The purity of the hydrogen produced at the outlet of the oxidizer indicates a trace of CO. From Figure 4.28, it is seen that the normalized H_2 concentration (moisture- and N_2-free basis) exceeds 99.95%. The results suggest that the purity of hydrogen produced from the SCL Process is such that it can be used directly for ammonia synthesis and oil-refining applications. Clearly, when hydrogen is to be used in polymer electrolyte membrane (PEM) fuel cell applications, further purification of the hydrogen product stream from the oxidizer using purification techniques such as PSA will be necessary.

Combustor Performance
The design of an entrained flow combustor is less challenging when compared with the reducer and the oxidizer as a result of much favored kinetics and thermodynamics for this oxidation reaction. The thermal stability of the composite particles under high operating temperatures in the combustor ($\sim$1,200°C), however, needs to be ascertained. A test using a quartz fixed bed reactor for the oxidation reaction of Fe_3O_4 to Fe_2O_3 at temperatures higher than 1,200°C reveals a full oxidation of Fe_3O_4 to Fe_2O_3 without identifiable loss in either reactivity or recyclability. The particle performance after extended high-temperature combustion operation, however, still needs to be studied.

Thermodynamic analyses exemplified in this section indicate that a countercurrent moving bed delivers the best overall performance for both the reducer and the oxidizer operations. The thermodynamic models constructed using the ASPEN simulation also are used to determine the optimum operating conditions for the SCL reactors as well as pollutant control strategies. Experimental studies in the bench reactor corroborate the reducer and the oxidizer operations predicted by the thermodynamic analysis, validating the proposed SCL Process concepts.

4.5   Process Simulation of the Traditional Gasification Process and the Syngas Chemical Looping Process

In this section, ASPEN simulation models are used to evaluate the overall performance of several representative coal gasification processes. These processes include the IGCC Process using the GE high efficiency quench (GE-HEQ) gasifier, the conventional Coal-to-Hydrogen Process using the Shell SCGP gasifier, and the SCL Process using the Shell SCGP gasifier.

4.5.1   Common Assumptions and Model Setup

To compare the performance of the conventional processes and the SCL process, a common set of assumptions and modeling parameters is defined as follows:

1. The CO_2 capture in the process is considered to be at least 90%.
2. The ambient temperature is 25°C, and the ambient pressure is 1 bar.
3. A feeding rate of 132.9 t/h of Illinois #6 coal is used (approximately 1,000 MW in HHV), and the properties of the Illinois #6 coal are shown in Table 4.7.
4. The GE-HEQ gasifier is used for the IGCC process, and the Shell gasifier with gas quench configuration is used for both the conventional coal-to-hydrogen process and the SCL Process.

TABLE 4.7. Physical and Chemical Properties of Illinois #6 Coal[56]

	Constituents (wt.%, As-Received)	Constituents (wt.%, dry)
Proximate Analysis		
Moisture	11.12	
Fixed Carbon	44.19	49.72
Volatiles	34.99	39.37
Ash	9.7	10.91
	100	100
HHV (MJ/kg)	27.14	30.53
Ultimate Analysis		
Moisture	11.12	
Ash	9.7	10.91
Carbon	63.75	71.72
Hydrogen	4.5	5.06
Nitrogen	1.25	1.41
Chlorine	0.29	0.33
Sulfur	2.51	2.82
Oxygen	6.88	7.75

5. Air consists of 21 vol.% O_2 and 79 vol.% N_2.
6. The solids circulating in the SCL consist of 70 wt.% Fe_2O_3 and 30 wt.% inert support.
7. The H_2 product is compressed to 3 MPa for subsequent transportation.
8. CO_2 is compressed to 15 MPa for sequestration.
9. The carbon conversion in the gasifier is 99%, and the heat loss in the gasifier is 0.6% of the HHV of coal.
10. The pressure level in a steam cycle is 12.4/3/0.2 MPa, and the HP and IP steam is superheated to 550°C, whereas the temperature of the flue gas in the stack is 130°C.
11. All compressors are designed using four stages, and the outlet temperature of the intercooler is 40°C.
12. The mechanical efficiency of pressure changers such as compressors and expanders is 1, whereas their isentropic efficiency is 0.8–0.9.
13. The gas and solids conversions in the SCL system are based on experimental results.
14. The energy consumption for CO_2 capture is 106 kWh/ton of CO_2 for traditional processes.[55]
15. The inlet firing temperature of the gas turbine is 1,250°C.

To simulate accurately the individual unit in the flow sheet, an appropriate ASPEN PLUS model (or models) for each unit are determined. These models are listed in Table 4.8.

4.5.2 Description of Various Systems

IGCC Process Using GE-HEQ Gasifier
The IGCC system illustrated in this case study uses a GE/Texaco slurry-feed, entrained flow gasifier, with a total water quench syngas cooler. The flow diagram of the process is shown in Figure 4.29, which is similar to that described in Section 4.2.1.

In this process, coal first is pulverized and mixed with water to form a coal slurry. The coal slurry then is pressurized and introduced to the gasifier to be oxidized partially at 1,500°C and 30 atm. The high-temperature raw syngas after gasification then is quenched to 250°C with water. The quenching step solidifies the ash. Moreover, most of the NH_3 and HCl in the syngas are removed during this step. After quenching, the syngas is sent to a venturi scrubber for further particulate removal. The particulate-free syngas, saturated with steam, then is introduced to the sour WGS unit. The syngas exiting from the WGS unit contains mainly H_2 and CO_2 with a small amount of CO, H_2S and mercury. This gas stream then is cooled to 40°C and passed through an activated carbon bed for mercury removal. The CO_2 and H_2S in the syngas

TABLE 4.8. ASPEN Models for the Key Units in the IGCC Process

Unit Operation	ASPEN PLUS Model	Comments/Specifications
Air Separation Unit	Sep	Energy consumption of the ASU is based on specifications of commercial ASU/compressors load
Coal Decomposition	Ryield	Virtually decompose coal to various components (prerequisite step for gasification modeling)
Coal Gasification	Rgibbs	Thermodynamic modeling of gasification
Quench	Flash2	Phase equilibrium calculation for cooling
WGS	Rstoic or Rgibbs	Simulation of conversion of WGS reaction based on either WGS design specifications or thermodynamics
MDEA	Sep or Radfrac	Simulation of acid gas removal based on design specifications
Burner	Rgibbs or Rstoic	Modeling of H_2/syngas combustion step
HRSG	MHeatX	Modeling of heat exchanging among multiple streams
Gas Compressors	Compr or Mcompr	Evaluation of power consumption for gas compression
Heater and Cooler	Heater	Simulation of heat exchange for syngas cooling and preheating
Turbine	Compr	Calculation of power produced from gas turbine and steam turbine

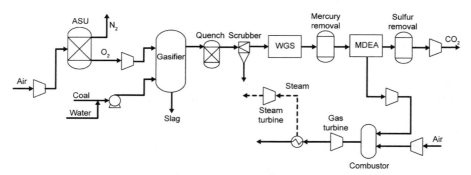

Figure 4.29. IGCC Process with CO_2 capture.

then are removed using an MDEA unit, resulting in a concentrated hydrogen stream with small amounts of CO_2 and CO. The hydrogen-rich gas stream then is compressed, preheated, and combusted in a combined-cycle system for power generation. The combined cycle system consists of a gas turbine and a two-stage steam turbine. The CO_2 obtained from the MDEA unit is compressed to 150 atm for sequestration.

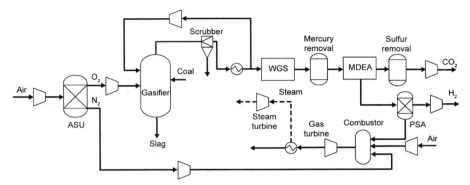

Figure 4.30. Conventional gasification—-water–gas shift Coal-to-Hydrogen Process.

Conventional Coal-to-Hydrogen Process
The Coal-to-Hydrogen system illustrated in this case study uses a Shell SCGP dry-feed, entrained flow gasifier with a gas quench configuration. The flow diagram of the process is shown in Figure 4.30, which is similar to that described in Section 4.2.2.

In this process, coal first is pulverized and dried. The coal powder then is pressurized in the lock hopper and introduced to the gasifier to be oxidized partially at 1,500°C and 30 atm. The high-temperature raw syngas after gasification then is quenched to 900°C with low-temperature syngas from the particulate removal unit. The quenched syngas then is introduced to a syngas cooler. After being cooled further to ~300°C, the syngas is introduced to a particulate removal unit where the entrained solids are removed from the syngas. The syngas then is saturated with steam and enters the sour WGS unit. In the WGS unit, nearly 96% of the CO is converted to H_2. This gas stream then flows to an activated carbon bed for mercury removal. Subsequently, acid gases, H_2S, and CO_2 in the raw hydrogen stream are removed using such solvents as MDEA. Further purification of hydrogen is made via a PSA unit. The tail gas from the PSA then is combusted for electricity generation.

Syngas Chemical Looping Process
To obtain data directly comparable with conventional processes, the SCL system illustrated here also uses a Shell SCGP dry-feed, entrained flow gasifier with a gas quench configuration. The flow diagram of the process is given in Figure 4.31, which is discussed in Section 4.3. The SCL system shown in Figure 4.31 uses a syngas generation, quenching, and cooling system identical to that used in the convention coal-to-hydrogen case. The syngas, however, is cooled only to 550°C. The cooled syngas is introduced to a hot gas cleanup unit to strip the sulfur level down to 50 ppm. The low sulfur syngas then is introduced to the SCL system for hydrogen and power cogeneration.

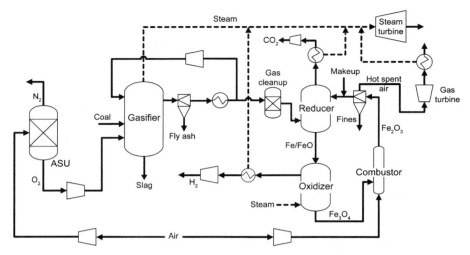

Figure 4.31. The Syngas Chemical Looping Process using Shell gasifier.

4.5.3   ASPEN PLUS Simulation, Results, and Analyses

Based on the parameters and process configurations described in Sections 4.5.1 and 4.5.2, ASPEN simulations are conducted based on the ASPEN flowsheets shown in Figure 4.32 for the three-coal gasification processes described. ASPEN PLUS has a comprehensive physical property database, as discussed in Section 4.4.2. Therefore, most of the chemical species involved in the process can be selected directly from this database. The nonconventional components such as coal and ash are specified conveniently using a general coal enthalpy modulus embedded in the ASPEN software. After the chemical species in the process are defined, the related physical property methods are selected among various choices for the simulator. In this simulation, the global property method used is PR-BM, and the local property methods are specified whenever required.

The ASPEN model is completed by establishing detailed operating parameters based on the operating conditions and design specifications of the individual units. The units then are connected in the same arrangement as shown in the flowsheet. An appropriate convergence setting is determined to ensure accurate simulation results. Table 4.9 compares the simulation results among the three systems. The case when all hydrogen generated in the SCL process is used for electricity generation also is investigated.

As is shown in Table 4.9, compared with conventional processes, the Syngas Chemical Looping Process is significantly more efficient, especially under a carbon-constrained situation. The advantage of the SCL Process results from an improved energy conversion scheme coupled with the integrated CO_2 capture capability.

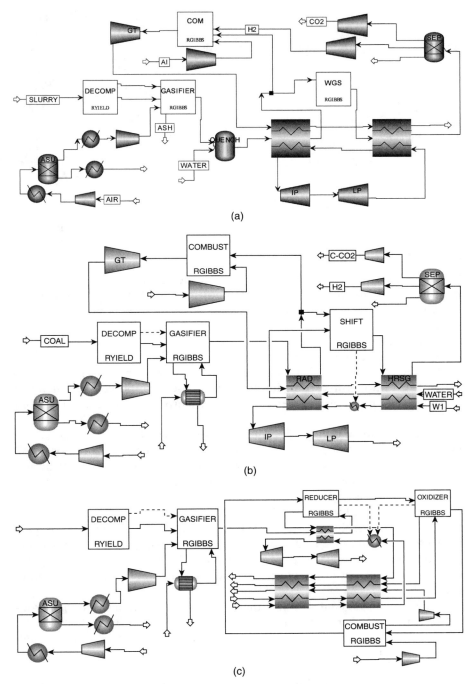

Figure 4.32. ASPEN simulation flowsheet: (a) IGCC system using GE-high-efficiency quench (HEQ) gasifier; (b) conventional Coal-to-Hydrogen system using Shell gasifier; and (c) SCL system using shell gasifier.

TABLE 4.9. Comparison of the Process Analysis Results

	IGCC Process	SCL Process Electricity	Conventional Coal-to-Hydrogen Process	SCL Process
Coal Feed (t/h)	132.9	132.9	132.9	132.9
Carbon Capture (%)	90	100	90	100
Hydrogen (t/h)	0	0	14.4	15.6
Net Power (MW)	348.1	422.0	57.6	57.4
Efficiency (%HHV)	34.8	42.2	62.7	66.5

4.6 Example of SCL Applications—A Coal-to-Liquid Configuration

4.6.1 Process Overview

The SCL Process can be integrated or retrofitted to existing processes to improve their overall energy conversion efficiencies. This section exemplifies a novel configuration that integrates the SCL Process with the traditional CTL process.

There are several different schemes to incorporate the SCL process into the CTL Process. In a conservative scheme, the SCL Process is used as a retrofit to the traditional CTL Process in which the role of the SCL Process is to generate additional hydrogen for Fischer–Tropsch synthesis. The schematic flow diagram for such a scheme is illustrated in a process configuration as given in Figure 4.33.

In a conventional CTL plant using cobalt-based catalysts, the syngas generated from the entrained flow gasifier has a hydrogen concentration of 30%–40% (dry basis), which is significantly lower than the required H_2 concentration (>60%) for the liquid fuel synthesis. This shortage in hydrogen concentration usually is compensated for by additional steps to shift the CO partially in the syngas stream in a traditional CTL process. Meanwhile, the Fischer–Tropsch reactor converts only part of the syngas (60%–80%) to a wide variety of hydrocarbons that range from methane to hard wax. The gaseous hydrocarbon fuels and unconverted syngas are considered to be by-products, and a significant portion of these gaseous compounds are combusted to generate electricity.

In the SCL–CTL configuration, the unconverted syngas and gaseous products from the Fischer–Tropsch reactor are introduced to the reducer of the SCL. These gaseous fuels are converted to carbon dioxide and water through the following reaction:

$$C_x H_y O_z + (2x + y/2 - z)MO \rightarrow (2x + y/2 - z)M + xCO_2 + y/2H_2O \quad (4.6.1)$$

where MO and M refer to the different iron oxide phases. Reaction (4.6.1) reduces the iron oxide from higher oxidation states to lower oxidation states.

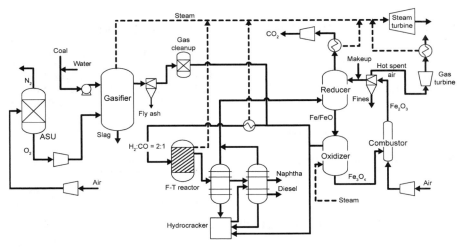

Figure 4.33. Syngas Chemical Looping enhanced coal-to-liquid (SCL–CTL) Process.

The reduced particles then are introduced to the oxidizer where they react with steam to produce pure hydrogen and regenerate iron oxide [Reaction (4.6.2)].

$$M + H_2O \rightarrow MO + H_2 \qquad (4.6.2)$$

The major feed gases for the SCL reducer are the by-products from the Fischer–Tropsch reactor. Meanwhile, the large amount of the medium-pressure (~25 atm) steam generated by the low-grade heat from the Fischer–Tropsch reactor provides ample steam supply required for the SCL oxidizer. Therefore, through the usage of the SCL, hydrogen—an essential feedstock for the CTL—is generated from the by-products of the Fischer-Tropsch synthesis. The liquid fuel yield of the CTL Process thus is improved. The integrated carbon capture capability of the SCL renders the SCL–CTL configuration even more attractive under a carbon-constrained situation.

4.6.2 Mass/Energy Balance and Process Evaluation

A system analysis based on ASPEN PLUS and flowsheet analysis is conducted on the SCL–CTL configuration to evaluate the performance of the SCL–CTL Process relative to the conventional CTL Process using assumptions identical to those in Section 4.5.1. Several additional assumptions used in the CTL Process analysis include the following:

1. Cobalt-based catalysts can achieve 75% per pass conversion on 2:1 $H_2:CO$ ratio syngas while having 85% selectivity toward liquid-phase products.

TABLE 4.10. Overall Energy Input/Output for the SCL-CTL Process

	Traditional CTL	SCL-CTL
Coal Feed In	3,190 t/day	3,190 t/day
Naphtha Out	2,579 bbl/day	2,811 bbl/day
Diesel Out	5,511 bbl/day	6,020 bbl/day
Net Electricity	5 MW	2 MW
Fuel Conversion Efficiency (%)	37.7	41.1
Overall System Efficiency (%)	38.2	41.3
Percent of CO_2 Captured (%)	0	100

2. All combustible gas in the traditional CTL Process is used for electricity generation.
3. The heating value of a naphtha product is 31.6 MJ/L, and the heating value of a diesel product is 35.5 MJ/L.
4. The steam of 200°C, 20 atm is generated from the high-temperature gas streams (200°C–500°C) and the exothermic reactions that occur in that range.

The mass and energy balances on both traditional CTL and SCL–CTL Process are shown in Table 4.10. The plant size is 1,000 MW$_{th}$.

Table 4.10 shows that the liquid fuel production efficiency for the SCL-CTL process is in excess of 40%. This production efficiency is a nearly 10% improvement over the traditional CTL Process, which is at 37.7%. When the extra electricity generation also is taken into account, the system efficiency for the SCL–CTL Process would be at 41.3% compared with 38.2% for the conventional CTL Process. In addition, the SCL–CTL process separates all the CO_2 generated, whereas the traditional CTL Process does not separate CO_2. Thus, the SCL-enhanced CTL Process yields a significantly improved performance for the liquid fuel production over the traditional CTL Process, which is illustrated further in terms of the improved energy usage scheme and the simpler process scheme given as follows:

1. Improved energy usage scheme. The SCL–CTL Process uses both the by-products (C_1–C_4) and the low-grade energy (steam) from the Fischer–Tropsch reactor and effectively converts them to H_2, which then is used to synthesize the desired liquid-fuel products. Thus, the SCL–CTL Process is thermodynamically more efficient compared with the traditional CTL Process for liquid fuel synthesis.
2. Simpler process scheme. The presence of the SCL system simplifies the traditional CTL Process in that the SCL system can perform multiple functions and effectively replace such units in the traditional CTL

Process as the WGS reactor, pressure swing adsorption unit, and multiple Fischer–Tropsch product upgraders. Furthermore, the load of CO_2 separation using Selexol units can be reduced drastically. The SCL system also can realize 100% carbon capture without additional economic burden.

An economic analysis of a commercial-scale CTL plant using either the conventional CTL or the SCL–CTL technology indicates that, although the capital requirement for a SCL–CTL plant is slightly higher than that for a traditional CTL plant with an identical coal-processing capacity, the normalized capital requirement based on the liquid fuel production capacity ($/daily barrel) for the SCL–CTL Process is notably lower. The normalized operating cost for the SCL–CTL Process also is reduced, which results from less coal input and from decreased parasitic energy consumption. The increase in liquid fuel yield, as well as the decrease in operating cost, renders the SCL–CTL Process more economical than the traditional CTL process. Thus, the integration of SCL to a conventional CTL plant has the potential to increase significantly the profitability of the plant. Such a configuration was explored by Noblis Systems, and more further information can be obtained from their report to the U.S. Department of Energy.[57]

4.7   Calcium Looping Process Using Gaseous Fuels

Enhancement in the production of high-purity hydrogen (H_2) from synthesis gas, obtained by coal gasification or steam methane reforming (SMR), is limited by the thermodynamics of the WGS reaction. This constraint, however, can be overcome by a strategy of concurrent WGS reaction and carbonation reaction to enhance H_2 production. The carbonation of calcium-oxide-forming calcium carbonate incessantly drives the equilibrium-limited WGS reaction forward by removing the CO_2 product from the reaction mixture. Calcium carbonate can be calcined separately to yield a pure CO_2 stream for subsequent compression, transportation, and sequestration with calcium oxide recycled back for the carbonation reaction.

There are several chemical-looping-based processes that enhance hydrogen production using calcium-based sorbents, as discussed in Chapter 1. These processes include those that introduce gaseous fuels into the looping reactor such as the Zero Emission Coal Alliance (ZECA) Process[58] and that introduce carbonaceous solid fuels directly into the looping reactor such as the Carbon Dioxide Acceptor Process, HyPr-Ring Process, the Alstom Hybrid Combustion-Gasification Process, and the GE Fuel-Flexible Process.[59–61] In the ZECA Process, hydrogen is used to gasify coal to produce a methane-rich stream, which then is reformed using steam in the presence of calcium-based sorbents to produce hydrogen. The sorbent removes the carbon dioxide produced during the reforming reaction, thereby shifting the equilibrium in the forward direction; because the carbonation reaction is exothermic, it also provides heat for the endothermic reforming reaction.[62]

In the HyPr-Ring Process, coal is gasified with high-pressure steam in the presence of a $CaO/Ca(OH)_2$ sorbent.[59] A high partial pressure of steam is maintained in the gasifier to hydrate the CaO so that the combined exothermic reactions of hydration and carbonation supply a sufficient amount of energy for the endothermic gasification reaction. These systems, however, require excess steam and produce only a 91%-pure hydrogen stream.[63] In addition, high temperatures and long residence time of the calcium-oxide sorbent in the gasifier have been found to increase considerably the solid–solid interaction between $CaO/Ca(OH)_2$ and the minerals in coal, hence, reducing the CO_2-capturing ability of the sorbent.[64] Steam also has been found to enhance the interaction of the sorbent with coal minerals at high temperatures and pressures.[65]

In a typical configuration of the ALSTOM Hybrid Combustion-Gasification Process for hydrogen production, calcium-based sorbents and bauxite ore are used to carry oxygen, CO_2, and heat in three loops. The first loop is the $CaSO_4$–CaS loop in which coal is gasified using $CaSO_4$—an oxygen-carrying agent—to produce CO. CO then is converted to CO_2 and H_2 by the WGS reaction. The CaS produced in this process is regenerated in air to produce $CaSO_4$ through an exothermic oxidation reaction. The second loop consists of the CaO-$CaCO_3$ loop in which the CaO sorbent is used to remove CO_2 during the WGS reaction, forming $CaCO_3$ while producing a pure stream of hydrogen. The third loop is a heat-transfer loop in which hot $CaSO_4$ or bauxite is used to transfer the heat from the exothermic CaS oxidation reaction to the calciner to support the endothermic calcination of $CaCO_3$.[60]

The GE Fuel-Flexible Process comprises two loops, an oxygen transfer loop and a carbon transfer loop, and it involves three reactors. In the first reactor, coal is gasified to produce CO and H_2, along with CO_2, which is removed constantly by the CaO sorbent. The reacted $CaCO_3$ product, along with the unconverted char, then is routed to the second reactor where hot oxygen transfer material from the third reactor is reduced while converting the char to CO_2. The hot solids also provide heat for the calcination of $CaCO_3$. In the third reactor, the reduced oxygen transfer material is reoxidized, releasing a considerable amount of heat that heats the solids and generates steam for power production. The GE Process obtains a hydrogen concentration of only 80%.[61] The calcium-based chemical looping process mentioned is discussed in detail in Chapter 5.

There also have been other applications for enhancing hydrogen production by coupling SMR with in situ CO_2 capture using sorbents.[66–69] Calcium-oxide-assisted SMR was attempted in earlier studies that examined the performance of a single-step, sorption-enhanced process using a Ni-based catalyst to produce hydrogen.[68,69] The studies of the SMR also probed the use of a mixed dolomite-CaO powder with Ni-based catalysts for the separation of CO_2 and the subsequent enhancement of H_2 concentration to 97%.

4.7.1   Description of the Processes

The Calcium Looping Process using gaseous fuels simplifies the production of H_2 by integrating into various functions of a single-stage reactor including the WGS reaction with *in situ* carbon dioxide, sulfur, and hydrogen halide removal from the synthesis gas at high temperatures. The unique features of the Calcium Looping Process are given as follows:

1. The amount of steam in the WGS reaction is reduced to close to stoichiometric level.
2. The process can be conducted with simultaneous removal of CO_2, as well as sulfur (H_2S + COS) and halide (HCl/HBr) impurities.
3. The process can be conducted with flexibility in the CO conversion to produce H_2:CO ratios, varying from 0.5 to 20, while reducing the sulfur and halide impurities in the product gas stream to levels of parts per billion, suitable for fuels/chemical synthesis by the Fischer–Tropsch reaction.
4. The process can be conducted with the production of a sequestrable CO_2 stream through spent sorbent regeneration at high temperatures (700°C–1100°C). The sorbent regeneration includes calcining the carbonated sorbent ($CaCO_3$) using H_2 and oxygen and/or steam to generate calcium oxide. The calcium sulfide sorbent can be regenerated partly to calcium oxide by treatment with steam and carbon dioxide.

The reactions in the process are given as follows.

Reaction phase

$$\text{WGS reaction:} \quad CO + H_2O \rightarrow H_2 + CO_2 \qquad (4.7.1)$$

$$\text{Carbonation:} \quad CaO + CO_2 \rightarrow CaCO_3 \qquad (4.7.2)$$

$$\text{Sulfur capture } (H_2S): \quad CaO + H_2S \rightarrow CaS + H_2O \qquad (4.7.3)$$

$$\text{Sulfur capture (COS):} \quad CaO + COS \rightarrow CaS + CO_2 \qquad (4.7.4)$$

$$\text{Halide capture (HCl):} \quad CaO + 2HCl \rightarrow CaCl_2 + H_2O \qquad (4.7.5)$$

Regeneration phase

$$\text{CaO regeneration:} \quad CaCO_3 \rightarrow CaO + CO_2 \qquad (4.7.6)$$

$$\text{CaS reaction:} \quad CaS + H_2O + CO_2 \rightarrow CaCO_3 + H_2S \qquad (4.7.7)$$

Hence, this process provides a "one-box" mode of operation for the production of high-purity hydrogen with integrated CO_2, sulfur, and chloride capture, along with the WGS reaction and hydrogen separation in one consolidated unit. Another advantage is that, in addition to generating a pure hydrogen stream, the Calcium Looping Process also is capable of adjusting the H_2:CO ratio to a required level at the reactor outlet while readily reducing sulfur in the syngas to a very low concentration for the catalytic Fischer–Tropsch synthesis of liquid fuels. This integrated one-box process yields higher system efficiencies and a lower overall footprint by combining different process units in one stage.

The performance of the Calcium Looping Process depends on the reactivity and recyclability of the calcium sorbent, as discussed in Chapter 2. Tests of various sorbents reveal the shortcomings of some of the calcium-based sorbents. The average calcium conversion achieved in dolomite sorbents is 82%. On a weight basis, the CO_2 capture capacity achieved by such a sorbent would be lower than 35%.[70] A lower solids conversion reflects a higher sorbent requirement and, hence, a larger reactor size. The recyclability of the sorbent, to a large extent is, dictated by the manner in which calcination is carried out in sorbent regeneration. The carrier gas for the calcination reaction can be pure CO_2, steam or other easily condensable gases. The heat for the endothermic calcination reaction may be provided by direct contact of the spent sorbent in the calciner with a hot CO_2 stream generated from the combustion of such fuels as syngas, natural gas, or coal. The heat also may be provided by indirect contact of the spent sorbent with heat sources. For $CaCO_3$ sorbent regeneration using the heat provided by a hot CO_2 stream, a high-purity CO_2 stream from regeneration can be obtained. Such regeneration, however, requires the calcination temperature to be greater than 950°C, which results in significant sintering of CaO and, hence, a loss in its reactivity. Moreover, when reforming catalysts such as Ni are used in the Calcium Looping Reforming process and the gaseous stream for sorbent regeneration contains oxygen, Ni catalysts can be oxidized to become NiO. Thus, the regeneration of the oxidized Ni catalysts to Ni will require a different step from the $CaCO_3$ regeneration step.[71] Similarly, the magnetite catalyst used to catalyze the WGS reaction can be oxidized to become hematite, and thus a separate regeneration step is required for converting hematite back to magnetite in the $CaCO_3$ regeneration step.[72] Furthermore, fuel gas impurities such as chlorine will be converted to a stable calcium chloride form and sulfur to calcium sulfide. A part of the calcium sulfide can be regenerated during calcination, and the remaining calcium sulfide as well as all the calcium chloride can be purged with the spent calcium sorbent.

4.7.2 Reaction Characteristics of the Processes

The creation of an effective Calcium Looping Process requires ensuring the reactivity of the calcium sorbent, and thus high reactivity calcium sorbents such as precipitated calcium carbonate (PCC) synthesized at the Ohio State University (OSU) are of value to the process operation. The PCC sorbent is synthesized by the wet precipitation technique using surface modifiers and has a reactivity higher than available sorbents,[73–77] as noted in Chapter 2.

Furthermore, it is capable of reducing the concentrations of sulfur and halides at the reactor outlet to the parts-per-billion level. Compared with PCC, CaO obtained from naturally occurring precursors [Linwood hydrate (LH), Linwood carbonate (LC), and dolomite] cannot react completely as a result of pore pluggage and pore-mouth closure. In addition, dolomite has a substantial amount of unreacted magnesium components (nearly 50 wt.%), which leads to lower sorbent conversions as observed through experimentation.[68] CaO obtained from the mesoporous PCC sorbent can achieve up to 95% conversion during carbonation and has a CO_2 capture capacity as high as ~700 g/kg of the sorbent. The high-CO_2 capture capacity ensures minimal sorbent usage and, hence, a smaller reactor. The life-cycle testing of the sorbent across multiple cycles of carbonation–calcination reactions indicates a capture capacity of 36–40 wt.% for 50–100 cycles for the PCC sorbent. In contrast, naturally occurring limestone such as LC generally exhibits a capture capacity of less than 20% after 50 cycles.

One of the drawbacks in the production of pure hydrogen using the conventional WGS reaction route is the requirement of excess steam. The excess steam varies from 7 to 50 times the stoichiometric value and is used to drive the equilibrium-limited WGS reaction forward for production of 99.5% pure hydrogen.[16] In contrast, the Calcium Looping Process achieves the same objective of shifting the equilibrium of the WGS reaction in the forward direction by the *in situ* removal of the CO_2 product. The high hydrogen yield makes it possible to operate at lower steam:CO ratios in this process. In addition to reducing the operating costs, lowering the steam addition assists the effective removal of sulfur and halide impurities (H_2S/HCl) through the sulfidation and chloridation of CaO. Similarly, it is evident from thermodynamics that the presence of CO_2 impedes COS capture by CaO. Because CO_2 is removed *in situ* by the CaO, the COS in the product hydrogen stream is reduced to the parts-per-billion level.

The results of the extent of the sulfidation reaction with three different CaO sorbents,[78] namely CaO obtained from Aldrich chemicals, PCC, and Linwood calcium carbonate, conducted at 800°C with a total pressure of 10 atm and P_{H_2S} of 0.03 atm (0.3%) are shown in Figure 4.34. The figure clearly points out the high reactivity of the PCC-CaO as compared with the other CaO sorbents.

The extent of chloridation of PCC–CaO and commercial limestone–CaO sorbent are shown in Figure 4.35. For HCl concentrations of 25 ppm and 1,500 ppm, it is shown that the reactivity of the PCC–CaO sorbent is higher than that of the limestone–CaO sorbent at a given temperature. Furthermore, for both sorbents, the reactivity peaks at temperatures between 500°C and 600°C. The results using the limestone–CaO sorbent similar to those given in Figure 4.35 also were reported in the literature.[79] This high reactivity of the PCC sorbent for sulfur and chloride capture again can be attributed to its desired sorbent morphology.

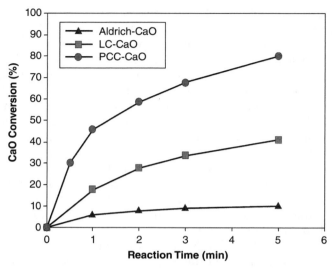

Figure 4.34. Reactivity of different CaO sorbents toward H_2S removal at 800°C and 10 atm[78].

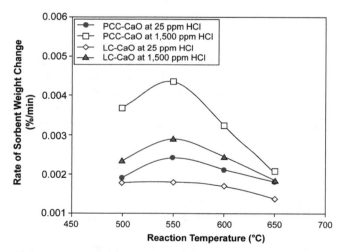

Figure 4.35. Reactivity of different CaO sorbents toward HCl removal at various temperatures and HCl concentrations at atmospheric pressure.

Catalytic Production of Hydrogen

Combined WGS and Carbonation Reaction without H_2S. The reaction results comparing LC and PCC in the combined carbonation and WGS reaction for enhanced H_2 production conducted in an integral fixed bed reactor are shown in Figure 4.36.[80] The high-temperature shift (HTS) iron oxide catalyst on chro-

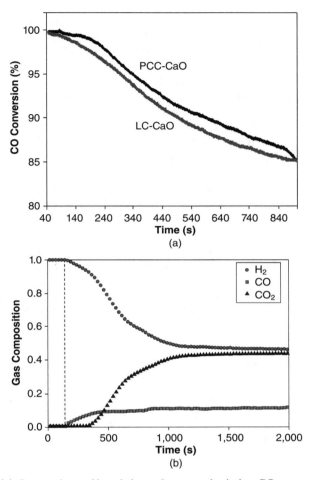

Figure 4.36. (a) Comparison of breakthrough curves depicting CO conversion for PCC and LC sorbents in the presence of a catalyst; (b) breakthrough curves for the gas composition using PCC sorbent in the presence of a catalyst at atmospheric pressure, at 600°C, and a steam:CO ratio of 3:1.

mium oxide support also is used in the reaction. Figure 4.36(a) illustrates the CO conversion breakthrough curves for both the PCC and LC sorbent-catalyst systems. It is evident that the presence of CaO enhances the CO conversion and, hence, the hydrogen purity. In both systems, a 100% initial conversion is observed. In addition, the PCC–CaO system performs better than the LC–CaO system at any given time, demonstrating the superior performance of the PCC sorbent for hydrogen production. Figure 4.36(b) describes the nitrogen- and steam-free product gas compositions for a PCC–HTS system at 600°C. As illustrated in the figure, during the initial breakthrough period, the system demonstrates the production of a 99.9% pure hydrogen stream, whereas the CO and CO_2 concentrations are negligible. As the system reaches steady state, the CO_2 and H_2 concentrations tend to converge.

Effect of Pressure on H_2 Yield. The effect of pressure on the combined WGS and carbonation reactions in a fixed bed reactor containing the calcined PCC sorbent and the HTS catalyst is shown in Figure 4.37. The combined reactions indicate the production of a 99.9% pure-hydrogen stream during the initial stage of the reaction at a pressure of 21 atm, which is significantly higher than the hydrogen purity obtained at atmospheric pressure.

Effect of Reducing Excess Steam Requirement. The effects of the steam:CO ratios on the CO conversion for the combined WGS and carbonation reaction are shown in Figure 4.38. The figures also show the effect of pressures at 1 atm [Figure 4.38(a)] and 21 atm [Figure 4.38(b)]. The steam:CO ratios considered are 3:1, 2:1, and 1:1. It is shown that, although at atmospheric pressure, a reduction in the steam:CO ratio results in a reduction in the CO conversion at a higher pressure of 21 atm, almost 100% conversion is achieved for all three steam:CO ratios. The high conversion achieved at high pressures clearly illustrates the benefit of using a smaller amount of steam at high pressures for a high CO conversion.

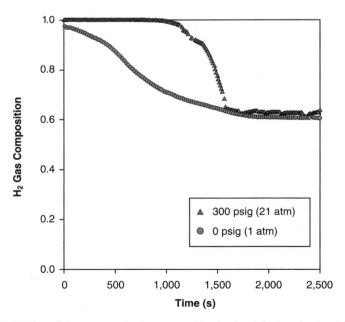

Figure 4.37. Effect of pressure on hydrogen purity obtained during the combined WGS and carbonation reaction in the presence of a catalyst at 650°C and a steam:CO ratio of 3:1.

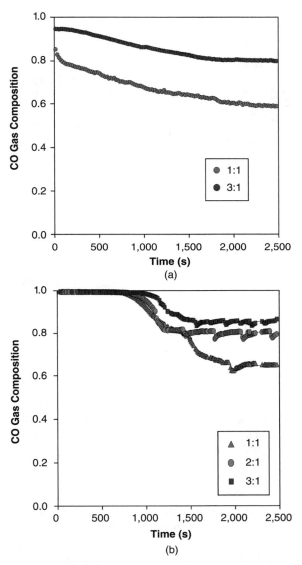

Figure 4.38. (a) Effect of the reduction in steam:carbon ratio on the CO conversion at atmospheric pressure; (b) effect of the reduction in steam:carbon ratio on the CO conversion at 21 atm in the presence of a catalyst at a temperature of 650°C.

Noncatalytic Production of H_2

The HTS iron catalysts can be poisoned in the presence of H_2S impurities. Thus, the reactions for producing pure hydrogen in the absence of the catalysts also are examined. Figure 4.39 shows that, at a pressure of 21 atm in the presence of sorbent, the WGS reaction can achieve almost complete conversion

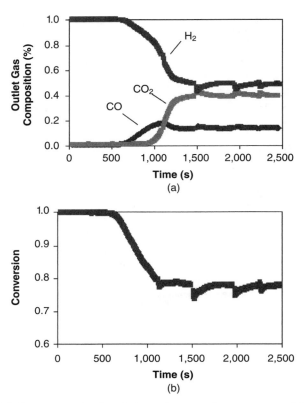

Figure 4.39. (a) Representative breakthrough curve for gas composition during the noncatalytic production of hydrogen at 21 atm; (b) breakthrough curve for CO conversion during the noncatalytic production of hydrogen at 21 atm, for a temperature of 600°C and a steam:carbon ratio of 3.

with a 99.9% pure hydrogen stream, even when WGS catalysts are not used. Hence, it is clear that the sorbent is effective in shifting the equilibrium of the WGS reaction to such an extent that pure hydrogen can be produced in the absence of a catalyst. Operating in this manner eliminates the complexities and costs involved in the separation of the sorbent and catalyst mixture and in the regeneration of the catalyst.

Effect of Pressure and Steam:Carbon Ratio on the Noncatalytic Production of Hydrogen. The CO conversions obtained for the combined WGS and carbonation reactions in the absence of catalysts at various steam:carbon ratios and pressures ranging from 1 atm to 21 atm are illustrated in Figures 4.40 and 4.41. With an increase in pressure, the purity of hydrogen increases and can reach greater than 99% at 21 atm. It is evident from Figure 4.41 that, at high pres-

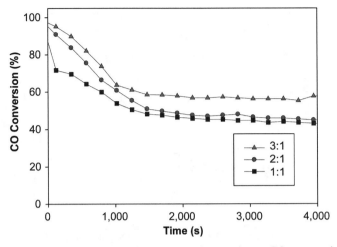

Figure 4.40. Effect of the reduction in steam:carbon ratio on CO conversion at atmospheric pressure and a temperature of 600°C.

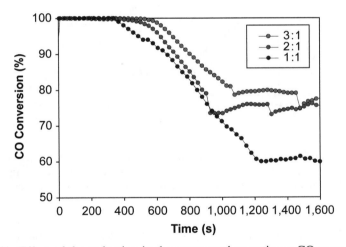

Figure 4.41. Effect of the reduction in the steam:carbon ratio on CO conversion at a pressure of 21 atm and a temperature of 600°C.

sures, the decrease in steam addition does not result in a significant change in the purity of hydrogen produced.

Combined Hydrogen Production, CO_2, and H_2S Capture. Figure 4.42(a) shows the breakthrough curves on the outlet H_2S composition for the combined carbonation, WGS reaction, and sulfidation for enhanced H_2 production from syngas in an integral fixed bed reactor assembly at 1 atm and at 21 atm. In the

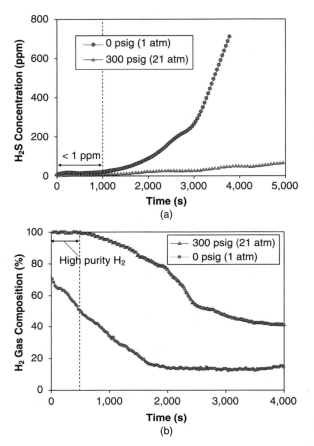

Figure 4.42. (a) Effect of pressure on the H₂S concentration observed in the outlet stream; (b) effect of pressure on the H₂ purity obtained during the noncatalytic production of hydrogen with sulfur capture at a temperature of 600°C and a steam:carbon ratio of 1:1.

prebreakthrough region of the curve, the calcium oxide sorbent undergoes sulfidation, removing H_2S to less than 1 ppm at 1 atm. At a higher pressure of 21 atm, lower levels of H_2S in the outlet of the fixed bed are observed. Because the stoichiometric quantity of steam is used, a very high H_2S removal is achieved in the reactor. Figure 4.42(b) illustrates the breakthrough curve of the H_2 purity, monitored during the same analysis. A very distinct prebreakthrough region is seen in the curve, which again shows that the sorbent is very effective in driving the WGS reaction forward. For 1 atm, the purity of H_2 obtained is 70%, whereas it is at 99.97% for 21 atm. It is clear that the high reactivity of the PCC sorbent results in rapid carbonation and high sorbent conversions (80%), which can reduce greatly the amount of the solids loading in the reactor.

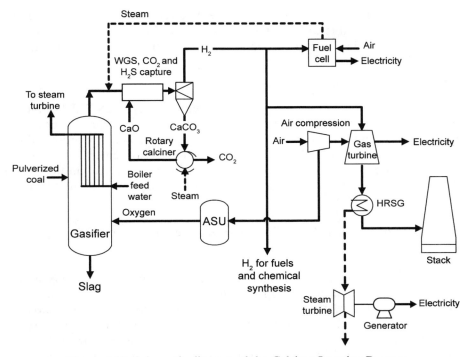

Figure 4.43. Schematic diagram of the Calcium Looping Process.

Integrated H_2 Production, CO_2 Capture, and Sulfur Removal in a Coal Gasification Process. Figure 4.43 (presented earlier as Figure 1.26) presents the flow diagram of the Calcium Looping Process for hydrogen generation using coal as the feedstock. The syngas from the gasifier flows into the combined one-box WGS carbonator reactor where a near stoichiometric amount of steam is injected along with CaO, leading to an enhanced WGS reaction coupled with CO_2, sulfur, and halide capture in an entrained or fluidized bed. In addition, the CO_2 concentration also can be minimized from the continuous removal of the CO_2 product via carbonation. Furthermore, the process favors sulfur (H_2S and COS) and halide removal using CaO at ~500°C–700°C from ~10 ppb to 20 ppm. Thus, the reactor system can achieve the CO_2, as well as H_2S, COS, and HCl removal while producing a pure H_2 stream. The Calcium Looping Process can be integrated in various configurations to apply to the following applications.

1. Air-blown gasification with sorbent (CaO) injection to produce electrical power from an advanced turbine.
2. Oxygen-blown gasification with sorbent (CaO) injection to produce greater than 90% purity of hydrogen by operating the looping reactor at atmospheric pressure.

3. Oxygen-blown gasification with sorbent (CaO) to produce hydrogen of purity compatible with solid oxide fuel cells or PEM fuel cells by operating the looping reactor at high pressures in conjunction with a PSA.

4.7.3 Analyses of the Processes

The Calcium Looping Process is analyzed based on the ASPEN PLUS simulation for the production of high-purity hydrogen from coal through two schemes. In the first scheme, the only product is hydrogen, and all energy produced in the process is used internally for the parasitic energy requirement. The second scheme is based on the cogeneration of hydrogen and electricity in the same facility. Both simulations are conducted for the production of 280 t/d of hydrogen from Illinois #6 coal.

Production of Fuel-Cell-Grade Hydrogen Having a Purity of 99.999% with a PSA

Production of Hydrogen with Internal Heat Integration. The ASPEN flowsheet for the Calcium Looping Process for fuel cell-grade hydrogen production is given in Figure 4.44. The entire flowsheet consists of five major blocks: (1) the gasifier and the ASU; (2) the calcium looping reactor where the production and purification of hydrogen and contaminant capture occur; (3) the calciner where the calcium sorbent is regenerated, and the concentrated CO_2 stream is produced; (4) the PSA where high-purity hydrogen is produced; and (5) the steam-generation block where the heat from various streams is used to generate steam for the process. For hydrogen production, a Shell gasifier is used to gasify 2,180 t/d of Illinois #6 coal, having an HHV of 29.2 MJ/kg, in the presence of oxygen supplied by the air separation unit. It is noted that the HHV value for coal used in the simulation is slightly lower than that for dry coal given in Table 4.7.

The Shell gasifier produces 847,200 m³/d of syngas at a temperature of 1,538°C and a pressure of 36 atm. Because of the high content of sulfur in the coal, the syngas contains 1.15% of H_2S and 848 ppm COS. Because the Calcium Looping Process is capable of *in situ* sulfur capture during the production of hydrogen, it can handle high-sulfur coal effectively. The composition of syngas produced at the outlet of the gasifier used in the simulation is given in Table 4.11. Of the syngas produced at the outlet of the gasifier, 88.7% is fed to the calcium looping reactor for the production of high-purity hydrogen, whereas 11.3% of the syngas is combusted in the calciner to provide the energy required for the endothermic calcination reaction.

The hot syngas is cooled in a radiant heater and is fed to the calcium looping reactor along with high-temperature and high-pressure steam and PCC-CaO sorbent. In the carbonation reactor, hydrogen production, purification, and sulfur removal are achieved by the integrated WGS reaction, carbonation and sulfidation of the calcium oxide sorbent at a temperature of 600°C and a pres-

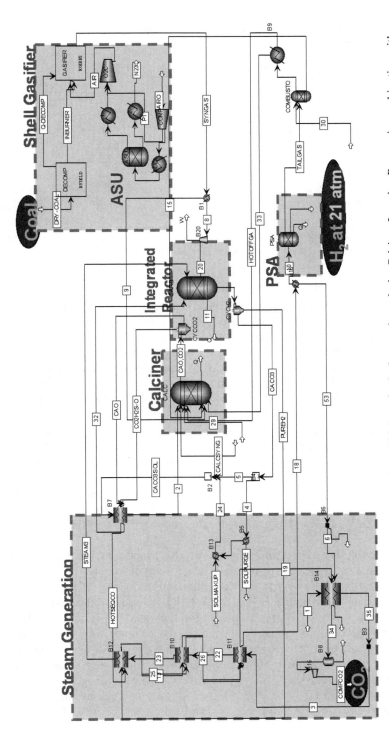

Figure 4.44. ASPEN simulation for the production of high-purity hydrogen through the Calcium Looping Process in combination with a PSA.

TABLE 4.11. Composition of the Syngas Exiting from the Shell Gasifier[a]

Syngas Component	Amount (mol%)
H_2O	2.5
N_2	4.1
O_2	0
H_2	27.6
CO	61.3
CO_2	2.2
Ar	0.8
COS (ppm)	848
H_2S	1.15
CH_4	0.1

[a]The Shell gasifier operating conditions were: temperature, 1,538°C; pressure, 36 atm; and mass flow rate, 177,038 kg/h.

sure of 21 atm. The hydrogen-rich product stream then is purified further in a PSA to produce up to 99.999% pure hydrogen, which can be used either in hydrogen fuel cells or for the production of fuels and chemicals. Because the purity of the hydrogen in the feed to the PSA is very high (94%–98%), the energy consumption in the PSA is reduced considerably.[20] The spent sorbent, which is separated from the hydrogen product in a cyclone, is regenerated in the calciner at 850°C to produce a sequestration-ready CO_2 stream. At this stage, 8% of the spent sorbent is purged, and a makeup of PCC sorbent is added to maintain the high reactivity of the sorbent mixture toward CO_2 and sulfur capture. In this process, the pure hydrogen stream is produced at a high pressure of 20 atm, and the CO_2 is compressed to a pressure of 150 atm for transportation to the sequestration site. A calcium:carbon ratio of 1.3 is used to achieve almost 100% carbon and sulfur capture and sequestration from coal. This process leads to the production of 280 t/d of hydrogen from coal having an HHV of 29.2 MJ/kg with an efficiency of 62.3% (HHV).

Cogeneration of Hydrogen and Electricity. The process layout for this mode of operation is similar to the previous case where hydrogen is the only product obtained. In this scenario, for the cogeneration of hydrogen and electricity, 2,405 t/d of coal having an HHV of 29.2 MJ/kg is used for the production of 280 t/d of hydrogen. All energy required for the calcination of the sorbent is supplied by the combustion of the syngas in the calciner. The heat produced in the carbonation reactor through the exothermic WGS and carbonation reactions is used to produce high-temperature, high-pressure steam, which is used to generate electricity. In this scenario, 280 t/d of hydrogen is produced, from coal having an HHV of 29.2 MJ/kg, with an efficiency of 56.5% (HHV), and 81-MW electricity is produced with an efficiency of 10%. The two sce-

TABLE 4.12. Summary of the Schemes Investigated for the Production of Hydrogen Alone and for the Coproduction of Hydrogen and Electricity with a PSA

	Hydrogen	Hydrogen and Electricity
Coal Feed (t/day)	2,180	2,405
Carbon Capture (%)	100	100
Hydrogen (t/day)	280	280
Hydrogen (MW, HHV)	457	457
Net Power (MW)	0	81
Overall Efficiency (%HHV)	62.3	66.5

narios for the production of hydrogen or cogeneration of hydrogen and electricity from coal by the Calcium Looping Process, followed by a pressure swing adsorption unit, are summarized in Table 4.12.

Production of Hydrogen Having a Purity of 94%–98% without a PSA

Production of Hydrogen with Internal Heat Integration. The ASPEN PLUS flowsheet for the Calcium Looping Process for the production of 94%–98% pure hydrogen without a PSA is given in Figure 4.45. The syngas obtained from the Shell gasifier is split into two streams, one fed to the calcium looping reactor and the other to the calciner. In the case of hydrogen production without a PSA, 19% of the syngas is required to supply the energy for the endothermic calcination reaction, and the remaining 81% is fed to the calcium looping reactor for the production of high-purity hydrogen. This process in the absence of a PSA also leads to the production of 280 t/d of hydrogen from 2,155 t/d of coal, having an HHV of 29.2 MJ/kg, with an efficiency of 63% from coal (HHV).

Cogeneration of Hydrogen and Electricity. In this scenario, for the cogeneration of hydrogen and electricity in the absence of a PSA, 2,350 t/d of coal having an HHV of 29.2 MJ/kg is used for the production of 280 t/d of hydrogen at an efficiency of 57.8% (HHV). In addition to the hydrogen, 67.56 MW of electricity is produced with an efficiency of 8.5% from coal.

The two scenarios for the production of hydrogen or cogeneration of hydrogen and electricity from coal by the Calcium Looping Process in the absence of a PSA are summarized in Table 4.13. Comparing the process with and without the PSA unit, it can be seen that the hydrogen generation efficiencies for these processes with internal heat integration are almost the same. For the cogeneration of hydrogen and electricity, the overall efficiencies of the processes with and without the PSA unit are also similar. Although with the PSA unit the hydrogen generation efficiency (56.5%) is lower than without the PSA unit (57.8%), more electricity is produced with the PSA unit (81 MW) than without the PSA unit (67.5 MW).

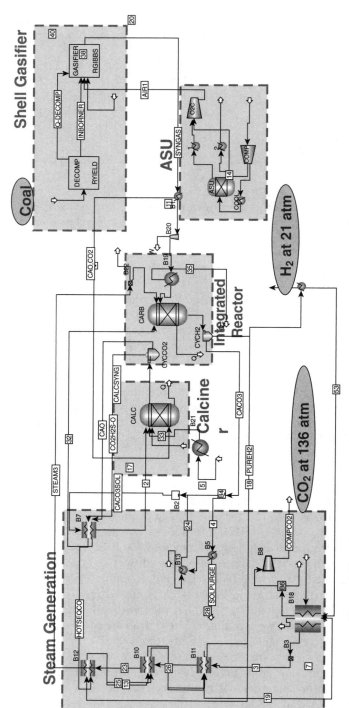

Figure 4.45. ASPEN simulation for the production of hydrogen using the Calcium Looping Process.

TABLE 4.13. Summary of the Schemes Investigated for the Production of Hydrogen Alone and for the Coproduction of Hydrogen and Electricity without a PSA

	Hydrogen	Hydrogen and Electricity
Coal Feed (t/day)	2,155	2,350
Carbon Capture (%)	100	100
Hydrogen (t/day)	280	280
Hydrogen (MW)	457	457
Net Power (MW)	0	67.56
Overall Efficiency (%HHV)	63	66.3

TABLE 4.14. Comparison of the Efficiency of the Hydrogen Production Process for Different Gasifiers

	Hydrogen	Hydrogen and Electricity
Shell	62.3%	66 (81 MW)
Lurgi (BGL)	55%	56 (32 MW)
GE	60%	63.6 (104.2 MW)

TABLE 4.15. Comparison of the Efficiency of the Hydrogen Production Process for Different Steam-to-Carbon Ratios (S:C Ratios)

S:C Ratio	Hydrogen	Hydrogen and Electricity
3:1	60%	63.6 (104.2 MW)
2:1	59.6%	61.5 (86 MW)
1:1	59%	60 (96.8 MW)

Comparison of the Efficiency of the Process for Different Gasifiers
The Calcium Looping Process is optimized for high-purity hydrogen production using syngas obtained from three different gasifiers, the Shell, Lurgi, and GE gasifiers. A comparison of the efficiencies obtained for the different gasifiers is given in Table 4.14. It is seen in Table 4.14 that the Calcium Looping Process in combination with the Shell gasifier has the highest efficiency because of the high efficiency of the dry-feed Shell gasifier.

Comparison of the Efficiency of the Process for Different Steam:Carbon Ratios
The effect of the reduction of steam in the Calcium Looping Process on the efficiency of the Coal-to-Hydrogen Process is given in Table 4.15. By decreasing the steam:carbon ratio in the calcium looping reactor, the efficiency in removing the sulfur and halide impurities by calcium oxide sorbent is improved.

The decrease in the efficiency of the process with a decrease in steam addition is very small and almost insignificant because of an insignificantly small decrease in the yield of hydrogen produced in the looping reactor. This conclusion is very similar to the experimental results observed in Figure 4.41, which shows that there is only an insignificant change in hydrogen purity with a decrease in the steam:carbon ratio.

4.7.4   Enhanced Coal-to-Liquid (CTL) Process with Sulfur and CO₂ Capture

The conventional CTL Process and the Syngas Chemical Looping CTL Process are described in Sections 4.2.3 and 4.6.2. When the Calcium Looping Process is used in the production of liquid fuels, a desired $H_2:CO$ feed ratio for the Fischer–Tropsch reactor is obtained through the conversion in the Calcium Looping Process of C_1–C_4 hydrocarbons and unconverted syngas generated from the Fischer–Tropsch reactor, and the syngas generated from the gasifier. This reaction scheme is illustrated in Figure 4.46, in which the unreacted syngas and light hydrocarbons from the Fischer–Tropsch reactor are mixed with the syngas from the gasifier and sent to the looping reactor system. Reforming the hydrocarbons [Reaction (4.7.8)] and shifting the syngas [Reaction (4.7.9)] in the presence of CaO lead to the adjustment of the ratio

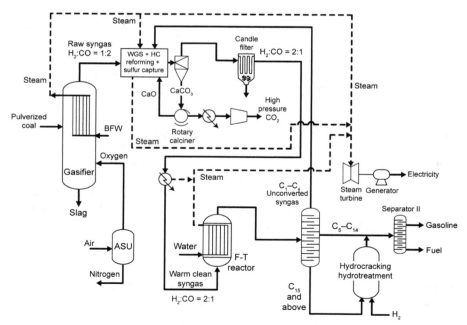

Figure 4.46. Calcium looping system integrated in a coal-to-liquid plant.

of the $H_2:CO$ in the syngas stream to 2. Various reactions occurring in this system are as given in the following reactions:

$$\text{Reforming:} \quad C_xH_y + xH_2O \rightarrow xCO + (y/2 + x)H_2 \qquad (4.7.8)$$

$$\text{WGS reaction:} \quad CO + H_2O \rightarrow H_2 + CO_2 \qquad (4.7.9)$$

$$\text{Carbonation:} \quad CaO + CO_2 \rightarrow CaCO_3 \qquad (4.7.10)$$

$$\text{Sulfidation:} \quad CaO + H_2S \rightarrow CaS + H_2O \qquad (4.7.11)$$

$$CaO + COS \rightarrow CaS + CO_2 \qquad (4.7.12)$$

$$\text{Chloridation:} \quad CaO + HCl \rightarrow CaCl_2 + H_2O \qquad (4.7.13)$$

$$\text{Calcination:} \quad CaCO_3 \rightarrow CaO + CO_2 \qquad (4.7.14)$$

4.8 Concluding Remarks

Coal gasification converts coal to syngas, which can be processed further to various products such as electricity, hydrogen, chemicals, and/or liquid fuels. This chapter describes the chemical looping processes using gaseous fuels in light of the well-developed coal gasification processes such as IGCC, coal-to-hydrogen, and indirect CTL processes. The gaseous fuels include syngas and C_1–C_4 hydrocarbons. The chemical looping processes possess significant advantages over the traditional processes from the viewpoint of energy conversion, process efficiency, carbon dioxide capture, process unit intensification, and, hence economics. There are two major chemical looping processes using gaseous fuels: iron-based and calcium-based chemical looping processes. The iron-based chemical looping processes have been practiced since the early 1900s, as demonstrated by the Lane and Messerschmitt steam-iron processes, the Bureau of Mines steam-iron processes, and the Institute of Gas Technology steam-iron processes. Renewed interest and recent efforts have been prompted by an important need to improve energy conversion efficiency and the CO_2 control techniques. Such efforts have been focused strongly on the design and synthesis of the oxygen-carrying particle, looping reactor design and operation, and process energy integration and optimization.

The SCL Process generates H_2 from syngas through reduction and oxidation reactions in a reducer, oxidizer, and combustor. The overall energy conversion scheme and control strategy of the pollutants such as H_2S, COS, and CO_2 in the SCL process can be simplified significantly and effectively over those in the traditional process. The SCL process is highlighted by a counter-current moving-bed contact mode of the gas and solid reactants; it exhibits optimum reactor operation under this contact mode for the reducer and the oxidizer in light of the reactant and product thermodynamic properties. The Calcium Looping Process, on the other hand, generates H_2 from syngas by integrating into various interlinking functions of a single stage reactor that are present in the multiple-stage reactors of the traditional processes. These functions include the WGS reaction, H_2S, COS, and HCl pollutant removal, and CO_2 capture. Furthermore, the Calcium Looping Process effectively can reduce excess steam consumption and produce a 99+% H_2 stream by the WGS reaction under high-pressure, catalyst-free conditions. Both the SCL and the Calcium Looping Processes readily can be integrated into the traditional CTL Process to produce liquid fuel at a substantially higher liquid fuel yield than does the traditional process. The process simulation and optimization based on the ASPEN PLUS further substantiate the effectiveness of the chemical looping processes using gaseous fuels.

References

1. Blankinship, S., "IGCC Rides a Regulatory Seesaw," *Power Engineering*, 111(11), 28–32 (2007).
2. Massachusetts Institute of Technology, "The Future of Coal: Options for a Carbon-Constrained World," http://web.mit.edu/coal/The_Future_of_Coal.pdf (2007).
3. Stiegel, G. J., and M. Ramezan, "Hydrogen from Coal Gasification: An Economical Pathway to a Sustainable Energy Future," *International Journal of Coal Geology*, 65(3–4), 173–190 (2006).
4. van der Ploeg, H. J., T. Chhoa, and P. L. Zuideveld, "The Shell Coal Gasification Process for the U.S. Industry," Proceedings of the Gasification Technology Conference, San Francisco CA, Oct 3–6 (2004).
5. Phillips, J., "Different Types of Gasifiers and Their Integration with Gas Turbines," U. S. Department of Energy NETL, http://www.netl.doe.gov/technologies/coalpower/turbines/refshelf/handbook/1.2.1.pdf. (2006).
6. Stultz, S. C., and J. B. Kitto, "Steam, Its Generation and Use," 40th edition, Babcock & Wilcox, Lynchburg, VA (1992).
7. Universal Industrial Gases, "Overview of Cryogenic Air Separation and Liquefier Systems," http://www.uigi.com/cryodist.html#Nitrogen. (2008).
8. Rhodes, C., S. A. Riddel, J. West, B. P. Williams, and G. J. Hutchings, "The Low-Temperature Hydrolysis of Carbonyl Sulfide and Carbon Disulfide: A Review," *Catalysis Today*, 59(3–4), 443–464 (2000).
9. Heaven, D., J. Mak, D. Kubek, M. Clark, and C. Sharp, "Synthesis Gas Purification in Gasification to Ammonia/Urea Plants," Proceedings of the Gasification Technologies Conference, San Francisco, CA, Oct 3–6 (2004).
10. Kanniche, M., and C. Bouallou, "CO_2 Capture Study in Advanced Integrated Gasification Combined Cycle," *Applied Thermal Engineering*, 27(16), 2693–2702 (2007).
11. Xue, E., M. OKeeffe, and J. R. H. Ross, "Water-Gas Shift Conversion Using a Feed with a Low Steam to Carbon Monoxide Ratio and Containing Sulphur," *Catalysis Today*, 30(1–3), 107–118 (1996).
12. Breckenridge, W., A. Holiday, J. O. Y. Ong, and C. Sharp, "Use of Selexol Process in Coke Gasification to Ammonia Project," Proceedings of the 50th LRGCC Conference, Norman, OK, February 27–March 1 (2000).

13. Bohlbro, H., "An Investigation on the Kinetics of Conversion of Carbon Monoxide with Water Vapour over Iron Oxide Based Catalysts," 2nd edition, Haldor Topsoe (1969).

14. Gerhartz, W., "Ullmann's Encyclopedia of Industrial Chemistry," 5th edition, VCH, Weinheim, Germany (1993).

15. Higman, C., and M. van der Burgt, "Gasification," 2nd edition, Gulf Professional, Burlington, MA, (2008).

16. David, N. S., "The Water-Gas Shift Reaction," *Catalysis Reviews—Science and Engineering*, 21, 275–318 (1980).

17. Li, Y. M., R. J. Wang, and L. Chang, "Study of Reactions over Sulfide Catalysts in CO-CO_2-H_2-H_2O System," *Catalysis Today*, 51(1), 25–38 (1999).

18. Andreev, A. A., V. J. Kafedjiysky, and R. M. Edreva-Kardjieva, "Active Forms for Water-Gas Shift Reaction on Nimo-Sulfide Catalysts," *Applied Catalysis A—General*, 179(1–2), 223–228 (1999).

19. Becker, P. D., H. Hiller, G. Hochgesa, and A. M. Sinclair, "Heavy Fuel Oil as Ammonia Plant Feedstock," *Chemical and Process Engineering*, 52(11) (1971).

20. Stöcker, J., M. Whysall, and G. Q. Miller, "30 Years of PSA Technology for Hydrogen Purification," UOP 2818, http://www.uop.com/objects/30YrsPSATechHydPurif.pdf (1998).

21. Fan, L.-S., "Gas-Liquid-Solid Fludization Engineering," Butterworth-Heinemann, Oxford, UK (1989).

22. Steynberg, A. P., and M. E. Dry, "Fischer-Tropsch Technology—Studies in Surface Science and Catalysis," Elsevier, Philadelphia, PA (2004).

23. Schulz, H., "Major and Minor Reactions in Fischer-Tropsch Synthesis on Colbalt Catalysts," *Topics in Catalysis*, 26(1–4), 73–85 (2003).

24. Hurst, S., "Production of Hydrogen by the Steam-Iron Method," *Journal of the American Oil Chemists' Society*, 16(2), 29–36 (1939).

25. Bleeker, M. F., S. R. A. Kersten, and H. J. Veringa, "Pure Hydrogen from Pyrolysis Oil Using the Steam-Iron Process," *Catalysis Today*, 127, 278–290 (2007).

26. Hacker, V., R. Fankhauser, G. Faleschini, H. Fuchs, K. Friedrich, M. Muhr, and K. Kordesch, "Hydrogen Production by Steam-Iron Process," *Journal of Power Sources*, 86(1–2), 531–535 (2000).

27. Lane, H., "Process for Production of Hydrogen," U.S. Patent 1,078,686 (1913).

28. Messerschmitt, A., "Process for Producing Hydrogen," U.S. Patent 971,206 (1910).

29. Teed, P. L., "The Chemistry and Manufacture of Hydrogen," Longmans, Green and Co. London, UK (1919).

30. Gasior, S. J., A. J. Forney, J. H. Field, D. Bienstock, and H. E. Benson, "Production of Synthesis Gas and Hydrogen by the Steam-Iron Process—Pilot Plant Study of Fluidized and Free-Falling Beds," Bureau of Mines, Washington, D.C. (1961).

31. Barr, F. T., "High-Pressure Hydrogen," U.S. Patent 2,449,635 (1948).

32. Reed, H. C., and C. H. Berg, "Hydrogen," U.S. Patent 2,635,947 (1953).

33. Institute of Gas Technology, "Development of the Steam-Iron Process for Hydrogen Production," U. S. Department of Energy EF-77-C-01-2435 (1979).

34. Velazquez-Vargas, L. G., T. Thomas, P. Gupta, and L.-S. Fan, "Hydrogen Production via Redox Reaction of Syngas with Metal Oxide Composite Particles," Paper Presented at the AIChE Annual Meeting, Austin, Texas (2004).

35. Gupta, P., L. G. Velazquez-Vargas, and L.-S. Fan, "Syngas Redox (SGR) Process to Produce Hydrogen from Coal Derived Syngas," *Energy & Fuels*, 21(5), 2900–2908 (2007).

36. Fan, L.-S., and C. Zhu, Principles of Gas-Solids Flows, Cambridge University Press, New York, NY (1998).

37. Chiesa, P., G. Lozza, A. Malandrino, M. Romano, and V. Piccolo, "Three-Reactors Chemical Looping Process for Hydrogen Production," *International Journal of Hydrogen Energy*, 33(9), 2233–2245 (2008).

38. Sanfilippo, D., F. Mizia, A. Malandrino, and S. Rossini, "Process for the Production of Hydrogen and the Co-Production of Carbon Dioxide," U.S. Patent 7,404,942B2 (2008).

39. Alcock, C. B., "Principles of Pyrometallurgy," Academic Press Inc., New York, NY (1976).

40. Gaskell, D. R., "Introduction to Metallurgical Thermodynamics," 2nd edition, Hemisphere Publications, Machias, NY (1981).

41. Elliott, J. F., R. M. Smailer, and R. L. Stephenson, "Direct Reduced Iron: Technology and Economics of Production and Use," Iron & Steel Society of AIME (1980).

42. ASPENTech, "ASPEN Physical Property System: Physical Property Methods and Models" (2006).

43. Perry, R. H., and D. W. Green, "Perry's Chemical Engineers' Handbook," 8th edition, McGraw-Hill, Columbus, OH (2008).

44. Barin, I., "Thermochemical Data of Pure Substances," VCH, Weinheim, Germany (1989).

45. Chase, M. W., "Nist-Janaf Thermochemical Tables," 4th edition, American Chemical Society, Washington, D.C. (1998).

46. Knacke, O., O. Kubaschewski, and K. Hesselmann, "Thermochemical Properties of Inorganic Substances," 2nd edition, Springer-Verlag, New York, NY (1991).

47. Mondal, K., H. Lorethova, E. Hippo, T. Wiltowski, and S. B. Lalvani, "Reduction of Iron Oxide in Carbon Monoxide Atmosphere—Reaction Controlled Kinetics," *Fuel Processing Technology*, 86(1), 33–47 (2004).

48. Ito, T., A. Tsutsumi, and T. Akiyama, "Coproduction of Iron and Hydrogen from Iron Carbide," *Journal of Chemical Engineering of Japan*, 36(7), 881–886 (2003).

49. Wu, S., M. A. Uddin, and E. Sasaoka, "Effect of Pore Size Distribution of Calcium Oxide High-Temperature Desulfurization Sorbent on Its Sulfurization and Consecutive Oxidative Decomposition," *Energy & Fuels*, 19(3), 864–868 (2005).

50. Cachu, S., and W. D. Gunter, "Acid Gas Injection in the Alberta Basin, Canada: A CO_2 Storage Experience," *Geological Society Special Publication*, 233, 225–234 (2004).

51. Thomas, T., L.-S. Fan, P. Gupta, and L. G. Velazquez-Vargas, "Combustion Looping Using Composite Oxygen Carriers," U.S. Provisional Patent Series No.11/010,648 (2004); U.S. Patent 7,767,191 (2010).

52. Gupta, P., L. G. Velazquez-Vargas, C. Valentine, and L.-S. Fan, "Moving Bed Reactor Setup to Study Complex Gas-Solid Reactions," *Review of Scientific Instruments*, 78(8), 0851061–0851067 (2007).

53. Simell, P. A., J. K. Leppalahti, and J. B. S. Bredenberg, "Catalytic Purification of Tarry Fuel Gas with Carbonate Rocks and Ferrous Materials," *Fuel*, 71(2), 211–218 (1992).

54. Gueret, C., M. Daroux, and F. Billaud, "Methane Pyrolysis: Thermodynamics," *Chemical Engineering Science*, 52(5), 815–827 (1997).

55. Shah, V. A., and T. L. Huurdeman, "Synthesis Gas Treating with Physical Solvent Process Using Selexol Process Technology," *Ammonia Plant Safety & Related Facilities*, 30, 216–224 (1990).

56. Shelton, W., and J. Lyons, "Shell Gasifier IGCC Base Cases," U. S. Department of Energy NETL, http://www.netl.doe.gov/technologies/coalpower/gasification/pubs/pdf/system/shell3x_.pdf (2000).

57. Tomlinson, G., and D. Gray, "Chemical Looping Process in a Coal-to-Liquids Configuration," U. S. Department of Energy NETL, U.S.DOE/NETL-2008/130 (2007).

58. Gao, L., N. Paterson, D. Dugwell, and R. Kandiyoti, "Zero-Emission Carbon Concept (ZECA): Equipment Commissioning and Extents of the Reaction with Hydrogen and Steam," *Energy and Fuels*, 22(1), 463–470 (2008).

59. Lin, S. Y., Y. Suzuki, H. Hatano, and M. Harada, "Developing an Innovative Method, Hypr-Ring, to Produce Hydrogen from Hydrocarbons," *Energy Conversion and Management*, 43, 1283–1290 (2002).

60. Andrus, H. E., G. Burns, J. H. Chiu, G. N. Liljedahl, P. T. Stromberg, and P. R. Thibeault, "Hybrid Combustion-Gasification Chemical Looping Power Technology Development," ALSTOM technical report DE-FC26-03NT41866 (2006).

61. Rizeq, R. G., J. West, A. Frydman, R. Subia, R. Kumar, and V. Zamansky, "Fuel-Flexible Gasification-Combustion Technology for Production of H_2 and Sequestration-Ready CO_2," Annual DOE Technical Progress Report (2002).

62. Ziock, H. J., K. S. Lackner, and D. P. Harrison, "Zero Emission Coal Power, a New Concept," http://www.netl.doe.gov/publications/proceedings/01/carbon_seq/2b2.pdf (2001).

63. Lin, S. Y., M. Harada, Y. Suzuki, and H. Hatano, "Process Analysis for Hydrogen Production by Reaction Integrated Novel Gasification (Hypr-Ring)," *Energy Conversion and Management*, 46(6), 869–880 (2005).

64. Lin, S., M. Harada, Y. Suzuki, and H. Hatano, "Continuous Experiment Regarding Hydrogen Production by Coal/CaO Reaction with Steam (I) Gas Products," *Fuel*, 83(7–8), 869–874 (2004).

65. Lin, S., M. Harada, Y. Suzuki, and H. Hatano, "Comparison of Pyrolysis Products between Coal, Coal/CaO, and Coal/Ca(OH)2 Materials," *Energy & Fuels*, 17(3), 602–607 (2003).

66. Hufton, J. R., S. Mayorga, and S. Sircar, "Sorption-Enhanced Reaction Process for Hydrogen Production," *AIChE Journal*, 45, 248–256 (1999).

67. Akiti, T. T., K. P. Constant Jr., L. K. Doraiswamy, and T. D. Wheelock, "A Regenerable Calcium-Based Core-in-Shell Sorbent for Desulfurizing Hot Coal Gas," *Industrial & Engineering Chemistry Research*, 41, 587–597 (2002).

68. Ortiz, A. L., and D. P. Harrison, "Hydrogen Production Using Sorption-Enhanced Reaction," *Industrial & Engineering Chemistry Research*, 40(23), 5102–5109 (2001).

69. Balasubramanian, B., A. L. Ortiz, S. Kaytakoglu, and D. P. Harrison, "Hydrogen from Methane in a Single-Step Process," *Chemical Engineering Science*, 54, 3543–3552 (1999).

70. Iyer, M. V., H. Gupta, B. B. Sakadjian, and L.-S. Fan, "Multicyclic Study on the Simultaneous Carbonation and Sulfation of High-Reactivity CaO," *Industrial & Engineering Chemistry Research*, 43(14), 3939–3947 (2004).

71. Johnsen, K., H. J. Ryu, J. R. Grace, and C. J. Lim, "Sorption-Enhanced Steam Reforming of Methane in a Fluidized Bed Reactor with Dolomite as CO_2 Acceptor," *Chemical Engineering Science*, 61, 1195–1202 (2006).

72. Iyer, M., S. Ramkumar, D. Wong, and L.-S. Fan, Enhanced Hydrogen Production with in-Situ CO_2 Capture in a Single Stage Reactor, Proceedings of the 23rd Annual International Pittsburgh Coal Conference, Pittsburgh, Pennsylvania, Septermber 25–28 (2006).

73. Agnihotri, R., S. K. Mahuli, S. S. Chauk, and L.-S. Fan, "Influence of Surface Modifiers on the Structure of Precipitated Calcium Carbonate," *Industrial & Engineering Chemistry Research*, 38, 2283–2291 (1999).

74. Ghosh-Dastidar, A., S. K. Mahuli, R. Agnihotri, and L.-S. Fan, "Investigation of High-Reactivity Calcium Carbonate Sorbent for Enhanced SO_2 Capture," *Industrial & Engineering Chemistry Research*, 35, 598–606 (1996).

75. Wei, S.-H., S. K. Mahuli, R. Agnihotri, and L.-S. Fan, "High Surface Area Calcium Carbonate: Pore Structural Properties and Sulfation Characteristics," *Industrial & Engineering Chemistry Research*, 36, 2141–2148 (1997).

76. Fan, L.-S., A. Ghosh-Dastidar, and S. K. Mahuli, "Calcium Carbonate Sorbent and Methods of Making and Using Same," U.S. Patent 5,779,464 (1998).

77. Gupta, H., and L.-S. Fan, "Carbonation-Calcination Cycle Using High Reactivity Calcium Oxide for Carbon Dioxide Separation from Flue Gas," *Industrial & Engineering Chemistry Research*, 41, 4035–4042 (2002).

78. Chauk, S. S., R. Agnihotri, R. A. Jadhav, S. K. Misro, and L.-S. Fan, "Kinetics of High-Pressure Removal of Hydrogen Sulfide Using Calcium Oxide Powder," *AIChE Journal*, 46, 1157–1167 (2000).

79. Weinell, C. E., P. I. Jensen, K. Dam-Johansen, and H. Livbjerg, "Hydrogen Chloride Reaction with Lime and Limestone: Kinetics and Sorption Capacity," *Industrial & Engineering Chemistry Research*, 31, 164–171 (1992).

80. Gupta, H., M. Iyer, B. B. Sakadjian, and L.-S. Fan, "Enhancing Hydrogen Production with *In-Situ* CO_2 Separation Using CaO/Catalyst Systems," Paper Presented at the AIChE Annual Meeting, Austin, Texas (2004).

CHAPTER 5

CHEMICAL LOOPING GASIFICATION USING SOLID FUELS

F. Li, L. Zeng, D. Sridhar, L. G. Velazquez-Vargas, and L.-S. Fan

5.1 Introduction

The gasification process converts carbonaceous fuels, such as coal and biomass, to syngas, which then can be processed to form various products including H_2, chemicals, heat/electricity, and liquid fuels. The ease of CO_2 separation using gasification, in addition to the variability in product generation, renders gasification a favorable coal-conversion process, particularly under a carbon-constrained situation.[1,2] Before gasification processes can gain wide application, however, challenges are associated with high capital and operating costs resulting from high-pressure and high-temperature gasification conditions, the use of oxygen, and syngas cleaning. Recent studies indicated that an integrated gasification combined cycle (IGCC) system required 6%–10% more capital investment than an ultra-supercritical PC plant (USPC).[1,3] Both plants yield similar energy conversion efficiencies. Although CO_2 capture from a coal gasification process can be conducted more efficiently as compared with that from a coal combustion process, CO_2 capture does consume a significant amount of parasitic energy. For example, employing the available CO_2 capture techniques will increase the cost of electricity by 25%–45% in a coal-based IGCC plant.[2,4–8] Furthermore, the efficiencies of traditional coal gasification technologies are far from optimized. The energy conversion efficiency for hydrogen production in a gasification-water–gas shift (WGS) hydrogen plant

Chemical Looping Systems for Fossil Energy Conversions, by Liang-Shih Fan
Copyright © 2010 John Wiley & Sons, Inc.

is reported to be at 64% (high heating value, HHV) without CO_2 capture.[9] This efficiency is much lower than the 70%–80% for a steam methane reforming process.[10] Clearly, the coal gasification processes could be more streamlined in light of process intensification, incorporating carbon capture to reduce the capital and operating costs and the carbon dioxide emission and to increase the energy conversion efficiency of the processes. Details of the traditional gasification processes using gasifiers to process carbonaceous fuels are discussed in Chapters 1 and 4.

The chemical looping gasification of carbonaceous solid fuels can be realized by using one of two approaches: (1) direct gasification using a high-temperature CO_2-reactive sorbent, (2) direct gasification using a high-temperature oxygen carrier. The first approach, chemical looping gasification processes using calcium-based sorbents, is discussed in Section 5.2. The second approach, iron-based chemical looping gasification processes, is discussed in Sections 5.3–5.5.

The processes for chemical looping gasification using calcium-based sorbents include the CO_2 Acceptor Process that was developed in the 1970s. Other processes that are being developed, considered, or have recently been developed include the HyPr-Ring Process, the Zero Emission Coal Alliance Process, the ALSTOM Hybrid Combustion–Gasification Process, and the Fuel-Flexible Advanced Gasification–Combustion Process.

In Section 5.3, the Coal-Direct Chemical Looping (CDCL) Process using iron-based oxygen carriers is illustrated. Two configurations for process operation are considered. The challenges to the CDCL Process and the strategy for improvements of several operational issues are discussed in Section 5.4. These issues include oxygen carrier particle reactivity and char reaction enhancement, reactant contact mode, conversions of the reducer, performance of the oxidizer and the combustor, fates of pollutants and ash, energy management, and heat integration. In Section 5.5, models used for the ASPEN simulation and the simulation results for the CDCL Process are presented. High-temperature oxygen carriers are discussed in Section 3.3.5 for applications in chemical looping combustion. In these applications, fluidized-bed reactors are used as the reducer. In this chapter, countercurrent gas–solid moving bed reactors are used as the reducer for applications in chemical looping gasification.

5.2 Chemical Looping Gasification Processes Using Calcium-Based Sorbent

The applications of the calcium-based chemical looping gasification process can be traced to early times. One notable application is the Carbon Dioxide Acceptor Process, developed in the 1960s and 1970s. This process ascertains the validity of the chemical looping concept using calcium-based sorbents and forms the basis of recent developments in calcium-based chemical looping gasification processes. The calcium-based looping processes to be covered are the Carbon Dioxide Acceptor, HyPr-Ring, ZECA, ALSTOM Hybrid Combustion–Gasification, and GE Fuel Flexible processes.

5.2.1 CO₂ Acceptor Process

The CO_2 Acceptor Process was developed by the Consolidation Coal Company and later Conoco Coal Development Company (Ogden, UT).[11] The American Gas Association, U.S. Department of Interior, and the Energy Research and Development Administration were among the major sponsors for the development of this process. The CO_2 Acceptor Process was designed to produce synthetic pipeline gas from pulverized lignite or subbituminous coal. There are five main operational blocks comprising this process, the feedstock preparation block, the gas synthesis block, the gas cleanup block, the methanation block, and the utility block. Figure 5.1 shows the gas synthesis block, which is the key to this process.

In the reactor system of the gas synthesis block shown in Figure 5.1, preheated coal is ground to ~150 μm before it is fed to the gasifier. The gasifier is a fluidized bed with steam as the fluidizing gas. It is operated at 800°C–850°C and 10 atm. The relatively low temperatures in the gasifier enable the use of high-sodium coal, which tends to fuse, forming silica and alumina solid

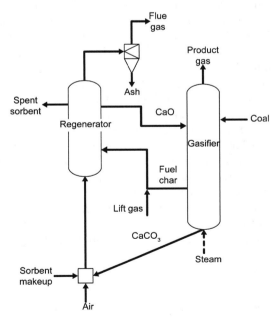

Figure 5.1. Schematic diagram of the reactor system in the gas synthesis block of the CO_2 Acceptor Process.[11]

aggregates at high temperatures (>870°C). The acceptor, which is the calcined limestone or dolomite sorbent, is fed from the top of the gasifier. The sorbent reacts with CO_2 produced from the combined steam–carbon reaction and WGS reaction. The exothermic carbonation reaction of the sorbent provides the heat for the endothermic steam–carbon reaction and also drives the water–gas shift reaction forward, thereby increasing the hydrogen content in the product gas. The major reactions that occur in the gasifier include

Steam–Carbon Reaction: $\quad C + H_2O \rightarrow CO + H_2$ $\qquad\qquad$ (5.2.1)

WGS Reaction: $\qquad\qquad CO + H_2O \rightarrow CO_2 + H_2$ $\qquad\qquad$ (5.2.2)

Carbonation: $\qquad\qquad CaO + CO_2 \rightarrow CaCO_3$ $\qquad\qquad$ (5.2.3)

$\qquad\qquad\qquad\qquad CaO \cdot MgO + CO_2 \rightarrow CaCO_3 \cdot MgO$ $\quad$ (5.2.4)

The hydrogen-rich gas obtained from the gasifier, containing ~20% CO and CO_2, is subsequently quenched, purified, and methanated, and can be transported by pipelines. The spent sorbent discharged from the bottom of the gasifier then is fed to the regenerator where the spent sorbent is regenerated at 1,010°C and 10 atm through the calcination reaction:

Calcination: $\qquad\qquad CaCO_3 \rightarrow CaO + CO_2$ $\qquad\qquad$ (5.2.5)

The heat for calcination is provided by combusting the residual char, which is discharged from the gasifier to the regenerator. The regenerator is also a fluidized bed. Air is used as both a fluidizing and an oxidizing gas for the regenerator. The regenerated sorbent is sent back to the gasifier to complete the loop. The CO_2-containing flue gas is generated in the regenerator.

The CO_2 Acceptor Process was proved technically feasible after being tested successfully in a pilot plant in Rapid City, SD. The designed capacity of the pilot plant was 40 t/d. During a period of six years that ended in 1977, an accumulative operation of 13,000 hours was achieved, with the longest continuous run of ~2,300 hours. A total of 6,500 t of dry coal of various ranks including three North Dakota lignites, one Texas lignite, and three subbituminous coals were tested in the facility. As noted, the gas-synthesis reactor system, which produces hydrogen-rich gas from coal with the aid of the calcium-based CO_2 sorbent or acceptor, characterizes the key innovation of the CO_2 Acceptor Process. The detailed description of a pilot-scale gas synthesis system, which includes the regenerator and the gasifier, is given in the following sections.

Regenerator
The regenerator decomposes the spent sorbent from the gasifier via coal char combustion. It was estimated that 2.04–2.34 MW_{th} was released by char combustion in the pilot regenerator.[11] The fluidized bed regenerator provides

some, though not full, mixing of the char and the spent sorbent. The heat released by char reactions, coupled with endothermic calcination, leads to a ~17°C gradient throughout the fluidized-bed regenerator. The feed to the regenerator contains CaS, part of which will be oxidized to $CaSO_4$ in the presence of CO_2 and O_2 as shown by

$$CaS + 4CO_2 \rightarrow CaSO_4 + 4CO \qquad (5.2.6)$$

$$CaS + 2O_2 \rightarrow CaSO_4 \qquad (5.2.7)$$

The concurrent presence of CaS and $CaSO_4$ in the regenerator generates calcium oxide and SO_2, as shown by

$$CaS + 3CaSO_4 \rightarrow 4CaO + 4SO_2 \qquad (5.2.8)$$

This reaction occurs through a series of intermediate steps. At temperatures above 955°C, a transient liquid of a eutectic mixture of $CaSO_4$ and CaS is formed as an intermediate, which solidifies and is deposited on the regenerator walls. Thus, a reducing environment with a CO concentration of 1%–5% is maintained in the regenerator to prevent the formation of the transient liquid.[11] Consequently, the gas exiting from the regenerator will contain a small amount of CO in the CO_2 stream. Further treatment of the CO-containing exit stream will be needed. The temperature in the regenerator is controlled by adjusting the air flow rate. During the regenerator operation, nearly all the char fed to the regenerator is consumed. The coal ash is entrained with the spent air and collected in the external cyclone-lock hopper located at the gaseous product outlet of the regenerator. It was determined that 99% of the carbon in coal was converted by the gasifier and the regenerator.

A minimum temperature of 471°C in the regenerator is required for the immediate combustion of char. At start-up, part of the flue gas from the regenerator, reheated in a natural gas furnace, was used to increase the temperature of the regenerator to ~538°C to initiate char combustion. After the initiation of combustion, the heat released from combustion increases the temperature of the regenerator. At steady state, the regenerator was operated at 1,010°C. Under such a high temperature, spent sorbent is decomposed, releasing CO_2 via the calcination reaction. The regenerated sorbent is discharged from the regenerator through two outlets: one outlet purges a predetermined amount of sorbent, whereas the other outlet discharges the remainder into the gasifier where the regenerated hot sorbent is used for coal gasification. The heat from the hot sorbent is partially used to balance the heat requirement of the endothermic steam gasification reaction in the gasifier.

Gasifier
The gasifier is operated using a fluidized bed with continuous solids feeding. The hot sorbent particles from the regenerator are fed to the upper part of

the gasifier, while coal and steam are injected to the middle and lower part of the gasifier, respectively. In the gasifier, the CaO sorbent is converted to calcium carbonate. The exothermic heat of the carbonation reaction is used to compensate for the endothermic gasification reaction. Examining the converted sorbent or the spent sorbent settled at the bottom of the gasifier before its transport by gravity to the engager pot (see Figure 5.1) shows that the extent of conversion of the sorbent in the gasifier from CaO to calcium carbonate is high. The spent sorbent in the engager pot then is pneumatically transported back to the regenerator by air. The fresh makeup sorbent also is provided to the engager pot.

The presence of the reaction products such as H_2, CO, CO_2, and CH_4 was found to limit the rate of gasification.[11] The rate of gasification at the lower part of the gasifier was much larger than at the higher part because of the difference in gas composition. For example, 62% of the char was gasified at the bottom section of the bed, whereas only 12% additional char was gasified in the middle section. To create a more uniform reaction rate throughout the gasifier, a portion of the product gas from the gasifier was recycled to the bottom of the gasifier. Such a product gas-recycling step assists in moderating the rate of the reaction in the gasifier where solids are not well mixed. Furthermore, the presence of the recycled gases decreases the partial pressure of the steam and, hence decreases the formation of $Ca(OH)_2$ in the gasifier. Thus, the formation of the eutectic mixture of $CaO–Ca(OH)_2–CaCO_3$ is minimized. It was found that lignite char was distinctively more reactive than subbituminous char.

Synthesis gas and synthetic natural gas are produced continuously in this process. A typical synthesis gas composition obtained from the gasifier is given in Table 5.1. The synthesis gas produced from the gasifier is used in the methanation system, which is not shown in Figure 5.1, for the production of synthetic natural gas. A typical composition of synthetic natural gas obtained from the CO_2 Acceptor Process is given in Table 5.2. As shown in the table, the heating value of the synthetic natural gas obtained from the pilot plant exceeds 900 Btu/SCF ($33.5 MJ/Nm^3$). The pilot plant studies also examined the factors that affected the activity of the sorbent and its environmental impact.

Process Testing
The activity of the CO_2 sorbent, which is expressed by the ratio of the weight of CO_2 absorbed to the weight of the fresh (unreacted) sorbent, was the key

TABLE 5.1. A Typical Composition of the Hydrogen-Rich Synthesis Gas from the Gasifier

Gas Type	CH_4	H_2	CO	CO_2	N_2	H_2O	HHV (Btu/SCF; MJ/Nm³)
Concentration (%)	11.4	65.6	15.7	4.7	0.7	1.9	379.1; 14.1

TABLE 5.2. A Typical Composition of the Synthetic Natural Gas from the Methanation System

Gas Type	CH_4	H_2	CO	CO_2	N_2	HHV (Btu/SCF; MJ/Nm^3)
Concentration (%)	92.6	4.7	0.01	0.5	2.2	>900; 33.5

parameter in determining the technical feasibility of this process. The average acceptor activity must exceed a certain level for the process to be in heat balance because the system heat requirements are met by the sensible and reaction heat released by the acceptor at a given CO_2 removal rate. The minimum activity for the CO_2 Acceptor Process was 0.26 for dolomite sorbent and 0.14 for limestone sorbent. Through the pilot plant testing, an activity of 0.35 was achieved using the dolomite sorbent, which exceeds the minimum activity requirement.

For the process to be economically attractive, the sorbent needs to maintain a high activity with a minimal purge rate. It was observed that the activity of the acceptor decreases when the number of carbonation–calcination cycles increases. Other important factors that affect the activity of the acceptor include the acceptor residence time and the reactor operating temperatures in the gasifier. The decrease in the acceptor reactivity was attributed to the CaO crystalline growth. It was hypothesized that the calcium atom is relatively mobile at the gasifier/regenerator operating conditions, especially when the operating temperature is high.[11] The high mobility of the calcium atom leads to fast CaO crystalline growth, which forms a "bridge" between closely placed CaO crystals during the carbonation reaction. The crystals formed in this bridging effect are highly stable. Such CaO crystals remain intact during the calcination reaction and tend to grow continuously in size and, hence reduce the gas diffusion rate. This observation is consistent with the recent experimental studies of the underlying mechanism of the declining carbonation rate caused by the sintering effect, as described in Chapter 2. Thus, the reactivity of the acceptor decreases across time, especially at high temperatures. The crystalline growth effect was more significant in the limestone acceptor than in the dolomite acceptor. To increase the recyclability of the acceptor, two approaches—acceptor reactivation and acceptor structure modifications—were tested in the pilot-scale facility. Satisfactory results for both approaches were reported. Beyond these attempts at improving acceptor reactivity and recyclability, several strategies were adopted to enhance the energy conversion efficiency of the CO_2 Acceptor Process. These strategies include the utilization of high-pressure exhaust gas for air compression and recovery of heat from exhaust gas for steam generation. Through the pilot testing, the metallurgical aspect of the reactor materials and their feasibility in usage were determined.

5.2.2 HyPr-Ring Process

The HyPr-Ring Process, currently under development in Japan as noted in Section 1.7, is similar to the CO_2 Acceptor Process. Both processes promote fuel conversion using CaO and/or $Ca(OH)_2$ sorbents.[12] While the CO_2 Acceptor Process was aimed at synthetic natural gas production, the goal of the HyPr-Ring Process is the production of high-purity hydrogen. The HyPr-Ring Process consists principally of two units, the gasifier and the regenerator, as shown in Figure 5.2.

Coal is introduced, along with CaO and steam, in the gasifier where the following reactions take place:

$$CaO + H_2O \rightarrow Ca(OH)_2 \qquad (5.2.9)$$

$$C + H_2O \rightarrow CO + H_2 \qquad (5.2.10)$$

$$CO + H_2O \rightarrow CO_2 + H_2 \qquad (5.2.11)$$

$$CaO + CO_2 \rightarrow CaCO_3 \qquad (5.2.12)$$

$$Ca(OH)_2 + CO_2 \rightarrow CaCO_3 + H_2O \qquad (5.2.13)$$

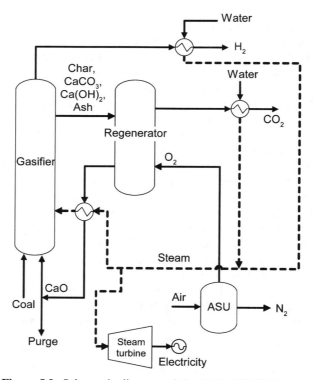

Figure 5.2. Schematic diagram of the HyPr-RING Process.

The solids mixture from the gasifier, which contains unreacted coal char, $CaCO_3$, $Ca(OH)_2$, and ash, is fed into the regenerator where the following reactions take place:

$$C + O_2 \rightarrow CO_2 \qquad (5.2.14)$$

$$CaCO_3 \rightarrow CaO + CO_2 \qquad (5.2.15)$$

The CaO sorbent and ash from the regenerator then are recycled back to the gasifier. Before re-entering the gasifier, a portion of the solids mixture is discharged, and fresh makeup is added. This purge step helps prevent ash accumulation and maintain the sorbent reactivity.

Figure 5.3 presents the pilot unit located at Japan's Coal Energy Center with a coal feeding rate of 3.5 kg/h. Figure 5.4 shows the typical concentrations of H_2, CH_4, C_2H_4, C_2H_6, CO and CO_2 in a typical continuous operation of the HyPr-Ring Process. The concentrations of H_2S, NH_3 and HCN in the syngas stream were reported to be 2.2 ppm, ~0 ppm, and 3.2 ppm, respectively.

The HyPr-Ring Process has been studied for hydrogen production. A comparison of the quantity of the synthesis gas produced from the pyrolysis of coal mixed with CaO and $Ca(OH)_2$ revealed that the extent of pyrolysis of coal is, in the order from high to low, coal/$Ca(OH)_2$, coal/CaO, and coal at pressures of 30–60 atm. This comparison reflects the advantages of using $Ca(OH)_2$ as a sorbent. Water is supplied at a high temperature so that calcium is in the form of $Ca(OH)_2$ and aids in the reforming reaction. Furthermore, CaO from the decomposition of $Ca(OH)_2$ enhances the pyrolysis reaction by

Figure 5.3. HyPr-Ring pilot-scale demonstration unit with 3.5-kg/h coal process rate. With permission from Dr. Shiying Lin of Japan Coal Energy Center; photo presented at the 3rd International Workshop on *In Situ* CO_2 Removal, ISCR, 2007.

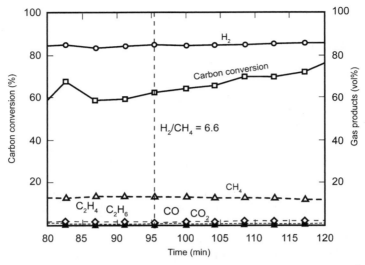

Figure 5.4. Typical gas concentration from the HyPr-Ring gasifier.[13]

the removal of CO_2 from the product gases. CaO also has a catalytic effect on the decomposition of tar, which further increases the gaseous product yield. The composition of hydrogen in the gaseous products was the highest in the temperature range of 650°C–700°C. This temperature range conforms to the optimal temperatures for the combined WGS and carbonation reactions. An increase in pressure results in an increase in the hydrogen purity because the combined WGS and carbonation reactions are favored at higher pressures.[14]

Studies of hydrogen generation from a mixture of pulverized coal and CaO with high-pressure steam in a fixed bed reactor reveal that, at a temperature of 700°C, the hydration of CaO occurs at a steam partial pressure higher than 30 atm. The yield of hydrogen was doubled with an increase in temperatures from 650°C to 700°C, and the yield of hydrogen increased by 1.5 times with an increase in the total pressure from 10 to 60 atm and in the steam partial pressure from 7 to 42 atm.[15] Gasification using pellets containing a mixture of coal and CaO in a fixed-bed reactor reveals that, although there is a decrease in the volume of the product gas, the composition of the product gas from pellet gasification is similar to that from gasification using the pulverized coal and CaO mixture. Furthermore, in the gasifier, pellets retain their size and morphology at a gasification temperature of 650°C. At 700°C, however, the pellets are separated into two distinct parts: a dark part containing carbon and a light part containing a mixture of CaO, $Ca(OH)_2$ and $CaCO_3$, which form a eutectic melting mixture of solids. Recycling the calcium oxide pellets between the gasifier and the regenerator results in a constant hydrogen yield over four cycles, beyond which, CaO is significantly deactivated because of the deposition of ash and inerts on the surface of the pellet.[16]

Studies of the HyPr-Ring Process in a fluidized bed reactor at 650°C and 50 atm revealed that the hydration of CaO and the carbonation of the $Ca(OH)_2$ occurred in series, resulting in a gaseous product containing 76% H_2, 17% CH_4, 2% C_2H_4, 3% C_2H_6, and 2% CO_2. As the time scale for the combined hydration, WGS, and carbonation reactions is 1–2 second, which is much shorter than that for the gasification reaction, the CO in the product gases is converted completely to CO_2, and almost all CO_2 is removed by CaO or $Ca(OH)_2$. An increase in the total pressure of the gasifier thermodynamically favors the formation of methane,[17] but the rates of the combined gasification, WGS, and carbonation reactions are higher than those of the methanation reactions, especially at high pressures. Thus, when the residence time of the reactants in the reactor is short, a high gasifier pressure leads to the enhancement of hydrogen production without inducing significant methane formation.[18] The carbon conversion was 60% near the entrance area of the fluidized bed reactor and 80% at the outlet of the reactor. The eutectic melting of the $Ca(OH)_2$, $CaCO_3$ and CaO mixture, which occurs at 700°C in the fixed bed experiments with the pelletized coal and CaO, was not present in the fluidized bed at 650°C. In the fluidized bed reactor, however, even at a low temperature of 650°C, particle growth occurs from crystallization and cohesion of calcium compounds.[18] Studies of the effect of various sorbents including $CaCO_3$, $CaOSiO_2$, MgO, SnO, and Fe_2O_3 on hydrogen production indicate that high-purity hydrogen is obtained only with $CaCO_3$ and $CaOSiO_2$ sorbents, and CO_2 cannot completely be removed from the product gas using the other sorbents.[17] For different Ca-based sorbents, the rate of hydration was found to decrease with an increase in CaO content. Furthermore, the initial rate of hydration increases with an increase in the surface area of the sorbent, whereas the final rate increases with an increase in the porosity.[19]

Studies of the regeneration of the spent calcium sorbent in a 100% CO_2 environment and the reactivity of the calcined sorbent for the hydration and carbonation reactions reveal that, for a calcination residence time of 70 minutes in a fluidized-bed reactor, 73% of $CaCO_3$ was calcined at 920°C, 95% was calcined at 1,020°C, and almost 100% was calcined above 1,020°C. The rates of the hydration and carbonation reactions decrease with an increase in the calcination temperature. Furthermore, the extent of carbonation of CaO decreases with an increase in the calcination temperature, from 60% at 950°C to 52% at 1,000°C, and 40% at 1,020°C. Thus, to improve the extent of carbonation, hydration of CaO is desirable to improve the porosity of the sorbent.[20] Calcination in the presence of steam yields a sorbent that requires only half the time for hydration compared with a sorbent obtained from calcination in the presence of 100% CO_2. The extent of carbonation of completely calcined CaO also is increased from 40% for 100% CO_2 calcination to 70% for steam calcination (0.4 atm CO_2 partial pressure and

30 atm total pressure).[21] Thus, by the combination of steam calcination and hydration, the sorbent loading in the process can be reduced significantly.

For generating hydrogen with high purity for fuel cell applications, extensive cleaning of H_2S, CH_4 and other pollutants or by-products from the hydrogen stream will be necessary. The energy efficiency, defined as the HHV of the H_2 produced divided by the HHV of the coal converted, for this process was reported to be 77%.[12] Note that the difference between the CO_2 Acceptor Process and the HyPr-Ring Process lies in the gasifier operating conditions. Comparing the operating conditions of the gasifier used in the CO_2 Acceptor Process (i.e., 800°C–850°C and 10 atm), the operating conditions of the gasifier used in the HyPr-Ring Process have a lower operating temperature (650°C) and a higher operating pressure (30 atm). The lower temperature and higher pressure in the HyPr-Ring Process gasifier thermodynamically favor the carbonation reaction, thereby further enhancing the hydrogen production. Moreover, an excess of steam is used in the HyPr-Ring Process to enhance the reactivity of the CaO sorbent by refreshing the pore structure of the particles.

5.2.3 Zero Emission Coal Alliance Process

The Zero Emission Coal Alliance Process, or ZECA Process, as noted in Section 1.7, was proposed by Klaus Lackner and H. Ziock.[22] Figure 5.5 outlines the process. In this process, coal first is converted to methane by reacting with hydrogen in a gasifier:

$$C + 2H_2 \rightarrow CH_4 \qquad (5.2.16)$$

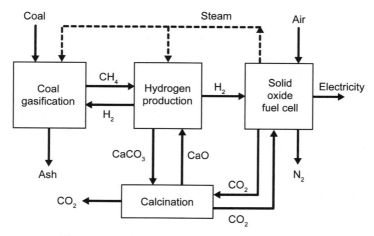

Figure 5.5. The schematic of the ZECA Process.

This hydrogasification step also produces light hydrocarbons. The methane and light hydrocarbons then are sent to the reformer. Steam and calcium oxide sorbents are introduced to the reformer to convert the hydrocarbons into hydrogen via sorbent enhanced reforming reactions:

$$CH_4 + H_2O \rightarrow CO + 3H_2 \qquad (5.2.17)$$

$$CO + H_2O \rightarrow CO_2 + H_2 \qquad (5.2.18)$$

$$CO_2 + CaO \rightarrow CaCO_3 \qquad (5.2.19)$$

The hydrogen gas generated in the reformer is split into two streams: one stream is recycled to the hydrogasifier, and the other is sent to a solid oxide fuel cell (SOFC) for power generation. The spent sorbent, consisting mainly of calcium carbonate, is regenerated in a calciner. Carbon dioxide readily is separated in this step:

$$CaCO_3 \rightarrow CaO + CO_2 \qquad (5.2.5)$$

The heat required for the calcination reaction is provided by the waste heat from the solid oxide fuel cell system, which is operated using hydrogen from the reformer. The fuel cell also generates steam, which is recycled back to the hydrogasifier and the reformer for methane and hydrogen generation, respectively. Stoichiometrically, to convert 1 mol of carbon, 2 mol of hydrogen gas (Reaction 5.2.16) is consumed in the gasifier, and 4 mol of hydrogen gas is generated in the reformer (Reactions 5.2.17 and 5.2.18). Therefore, there is a net gain of 2 mol of hydrogen gas per mol of carbon converted. The excess hydrogen stream is used to meet the process heat requirement and to generate the electricity product.

The ZECA Process has not been tested experimentally. ZECA studies have focused on the conceptual analysis of the process based on ASPEN PLUS (Aspen Technology Inc., Houston TX) simulation. The performance of the ZECA Process was highly dependent on that of the SOFC, which is difficult to predict for the scale required of the ZECA Process applications.[23–25] The ASPEN PLUS simulation and the economic analysis showed an efficiency of 68.9% (HHV) for a 600 MW ZECA plant,[24] when the SOFC efficiency was assumed to be 59% (LHV). It can be significantly lower (e.g., ~60%, HHV) when the SOFC efficiency is lower (e.g., 50%, LHV).[24] The use of the 50% efficiency value for SOFC in the estimation of the ZECA plant efficiency is more reasonable, considering the current SOFC performance and the heat integration requirement between the SOFC and the calciner in the ZECA Process. The estimated cost for a ZECA plant (in terms of year 2000 dollars) was 8.6% higher than a comparable IGCC plant with 90% CO_2 capture.[25] Due to a lower projected operating cost, however, the estimated cost of electricity for the ZECA plant is 3.4% lower than that for the IGCC plant.[25] Hurdles need to be overcome for the ZECA Process to be used, such as the heat transfer from the SOFC to the calciner.[24] Because the heat required in the calciner is provided by the SOFC, the operating temperature of the SOFC should be significantly higher than in the calcination reactor, which is at ~1,000°C. The high operating temperature required for the SOFC will be difficult to achieve because of the materials constraint and anode stability. The sulfur in coal also can be detrimental to the activity of the reforming catalyst and the SOFC stability.[24]

5.2.4 ALSTOM Hybrid Combustion–Gasification Process

As described in Section 1.7, the ALSTOM hybrid combustion–gasification chemical looping coal power technology encompasses three different configurations that produce: (1) heat for power generation, (2) coal-derived syngas, and (3) hydrogen.[26] The flow diagram for the first configuration is shown in Figure 5.6. In this configuration, two fluidized bed reactors, a reducer and an oxidizer, are used to complete one chemical loop that converts coal to heat using $CaS/CaSO_4$ as the oxygen-carrier particle. This is similar to the other chemical looping combustion processes discussed in Chapter 3. In the reducer, which operates at 6 atm and 880°C–980°C, high-temperature $CaSO_4$ particles (~1,100°C) from the oxidizer react with coal to form CaS and an exhaust gas stream with a high CO_2 concentration. Steam also is introduced to the reducer to enhance the reaction rates. The heat needed for this endothermic reaction is provided by the high-temperature $CaSO_4$ particles. The reduced CaS particles discharged from the reducer then are transported to the oxidizer where they are oxidized back to $CaSO_4$ with air. The oxidizer operates at ~1,100°C and 1 atm. The oxidation of CaS using air is highly exothermic. The heat released in the oxidizer is used to generate high-temperature, high-pressure

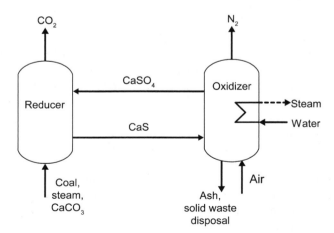

Figure   5.6. ALSTOM's   Hybrid   Combustion–Gasification   Process   for   heat generation.

steam, as well as hot $CaSO_4$. The steam produced in the oxidizer is used for power production, whereas the hot $CaSO_4$ particles are transported back to the reducer to complete the chemical loop. The spent $CaSO_4$ particles are discharged periodically along with ash to maintain particle reactivity and to prevent ash buildup. The reactions involved in the first configuration are as follows:

$$\text{Reducer:} \quad CaSO_4 + 2C \rightarrow CaS + 2CO_2 \qquad (5.2.20)$$

$$\text{Oxidizer:} \quad CaS + 2O_2 \rightarrow CaSO_4 \qquad (5.2.21)$$

$$\text{Overall reaction of the first configuration:} \quad C + O_2 \rightarrow CO_2 \qquad (5.2.22)$$

The major advantage of the first configuration compared with the traditional coal-fired power plant is its integrated CO_2 separation feature, which is common among chemical looping combustion technologies.

In the second configuration, as shown in Figure 5.7, syngas is produced without the use of an air separation unit (ASU). The major difference from the first configuration lies in the reducer operation, in which a higher coal/$CaSO_4$ ratio and an increased steam feed rate are used. As a result, most of the coal in the reducer is gasified to syngas by steam and $CaSO_4$. CaS also is produced in this step. Because of the highly endothermic nature of the steam–carbon reaction, a significant amount of heat is needed in the reducer. This again is provided by the heat transfer from high-temperature $CaSO_4$ particles circulated from the oxidizer. Thus, in this configuration, the heat generated in the oxidizer mainly is used to produce high-temperature $CaSO_4$ particles as opposed to producing steam. The key reactions in the second configuration can be given by

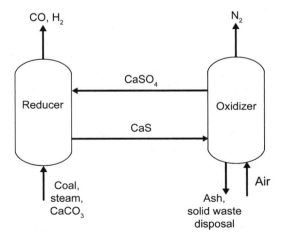

Figure 5.7. ALSTOM's Hybrid Ccombustion–Gasification Process for syngas production.

Reducer:

$$CaSO_4 + 4C \rightarrow CaS + 4CO \tag{5.2.23}$$

$$H_2O + C \rightarrow H_2 + CO \tag{5.2.24}$$

$$CO + H_2O \rightarrow CO_2 + H_2 \tag{5.2.25}$$

Oxidizer:

$$CaS + 2O_2 \rightarrow CaSO_4 \tag{5.2.26}$$

Overall reaction of the second configuration:

$$(m + 2n)C + mH_2O + nO_2 \rightarrow (m + 2n)CO + mH_2 \tag{5.2.27}$$

As is shown in the overall reaction, less than a stoichiometric amount of oxygen is used in the second configuration to oxidize coal partially into syngas in an indirect manner. For comparison, a stoichiometric amount of oxygen is used in the first configuration to oxidize the coal fully into CO_2. In the second configuration, the carbon in the coal exits from the reducer in the form of syngas; therefore, no sequestrable carbon dioxide is generated and, hence is captured.

In the third configuration, as shown in Figure 5.8, hydrogen is produced using two chemical loops with three reactors: a reducer, an oxidizer, and a calciner. The reducer is used to produce hydrogen, and the oxidizer is used to produce the heat for gasification and calcination reactions, both of which are operated under conditions similar to those shown in the first configuration. The calciner, operated at ~880°C and 0.5 atm, is used to regenerate the spent calcium sorbent from the reducer. In this configuration, both $CaSO_4$ and CaO are introduced into the reducer. A typical solids mixture circulating in the system, known as the fluid-bed heat exchanger (FBHE) sorbent, has a composition by mass of 35.6% $CaSO_4$, 18.4% CaO and 44.6% coal ash. The $CaSO_4$ acts as the oxygen carrier for indirect oxidation, whereas the CaO sorbent captures CO_2 produced from the WGS reaction to enhance hydrogen production. Thus, most of the CO formed in the gasification step will be converted to H_2.

The heat integration of this configuration includes the utilization of the heat generated from calcium-sulfide combustion in the oxidizer.[27] Bauxite, a naturally occurring, heterogeneous material that predominantly contains aluminum hydroxide, is used to transfer the heat from the oxidizer to the calciner, forming a thermal loop that is independent of the chemical loops. During the regeneration of the spent CaO sorbent, a concentrated CO_2 gas stream is produced. This stream later can be compressed and sequestrated. Note that the carbonation reaction of CaO in the reducer also provides heat for the steam–carbon reaction in the gasification step. The main reactions in this configuration are

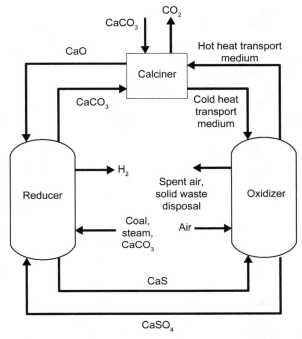

Figure 5.8. ALSTOM's Hybrid Comubstion–Gasification Process for hydrogen generation.

Reducer:

$$CaSO_4 + 4C \rightarrow CaS + 4CO \qquad (5.2.28)$$

$$C + H_2O \rightarrow CO + H_2 \qquad (5.2.29)$$

$$CO + H_2O \rightarrow CO_2 + H_2 \qquad (5.2.30)$$

$$CaO + CO_2 \rightarrow CaCO_3 \qquad (5.2.31)$$

Oxidizer:

$$CaS + 2O_2 \rightarrow CaSO_4 \qquad (5.2.32)$$

Calciner:

$$CaCO_3 \rightarrow CaO + CO_2 \qquad (5.2.33)$$

Overall reaction of the third configuration:

$$(m+n)C + 2mH_2O + nO_2 = (m+n)CO_2 + 2mH_2 \qquad (5.2.34)$$

Process Development Unit Testing

A process development unit (PDU) was constructed by ALSTOM to conduct the feasibility studies of the chemical looping process. The main components of the PDU are the riser reducer and the riser oxidizer. A dedicated calciner, described in the third configuration of the process, was not included in the PDU. Both the reducer and the oxidizer are 15 ft (4.6 m) in length with a diameter of 1 in. (2.54 cm) and 0.75 in. (1.91 cm), respectively. These two risers are fast fluidized-bed reactors[28] operated with a gas residence time of ~0.57 seconds. In the reducer, the high-temperature $CaSO_4$ particles, obtained from the oxidizer, pass through the riser to convert coal to CO_2 and H_2O or CO and H_2 depending on the configuration of the process. In the hydrogen generation configuration, the FBHE sorbent, which is a solid mixture of $CaSO_4$, CaO, and ash, as mentioned earlier, is used in the reducer. The reacted solids then enter a seal pot control valve, which is a nonmechanical valve that guides the solids to three paths: the steam activation path, the recirculation path, or the oxidizer riser path. A portion of the reacted solids enters the steam activation path where it is hydrated and reactivated. Another portion of these solids is recycled back to the reducer to improve the solids conversion. The rest of the solids is sent to the oxidizer where they enter the oxidation cycle. CaS formed in the reducer is oxidized in the oxidizer to $CaSO_4$ while releasing heat. Similar to the reducer, a seal pot control valve distributes the solids in the oxidizer. Figure 5.9 shows the various paths for the solids flow. A thermal loop also is present that utilizes bauxite to carry the heat from the oxidizer to the calciner. The thermal loop is not included in the PDU. Thus, the complete operation of the PDU involves simultaneous circulation of solids among multiple units.

ALSTOM conducted the looping process testing on the PDU at ALSTOM's Power Plant Laboratories in Windsor, CT for indirect combustion and syngas production from coal char.[27] Two main sets of experiments were conducted with respect to the $CaS/CaSO_4$ loop. The first set of experiments was conducted to validate the ability of $CaSO_4$ to gasify coal. These experiments were carried out in two steps. In the first step, methane was used as the fuel. The results indicate that a higher temperature favors methane conversions. For example, a 50% methane conversion was observed at 927°C and 17% methane feed compared to a 10–15% methane conversion at 870°C and 5% methane feed.[29] This was followed by char experiments using Pittsburgh #8 coal char (a less reactive char) and hardwood charcoal (a highly reactive char). An increase in CO concentration was observed with an increase in the feed ratio between char and $CaSO_4$. The char gasification rates, expressed in terms of the percentage of carbon conversion per second under one absolute atmosphere of pure CO_2 (ata), were higher than 10%/s·ata and 100%/s·ata for Pittsburgh #8 char and hardwood charcoal, respectively.[29] These values exceed the minimum char gasification rate required for feasible commercial operation estimated by ALTSOM. The extent of char conversion, however, was not reported.

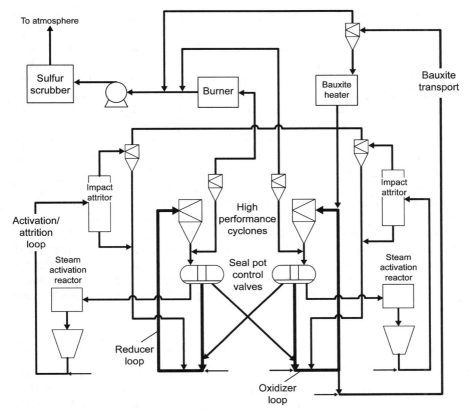

Figure 5.9. Schematic of the process development unit for ALSTOM's Hybrid Combustion–Gasification Process. Adapted from ALSTOM 2006 annual technical report to U.S. DOE Report No. PPL-06-CT-27. Project title "Hybrid Combustion–Gasification Chemical Looping Coal Power Technology Development," under DOE contract #DE-FC26-03NT41866.

The second test was to verify the oxidation behavior of CaS to $CaSO_4$. The formation of SO_2 is a major concern in the oxidizer operation:

$$CaS + 2O_2 \rightarrow CaSO_4 \text{ (desired reaction)} \qquad (5.2.35)$$

$$CaS + 1.5O_2 \rightarrow CaO + SO_2 \text{ (undesired side reaction)} \qquad (5.2.36)$$

The experiments performed in a 4-in. (10-cm)-ID fluidized bed reactor showed SO_2 formation at the initial stage of the reaction as a result of the relatively low air-to-CaS ratio. The experiments also showed that the SO_2 formation could be prevented by reacting CaS with an excessive amount of air or oxygen.[29] Enhanced H_2 production from syngas using calcium oxide also was tested in a 4-in. (10.16-cm)-ID fluidized-bed reactor in addition to the PDU. The test results in the 4-in. (10.16-cm)-ID fluidized-bed reactor indicted that by adding calcium-oxide sorbent, the CO conversion increased from 25% to as high as 95%. These results suggest that the presence of calcium oxide is effective in WGS reaction enhancement. The sorbent also was able to capture up to 83% of the CO_2 generated; however, less than 50% of the CaO sorbent was used throughout the tests.[26] The PDU tests for hydrogen production achieved 54% CO conversion and a 60% CO_2 capture rate. Although both the CO conversion and the CO_2 capture rate are less than 100%, ALSTOM stated that, by using a finer CaO sorbent and a longer residence time, a very high-purity H_2 can be produced in a commercial system.[26]

Thus, in the ALSTOM process, two types of solids, $CaS/CaSO_4$ and $CaO/CaCO_3$, are used for hydrogen production. $CaS/CaSO_4$ particles are used as an oxygen carrier for indirect combustion of coal, avoiding the energy-intensive oxygen separation scheme, while providing the heat needed for coal gasification and spent sorbent regeneration. $CaO/CaCO_3$, on the other hand, is used as a CO_2 capture sorbent as well as a H_2 production enhancer. Therefore, this process uses two chemical loops [a combustion loop ($CaS/CaSO_4$ loop) and a CO_2 capture loop ($CaO/CaCO_3$ loop)] and one thermal loop (bauxite loop).

5.2.5 Fuel-Flexible Advanced Gasification–Combustion Process

The Fuel-Flexible Advanced Gasification–Combustion (AGC) Process was developed by the General Electric Energy and Environmental Research (GE EER) Corporation. The process resembles the HyPr-Ring Process with a key exception in that iron oxide is used as the oxygen carrier, instead of having a stream of pure oxygen. The process also uses calcium oxide as the hydrogen production enhancer, whereas Fe_2O_3 is used as the oxygen carrier for indirect fuel combustion. Thus, this process requires two chemical loops in the operation. As shown in Figure 5.10, which previously was shown as Figure 1.27, the Fuel-Flexible AGC Process consists of three interconnected fluidized-bed reactors.[30] In the first reactor, or the gasifier, coal partially is gasified with steam to form a mixture of H_2, CO and CO_2. The CO_2 generated during the coal gasification is captured *in situ* by CaO under operating conditions of 750°C–850°C and 17–20 atm. This carbonation reaction provides the heat needed for coal gasification with steam. The carbonation reaction also enhances H_2 formation, as discussed in prior techniques. Moreover, the calcium-based sorbents also capture sulfur in the coal. As a result, a concentrated H_2 stream is obtained from the gasifier. The main reactions in the gasifier include

$$C + H_2O \rightarrow CO + H_2 \qquad (5.2.37)$$

$$CO + H_2O \rightarrow CO_2 + H_2 \qquad (5.2.38)$$

$$CO_2 + CaO \rightarrow CaCO_3 \qquad (5.2.39)$$

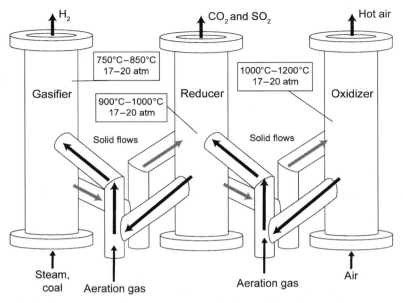

Figure 5.10. Schematic diagram of the Fuel-Flexible Gasification–Combustion Process.[30]

Because the calcium sorbent and iron oxide particles are in mixed form, the partially reduced iron oxide particles unavoidably will be carried from the reducer to the gasifier and be further reduced:

$$Fe_3O_4 + H_2 \rightarrow 3FeO + H_2O \qquad (5.2.40)$$

As discussed subsequently, the presence of iron oxide can enhance hydrogen production.

The solids discharged from the gasifier, which mainly consist of spent sorbents and unconverted char, are introduced to a second reactor, known as the reducer, operated at 900°C–1,000°C and 17–20 atm. In the reducer, the unconverted char is gasified with steam to form CO and H_2. CO and H_2 are oxidized *in situ* by Fe_2O_3 from the oxidizer, forming reduced iron oxide mainly in the form of Fe_3O_4 and also FeO, H_2O, and CO_2. There is no Fe formed because of the use of a fluidized bed and the presence of excess steam. The high operating temperature of the reducer also decomposes the spent sorbent from the gasifier, producing CaO and CO_2. Thus, a concentrated CO_2 stream is generated from the reducer once the water is condensed out. The regenerated sorbent particles and the reduced iron oxide particles produced from the reducer are recycled back to the gasifier and the oxidizer, respectively.

The main reactions involved in the reducer include

$$C + H_2O \rightarrow CO + H_2 \tag{5.2.41}$$

$$CO + 3Fe_2O_3 \rightarrow 2Fe_3O_4 + CO_2 \tag{5.2.42}$$

$$H_2 + 3Fe_2O_3 \rightarrow 2Fe_3O_4 + H_2O \tag{5.2.43}$$

$$Fe_3O_4 + CO \rightarrow 3FeO + CO_2 \tag{5.2.44}$$

$$Fe_3O_4 + H_2 \rightarrow 3FeO + H_2O \tag{5.2.45}$$

$$CaCO_3 \rightarrow CaO + CO_2 \tag{5.2.46}$$

$$CaSO_3 \rightarrow CaO + SO_2 \tag{5.2.47}$$

Note that both the steam–carbon reaction and the calcination reaction are highly endothermic, whereas other reactions are nearly heat neutral. GE stated that the heat required by the reducer can be provided by the high-temperature Fe_2O_3 particles from the oxidizer; however, meeting the reducer heat requirements solely from the high-temperature particles can be extremely challenging.

The third reactor, the oxidizer, regenerates the reduced oxygen carrier particles fed from the reducer by oxidizing the particles using air. The operating temperature for this reactor (1,000°C–1,200°C) is significantly higher than the first two reactors as a result of the highly exothermic oxidation reaction. Part of the heat generated in this reactor is carried by the hot solids to the reducer. The rest of the heat carried by the hot exhaust gas stream is recovered and used to drive the gas turbine for electricity generation.[31] The reaction in the oxidizer is as follows:

$$4FeO + O_2 \rightarrow 2Fe_2O_3 \tag{5.2.48}$$

The products from this process, thus, are pure hydrogen from the gasifier, carbon dioxide from the reducer, and high-temperature exhaust gas from the oxidizer, which then is converted to electricity. The spent particles along with ash are discharged from the beds periodically, while fresh particles are added to the system to prevent ash buildup and to maintain particle activity.[32]

Lab- and Bench-Scale Tests
To examine the feasibility of the AGC process and to determine the optimal operating conditions, tests first were carried out in a laboratory- and a bench-scale fluidized-bed reactor. These tests studied the individual reactor operation in the AGC process. The laboratory-scale reactor, as shown in Figure 5.11, is a 1-in. (2.54-cm)-ID reactor vessel enclosed within a 4-in. (10.16-cm) double, extra-heavy-grade shell for high-pressure operation. The feed side, which extends beyond the shell, is fitted with a flange that is sandwiched between two 900-lb (408 kg) flanges. The outlet tube was attached to one 900-lb (408-kg) flange, and the inlet is attached to the other side of the reactor

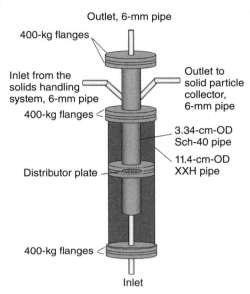

Figure 5.11. Laboratory-scale fluidized bed for testing individual reactors.[31]

pipe. A distributor is installed at the feed side of the reactor pipe. The maximum operating temperature of the reactor is 860°C.

The bench scale reactor, as seen in Figure 5.12, has a design similar to the laboratory-scale unit. The outer vessel still has an outer diameter of 4-in. (10.16-cm), whereas the inner reactor is a 2-in. (5.08-cm) pipe. This reactor also has been outfitted with coal-injection ports. This bench-scale system can be operated at up to 1,000°C and 20 atm.

Gasifier Test
The gasifier conducts coal gasification while producing hydrogen in the presence of CaO. Experiments performed in the laboratory- and bench-scale reactors showed that the H_2 yield and purity are both at a maximum at 800°C. The purity of the hydrogen produced was ~80%. A temperature lower than 800°C will lead to reduced coal gasification kinetics, while a higher temperature (~900°C) can lead to the decomposition of calcium carbonate. Experiments conducted in the labatory-scale reactor further suggested that a high calcium sorbent to coal ratio (~16:1 by weight) was required to avoid CO_2 emission from the reactor.[32] The experimental results also suggest that a 4:1 ratio by weight of calcium sorbent to Fe_2O_3 enhances H_2 production.[32] A further increase in Fe_2O_3, however, will lead to a reduction in H_2 yield.

Reducer and Oxidizer Tests
The reducer or the CO_2 separator decomposes spent calcium sorbents while reducing the oxygen carrier (primarily Fe_2O_3) with unconverted char from the

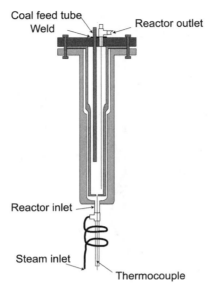

Figure 5.12. Bench-scale fluidized bed with expansion zone and steam superheater.[31]

gasifier. To determine the reducer operating conditions, isolated experiments were performed for the reducer using a bench-scale fluidized bed reactor. It was found that at ~920°C, a significant amount of CO_2 was generated from a mixture of spent $CaCO_3$ sorbent, char, and the oxygen carrier. Further experiments at 920°C determined that the highest CO_2 concentration at the reactor outlet was achieved with a 3:1 weight ratio between the oxygen carrier and the coal. Note that the oxidizer and reducer are operated in a highly integrated manner with the gasifier. Thus, the optimum operating conditions obtained for these reactors conducted in an integrated manner will be different from the isolated condition in which these tests were done.

The reducibility of the oxygen carrier was examined using gaseous fuels such as H_2 and/or CO in a bench-scale reactor. A maximum reduction of ~20% of the oxygen carrier was observed. Thus, the reduction reaction between the oxygen carrier and the reducing gas substantiates the oxygen carrier–char reaction when steam is present. The oxidizer provides the heat for the reducer and for electricity generation. The operation of the oxidizer was conducted by oxidizing the reduced oxygen carrier with air, giving rise to an increase of the solids temperature by more than 200°C, which was deemed satisfactory.

ASPEN Simulation
ASPEN PLUS simulations were carried out to simulate the performance of both a pilot-scale AGC system and a commercial AGC plant. The parameters used in the pilot-scale system were those obtained in laboratory-scale and bench-scale tests.[31] The simulation under these constraints opened up the

possibility of obtaining >85% H_2 concentration from the gasifier. Also, more than 75% of the CO_2 generated from coal can be separated in the reducer. However, because of incomplete conversion of the coal char and the reducing gas in the reducer, however, a leakage of H_2 (~35%) from the reducer and CO_2 (<3%) from the oxidizer were predicted by the simulation. The leakage was attributed primarily to the size of the pilot-scale plant. Rizeq et al.[31] suggested that a higher solids circulation rate in a commercial plant would prevent the leakage from the reducer by providing more oxygen-carrier particles in the reducer.

ASPEN simulations for a commercial-scale AGC plant also were performed based on the RGIBBS module in ASPEN PLUS. Rizeq et al.[31] found a 6%–10% improvement in efficiency compared with the IGCC process. Furthermore, a cost of electricity for the AGC process at $0.053 per kWh[32] was obtained in the preliminary economic analysis, which was lower than the cost for the IGCC process at ~$0.067/kWh. However, key assumptions in obtaining these results such as the sorbent and the oxygen carrier reactivities and recyclabilities, purge rate, and solid residence time were not specified in their report.[32]

Pilot-Scale Tests
A pilot-scale system with a designed capacity of 25 lb/h (11.3 kg/h) also was constructed. The outer diameter of each reactor was 18-in (45.72-cm), and the inner wall was coated with a 2 1/8-in. (5.40-cm)-thick layer of low thermal conductivity material. Inside the low thermal conductivity material was another layer of highly abrasion resistant material that was 1 3/8-in. (3.5-cm)-thick. Therefore, the resulting cylindrical reactor had a diameter of 10-in. (25.4-cm), as shown in Figure 5.13(a). The solid transfer ports were designed with a similar refractory lining to prevent heat loss. These ports were connected according to the scheme shown in Figure 5.10, using multiple flanged ducts. The reactors were designed to operate at 20 atm and between 750°C and 1,300°C. The coal is prepared in the form of a slurry and is delivered using a pump. The steam is generated in a boiler and then superheated to the reactor inlet temperature. An air compressor is used to deliver the high-pressure air to the oxidizer. The entire pilot-scale plant is shown in Figure 5.13(b).

Two sets of semibatch experiments were carried out in the reactors shown in Figure 5.13(b). The first experiment tested the operation of the gasifier and the reducer, whereas the second experiment tested the operation of the oxidizer. The first experiment was done in two steps. In the first step, bed materials containing a 1:1 ratio (by weight) of calcium sorbents to oxygen carrier material (in its fully oxidized form) were fluidized by steam at 2.4 atm. Coal slurry was introduced simultaneously for 6 minutes. In the second step, the coal feed was stopped, but the steam supply continued. The gases exiting from the reactor were monitored throughout the experiments. The first step was intended to mimic the operation of the gasifier, whereas the second step was intended to simulate the reducer operation. A product stream with relatively low H_2 concentration (~60%) was obtained during the first step (gasifier

(a)

(b)

Figure 5.13. (a) Reactor showing refractory lining; (b) pilot-plant layout.[33] With permission from Francis T. Coppa and George Rizeq of GE; photos from GE 2004 Annual Report to U.S. DOE on Project "Fuel-Flexible Gasification-Combustion Technology for Production of H_2 and Sequestration-Ready CO_2," under DOE Contract #DE-FC26-00FT40974.

operation). The by-products in the stream included CO, CH_4, CO_2, and other contaminants. The low hydrogen product purity was attributed to the use of an Fe_2O_3-to-calcium sorbent ratio of 1:1, which led to the reduction of Fe_2O_3 and, hence, the consumption of H_2 in the first step. About a 40% reduction of Fe_2O_3 was observed after both steps were completed.[33]

In the second experiment, the solids obtained from the first experiment were fluidized with air. The experiment was set up to mimic oxidizer operation. The parameters observed in the experiment were bed temperature and O_2

consumption. An increase in reactor temperature of ~120°C was observed. Such a temperature increase was considered sufficient for the heat integration of the process. As noted earlier for the bench-scale tests, the semibatch experiments cannot represent the complexity of the actual AGC operation. Specifically, in the AGC operation, the iron oxide that enters the gasifier from the reducer is in partially reduced form, and thus the gasifier test using iron oxide in fully oxidized form deviates from the actual gasifier operating conditions. Furthermore, the Fe_2O_3 reduction rate of 40% resulting from the sequential gasifier and reducer tests conducted in a semibatch mode is higher than that seen in actual AGC operation.

The continuous operation of all three reactors and the flexibility of using various fuels have not yet been demonstrated. With three interconnected fluidized-bed reactors simultaneously exchanging a mixture of two-looping particles, that is, the CO_2 sorbent and the oxygen carrier, the operation of the GE AGC process can be very challenging. Issues that need to be addressed further include the simultaneous handling of well-mixed sorbent and oxygen carrier, sorbent and oxygen carrier reactivity and recyclability, heat transfer and solid/gas leakage between the reactors, and eutectic formation among solids. Thus, the viability of the process is uncertain.

5.2.6 General Comments

Various calcium-based chemical looping gasification processes share a common approach in which $CaO/Ca(OH)_2$ sorbents are used to enhance the fuel conversion to H_2 by *in situ* CO_2 removal. These processes differ, however, in the manner under which the spent $CaCO_3$ sorbent is regenerated. The regeneration of sorbents for effective applications in a looping system continues to pose a challenge, particularly when calcium-based sorbents are used. Because the carbonation–calcination reactions involve a highly endothermic calcination reaction that typically takes place at much higher temperatures than a highly exothermic carbonation reaction, this system imposes an inevitable heat penalty for looping process operation. The heat penalty can be mitigated partially by using an optimized heat integration scheme while avoiding the energy-intensive CO_2 separation scheme.

The heat sources for sorbent regeneration can originate from two approaches: (1) direct combustion of leftover fuels or product fuels; (2) indirect combustion of fuels using an oxygen carrier. The first approach is used by the CO_2 Acceptor Process, the HyPr-Ring Process, and the ZECA Process, whereas the second approach is employed by the ALSTOM Hybrid Combustion–Gasification Process and the GE Fuel-Flexible Process. The first approach uses fuels such as coal, coal char, and hydrogen that are oxidized either with air or oxygen to generate heat for the calcination reaction. If carbon emissions control is not considered, then air can be used directly for fuel combustion, generating a flue gas stream that contains N_2 and CO_2. If carbon emissions control is considered, however, then a stream of oxygen should be used so that combustion of the fuel yields steam and CO_2, from which CO_2 can be readily separated. This approach would require a capital- and an energy-intensive air-separation unit to provide oxygen. An alternative method, proposed by the ZECA Process, involves oxidizing the hydrogen product in a SOFC with air. The heat then is used indirectly for the calcination reaction. Overall, the high cost for the oxygen production and that for the fuel cell operation lead to methods for calcination either through indirect combustion/oxidation or through direct air combustion with an integrated CO_2 capture scheme, as described in Chapter 6.

In the second approach, the indirect combustion approach adopted by ALSTOM and GE, which also is discussed in Chapter 3 in connection with the chemical looping combustion processes, oxygen-carrier particles first are reduced with the fuel/leftover fuel and then combusted to provide the heat for the calcination reaction. In this manner, direct contact between the air and the carbonaceous fuel is avoided to prevent the mixing of CO_2 with N_2 from the air. Here, the oxygen-carrier particles serve the role of a separator in separation of oxygen from the air. Although different oxygen-carrier particles are used in the ALSTOM ($CaSO_4$) and GE (Fe_2O_3) Processes, their functions are fundamentally identical. Indirect combustion avoids the costly ASU; however, at least one additional reactor, compared with the direct combustion approach, will be required for the indirect combustion step. Moreover, as noted earlier, the circulation of two types of looping particles in a reactor system can be highly challenging.

Other important issues for calcium-based gasification processes include the sorbent reactivity, sorbent recycleability, and sorbent reaction kinetics. These factors will affect the sorbent loading in the reactors, the CO_2 capture rate, the fuel/sorbent ratio, the makeup sorbent amount and, hence, the reactor size. Thus, desirable calcium-based gasification processes will need to use highly reactive, highly recyclable sorbents and to adopt a heat integration system that minimizes the heat penalty involved in the calcination step.

5.3 Coal-Direct Chemical Looping (CDCL) Processes Using Iron-Based Oxygen Carriers

Unlike calcium-based chemical looping gasification processes, iron-based chemical looping gasification processes convert solid fuels directly through steam–iron reactions:

$$FeO_x + R \rightarrow FeO_{x-1} + RO \qquad (5.3.1)$$

$$FeO_{x-1} + H_2O \rightarrow FeO_x + H_2 \qquad (5.3.2)$$

where x can vary between 0 (Fe) and 1.5 (Fe_2O_3), and R can be any solid fuel, including coal and biomass. As indicated in Chapter 2, iron oxide is a suitable oxygen carrier for solid fuel gasification. Chapter 4 discusses several challenges for iron-based looping processes using gaseous fuels. These challenges include iron oxide conversion and recyclability, carbonaceous fuel conversions in the reducer, and steam-to-hydrogen conversion in the oxidizer. The iron-based looping processes using solid fuels, however, impose additional challenges such as solid fuel conversion enhancement, ash separation, and pollutant control evolved *in situ* from solid fuels. The Coal-Direct Chemical Looping (CDCL) Process features the iron oxide-based looping process using coal as the solid fuel with a gas–solid contact pattern similar to the Syngas Chemical Looping (SCL) Process discussed in Chapter 4. The CDCL is presented in detail in Sections 5.3–5.5. A similar process concept can be extended to the use of other solid fuels such as petroleum coke and cellulosic biomass. The CDCL Process can be configured differently depending on the heat integration schemes in the reactor system. Two representative configurations are presented in this section. Note that the chemical looping operation using direct feeding of coal to the reducer also appears in chemical looping combustion applications described in Section 3.3.5. In those applications, fluidized bed reactors are used. In the following discussion, countercurrent gas–solid flow reactors are used as the reducer in chemical looping gasification applications.

5.3.1 Coal-Direct Chemical Looping Process—Configuration I

Figure 5.14 shows the simplified flow diagram of Configuration I for the CDCL Process. Similar to the SCL Process, the CDCL Process also consists of three reactors: the reducer, the oxidizer, and the combustor. The reducer converts carbonaceous fuels to CO_2 while reducing Fe_2O_3 to a mixture of Fe and FeO; the oxidizer oxidizes the reduced Fe/FeO particles to Fe_3O_4 using steam, producing a H_2-rich gas stream; and the combustor pneumatically transports the Fe_3O_4 particles from the oxidizer outlet to the reducer inlet by air. The Fe_3O_4 particles are reoxidized to Fe_2O_3 during the conveying process.

Reducer
The reducer is a countercurrent gas–solid reactor operated at 750°C–950°C and 1–30 atm. The countercurrent operational mode is intended to maximize the solids and gas conversions. The solids flow can be in a moving bed or in a series of fluidized beds. Note that the moving bed contact mode is highlighted in this design, as it represents the fundamental countercurrent solids contact pattern with gases that is preferred in this reactor system. The desirable reaction in the reducer is

$$C_{11}H_{10}O\,(coal) + 8.67Fe_2O_3 \rightarrow 11CO_2 + 5H_2O + 17.34Fe \qquad (5.3.3)$$

The coal used here is Pittsburgh #8 and is represented as $C_{11}H_{10}O$ given the elemental composition.[34] The reaction is highly endothermic with the heat of

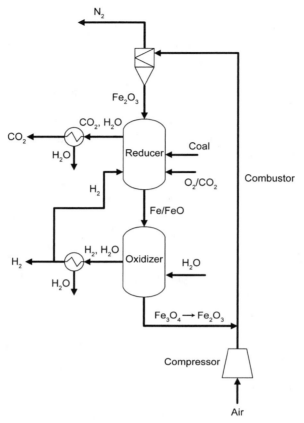

Figure 5.14. A simplified flow diagram for Coal-Direct Chemical Looping Process—Configuration I.

reaction equal to 1,794 kJ/mol at 900°C. Therefore, a significant amount of heat needs to be provided to the reducer.

One option for balancing the heat is to combust coal partially *in situ* by sending a substoichiometric amount of O_2 into the reducer. The overall reaction then would be

$$C_{11}H_{10}O + 6.44Fe_2O_3 + 3.34O_2 \rightarrow 11CO_2 + 5H_2O + 12.88Fe \qquad (5.3.4)$$

This reaction takes place with zero heat of reaction at 900°C. Because the amount of oxygen required for this reaction is significantly less than that for the coal gasification reactions, the size of the ASU is smaller than in traditional gasification processes. The reduction in oxygen demand leads to savings in both operating cost and capital investment of the coal to hydrogen plant. Another option for the heat balance in the reducer is to combust a portion of the reduced particles and use them as the heat source. This option is similar to the steam methane reforming process where heat is provided by combusting a portion of the methane outside the reforming tubular reactors. Such an option will be discussed further in Sections 5.3.2, 5.4 and 5.5.

Oxidizer
The oxidizer is a moving bed reactor operating at 500°C–850°C and 1–30 atm. In the oxidizer, the Fe and FeO mixture from the reducer reacts with steam countercurrently. The reactions in the oxidizer are as follows:

$$Fe + H_2O \rightarrow FeO + H_2 \quad (5.3.5)$$

$$3FeO + H_2O \rightarrow Fe_3O_4 + H_2 \quad (5.3.6)$$

This reaction is slightly exothermic. To maintain adiabatic operation, steam at a moderately low temperature is introduced into the oxidizer to modulate the reactor temperature.

Combustor
Fe_3O_4 from the oxidizer is regenerated fully in the combustor. The combustor is an adiabatic entrained-bed reactor operated at a pressure similar to the reducer and the oxidizer. If necessary, a dense bed operated in the turbulent fluidization regime can be installed below the entrained bed to provide a sufficient reaction time to oxidize the oxygen carrier. Air is used to convey pneumatically the Fe_3O_4 particles from the oxidizer outlet to the reducer inlet while fully regenerating the particles by the following reaction:

$$4Fe_3O_4 + O_2 \rightarrow 6Fe_2O_3 \quad (5.3.7)$$

This exothermic reaction heats up the solids and the gas. The hot solids are subsequently fed into the reducer partially to compensate for the heat needed for the coal conversion. The hot gas then is used for power generation. The coal ash, which is significantly smaller in size than the Fe_2O_3 composite particles, is separated out along with the fine particles using a cyclone on the top of the reducer. Fresh makeup particles also are introduced to the reducer to maintain reactivity.

5.3.2 Coal-Direct Chemical Looping Process—Configuration II

Configuration II of the Coal-Direct Chemical Looping Process is presented briefly in this section, with details given in Sections 5.4 and 5.5 in the context of process analysis and ASPEN PLUS simulation.

As in CDCL—Configuration I, there are three major units involved in CDCL—Configuration II: the reducer, the oxidizer, and the combustor. A simplified diagram for Configuration II of the CDCL Process is shown in Figure 5.15. The key difference between Configurations I and II lies in the heat

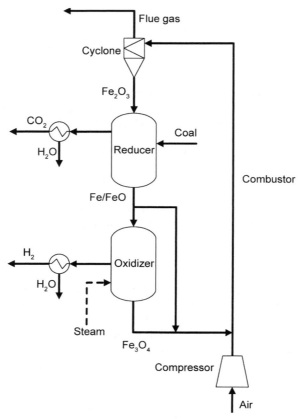

Figure 5.15. A simplified flow diagram for the Coal-Direct Chemical Looping Process—Configuration II.

integration strategy. As indicated in Section 5.3.1, the reaction between coal and iron oxide in the reducer is highly endothermic. Unlike Configuration I where the heat requirement of the reducer is met by partial oxidation of coal, in Configuration II, reduced iron oxide particles are combusted to compensate for the heat deficit of the reducer.

In Configuration II, composite Fe_2O_3 particles with an inert support are used to oxidize coal in the reducer while being reduced to a mixture of metallic iron and FeO. The gaseous product of this reactor is mainly CO_2 mixed with a small amount of H_2O. The reduced Fe/FeO particles from the reducer are split into two streams. The first stream, which includes most of the reduced Fe/FeO particles, is sent to the oxidizer to perform the steam-iron reaction. The oxidizer produces H_2 while oxidizing the reduced Fe particles to Fe_3O_4 with steam. The rest of the reduced particles from the reducer, along with the Fe_3O_4 particles discharged from the oxidizer, are burned in the combustor with air. As a result, high-temperature solid and gas streams are generated from the combustor. The sensible heat carried by the high-temperature solids is used to support the heat requirement in the reducer. By increasing the amount of particles being combusted, excess heat also can be produced from the combustor for electricity generation at the expense of the hydrogen yield.

5.3.3   Comments on the Iron-Based Coal-Direct Chemical Looping Process

Similar to the calcium-based chemical looping gasification processes, the goal of iron-based chemical looping gasification processes is to produce efficiently hydrogen or syngas from coal in a cost-effective and environmentally friendly manner. The underlying coal gasification strategy of the calcium and the iron chemical looping processes, however, are quite different. Unlike calcium-based chemical looping gasification processes where $CaO/Ca(OH)_2$ sorbents are used to enhance coal gasification, coal is gasified indirectly in the iron-based processes. Therefore, the WGS reaction, that commonly appears in the carbonaceous fuel conversion processes, is not used in the iron-based chemical looping gasification processes. The successive CO_2 and H_2 separation step can thus, be avoided.

Iron-based chemical looping gasification processes are characterized by the unique properties of the iron-based particles (see Chapter 2). Unlike calcium, iron has multiple oxidation states that can be effectively utilized. For example, iron oxides can be reduced to a lower oxidation state by various types of fuels. Furthermore, the reduced iron oxide can be regenerated fully or partially with oxygen/air or steam, respectively. Various reduction and oxidation reactions along with different kinetic/thermodynamic behavior, which is associated with these reactions, allow for numerous possible schemes of heat and energy management when an iron-based chemical looping medium is used. These schemes, once optimized dramatically can improve the energy conversion efficiencies of the coal gasification processes. The design criteria, practical issues and the energy management schemes of the chemical looping gasification processes will be discussed in the following sections.

5.4   Challenges to the Coal-Direct Chemical Looping Processes and Strategy for Improvements

This section illustrates several critical issues challenging the Coal-Direct Chemical Looping Process and the strategies that can be adopted to overcome them. These issues include oxygen carrier particle reactivity, char reaction enhancement, gas and solid conversion, the fate of pollutants and ash, and heat management and integration. The results from both experiments and theoretical analysis using ASPEN PLUS simulations are presented.

5.4.1   Oxygen-Carrier Particle Reactivity and Char Reaction Enhancement

As discussed in Chapter 2, the reactivity of oxygen-carrier particles is the most widely examined subject in the research and development of chemical looping processes employing gaseous fuels such as methane and/or syngas. For chemical looping processes employing solid fuels additional issues must be considered that are associated with the oxygen carrier interaction *in situ* with gaseous and solid pollutants and coal ash.

Particle Reactivity
NiO often is considered as an attractive oxygen carrier because its reactivity with gaseous fuels is higher than other oxygen carriers.[35] However, it is noted that for the solid fuel chemical looping reactor, the solid–solid reaction between coal char and metal oxide is very slow. Therefore, the reactions of the oxygen carrier and the char are mainly through the following solid–gas reactions in the presence of CO_2 or H_2O:

$$H_2O/CO_2 + C \rightarrow CO + H_2/CO \qquad (5.4.1)$$

$$MeO + H_2/CO \rightarrow Me + H_2O/CO_2 \qquad (5.4.2)$$

The overall reaction is

$$2MeO + C \rightarrow 2Me + CO_2 \qquad (5.4.3)$$

In this reaction scheme, the overall reaction rate is controlled by the rate of gasification of char.[36] Thus, CO_2 and H_2O act as char reaction enhancers that promote the overall reaction in the reducer. For most metal oxide based oxygen carriers such as NiO and Fe_2O_3, Reaction (5.4.2) is significantly faster than Reaction (5.4.1). Thus, in the presence of CO_2/H_2O, the overall rate of reaction, Reaction (5.4.3), between coal char and most metal-oxide-based oxygen carriers will be comparable irrespective of the metal oxide oxygen carriers used. For solid carbonaceous fuel conversions, there are advantages of using an iron-based oxygen carrier, including low raw material cost, favorable thermodynamic properties, high oxygen-carrying capacity, high melting points for all oxidation states, high particle strength, and low environmental and health concerns (see Chapter 2). These advantages render iron based oxygen carriers desirable to be used as looping particles for solid carbonaceous fuel applications.

Coal Char Reaction Enhancement
Char reaction enhancement schemes that improve the extent of the conversion for coal char and oxygen carrier particles are illustrated in Figure 5.16. Figure 5.16(a) shows Scheme A, where hydrogen is used as a char reaction

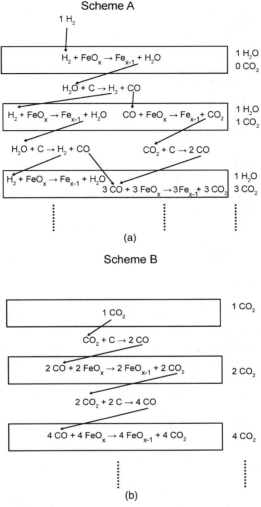

Figure 5.16. Char reaction enhancement schemes: (a) using recycled hydrogen from the oxidizer; (b) using recycled CO_2 from the reducer exhaust.

enhancer. Under such a scheme, a small portion of the hydrogen ($<5\%$) produced in the oxidizer is recycled and introduced to the bottom of the reducer, and flows countercurrently to the solids flow. As shown in Figure 5.16(a), the hydrogen reacts with iron oxide to form steam (Reaction [5.4.2]) as it enters the reactor. The steam then will react with carbon in coal char to form hydrogen and carbon monoxide via the steam carbon reaction (Reaction [5.4.1]).

Because 1 mol of steam will generate 2 mol of reducing gas, that is, CO and H_2, more iron oxides will be reduced, producing steam and CO_2. As a result, the amount of char reaction enhancer, which is represented by compounds that can enhance char gasification, is doubled. Thus, introducing a small amount of hydrogen into the reducer initiates a chain reaction, which produces a large amount of steam and CO_2, which enhances the conversion of coal char and metal oxide particles.

As illustrated in Figure 5.16(b), Scheme B introduces CO_2 or steam at the bottom of the reducer to enhance the char conversion in a manner similar to that explained in Scheme A for the iron oxide conversion. The exiting CO_2 can be obtained from the reducer exhaust, and then is recycled to the reducer inlet. Although CO can trigger an effect similar to that of H_2, CO_2, or steam, it is not attractive because (1) CO is not produced in the standard CDCL configuration and, hence is not readily available, and (2) high CO concentration at the CO entrance to the reducer can cause carbon deposition, as noted in Chapter 2.

Schemes A and B have their respective advantages and disadvantages. From the thermodynamic analysis using ASPEN PLUS, Scheme A leads to a slightly higher iron oxide conversion in the reducer than Scheme B; however, a portion of the valuable hydrogen product must be recycled under Scheme A. However, although CO_2 and steam negatively may affect the iron oxide conversion in the reducer, they are by-products with a low economic value. Clearly, the choice of one reaction enhancement scheme over the other will depend on the process economics, which is affected by several factors including hydrogen output and price, coal/coal char properties, and the performance of the enhancing agent under the specific reducer design and operating conditions. Simulations and bench-scale tests reported in Section 5.4.2 further illustrate the effects of the addition of a char reaction enhancer.

5.4.2 Configurations and Conversions of the Reducer

The reducer using fluidized bed reactors is discussed in Section 3.3.5 for chemical looping combustion applications. From the discussion in Section 5.4.1, note that the reducer in the CDCL is a countercurrent gas–solid flow reactor. The operation of the oxidizer and the combustor in the CDCL Process is similar to the SCL Process, with a key difference in the reducer where coal and metal oxide are converted.

At the reducer operating temperature, which is between 750°C and 950°C, coal is decomposed to volatiles and char. To utilize fully the chemical energy in coal and to generate sequestration-ready CO_2, both volatiles and coal char need to be fully oxidized in the reducer. As discussed in Chapter 4, a higher conversion of the oxygen carrier is also desirable to increase the steam-to-hydrogen conversion in the oxidizer and to reduce the solids circulation rate. Therefore, a well-conceived reducer design and gas–solid contacting pattern are essential.

Overview of the Reducer Configurations

Figure 5.17 shows the reducer design for CDCL—Configuration II given in Figure 5.15. In this configuration, fresh Fe_2O_3 composite particles are fed from the top of the reducer, whereas pulverized coal is conveyed pneumatically to the middle section of the reactor using CO_2. A small amount of CO_2 or steam also is introduced at the bottom of the reducer to enhance char conversion. The coal injection port divides the reducer into two sections. The function of the upper section (Stage I) is to ensure full conversion of gaseous species to CO_2 and H_2O, whereas the lower section (Stage II) is used to maximize the char and iron oxide conversions.

Because of the high reducer operating temperature, coal will be devolatilized in the pneumatic injection zone. The coal volatiles will move upward, along with other gases such as CO_2, H_2O, CO, and H_2. Fresh iron oxide particles that enter from the top of the reducer will interact with these gases in a countercurrent manner. The countercurrent interaction between the coal volatiles and iron oxide particles ensures the complete conversion of gaseous carbonaceous fuels. As shown in Figure 5.17, an annular region is present for the coal flow in the reducer around the internal hopper in Stage I. The hopper is designed to allow an upward flow of gas through it while iron oxide solids are being discharged downward. Coal char and iron oxide particle mixing is initiated in the annular region where coal devolatization begins. The mixing of the devolatilized coal char and partially reduced iron oxide particles continues to occur as particles descend from Stage I to Stage II. In Stage II, the ascending gaseous species, which contain mainly H_2O, CO_2, CO, and H_2, will react with

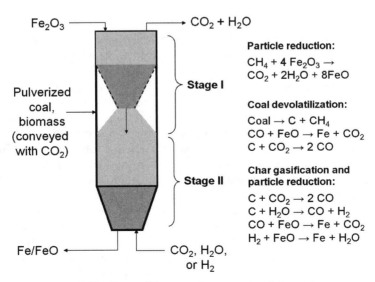

Figure 5.17. Gas–solid contacting pattern of the reducer.

the descending solids. During this contact, the coal char is gasified progressively by CO_2 and H_2O formed at the lower portion of the reducer. Provided that an adequate residence time is given, which is estimated at 30–90 minutes, coal char can be converted fully. Furthermore, the Fe_2O_3 particles can be reduced to a mixture of metallic Fe and FeO. Coal ash will exit from the bottom of the reducer along with the reduced particles. Thus, reaction product streams from the reducer include a solid particles stream that exits from the bottom of the reducer, containing Fe, FeO, and coal ash as well as an exhaust gas stream that exits from the top of the reducer, containing mainly CO_2 and H_2O. By condensing out the H_2O in the exhaust gas stream, a pressurized CO_2 stream can be obtained from the reducer. The following section further illustrates the reducer performance using ASPEN PLUS simulation.

ASPEN PLUS Simulation on the CDCL Reducer

Model Setup. The ASPEN PLUS thermodynamic models described in Chapter 4 are shown to predict the theoretical performance of a reducer that converts gaseous fuels. To extend the models for simulating the performance of a CDCL reducer that converts solid fuels, additional model simulation information needs to be provided. This includes a definition of the solid fuel, changes in the physical property method, and designation of solid unit operation. In this simulation, solid fuels are handled with ASPEN PLUS built-in methods.[37] The choice of a suitable method is dependent on the fuel type. For example, simple solid fuels such as pure carbon are defined as conventional solids, whereas complicated mixtures such as coal and biomass are defined as nonconventional solids. Because the analysis on conventional solids directly uses the built-in physical/chemical property databank in ASPEN PLUS, simulation of pure carbon is rather straightforward. When a nonconventional solid fuel is used, however, the physical and chemical property information of the solid is necessary for accurate simulation. Such information includes proximate analysis, ultimate analysis, and heat of combustion.

In the following sections, Illinois #6 coal with properties, as given in Chapter 4, is used as the fuel, unless otherwise noted. Because the moisture in coal can affect the oxygen-carrier conversion, an extra coal-drying step using high-temperature nitrogen gas is incorporated prior to the coal entering the reducer. The drying step is modeled using an RStoic block and a Flash2 block. Because coal is defined as a nonconventional solid, it cannot directly "react" with other reactants in an ASPEN module. Instead, an RYield block with a FORTRAN subroutine is used to decompose coal into water, inert ash, and elements such as oxygen, hydrogen, and carbon. The decomposed components, all of which are conventional components except for ash, then are sent to an RGibbs block to perform the desired chemical reactions. The heat of reaction for coal conversion is equal to the sum of the heats of conversion of the RYield (decomposition) block and the RGibbs (reaction) block. If carbon conversion is not completed fully according to experimental results, then unreacted carbon will be split out from the interblock stream between RYield and RGibbs and mixed with the solid stream after the reducer.

Figure 5.18 illustrates the block diagram of the ASPEN PLUS models for both the fluidized-bed and moving-bed reducers. Similar to the models used for the looping process with gaseous fuels given in Chapter 4, a model of one RGibbs block is used to represent a fluidized-bed reducer as shown in Figure 5.18(a), whereas a model of five RGibbs blocks in series is used to represent

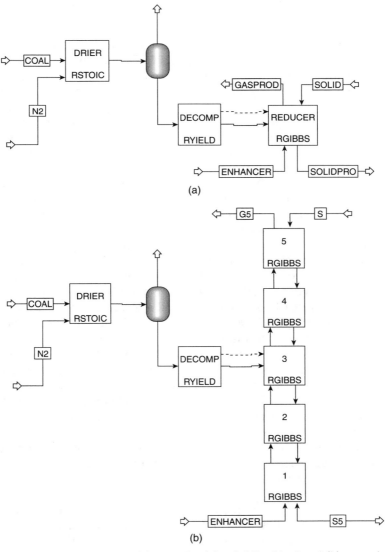

Figure 5.18. ASPEN PLUS model setup for (a) a fluidized bed and (b) a moving bed.

a moving-bed reducer as shown in Figure 5.18(b). For the moving-bed model, dry coal is injected into the middle block (Block 3) in the five-block moving-bed model to mimic the reducer design given in Figure 5.17. In addition, char reaction enhancers such as H_2, CO_2, and/or steam can be introduced from the bottom block (Block 1). When there is an excess of Fe_2O_3, coal can be gasified fully in Block 3 of the moving bed model. Thus, the syngas from coal gasification then will move upward and be converted in Blocks 4 and 5. In this case, no reaction will take place in Blocks 1 and 2. In the case where coal cannot be gasified fully in Block 3, further conversion of coal char will take place in Blocks 1 and 2.

Coal and Oxygen-Carrier Conversions. An ideal reducer should be configured such that coal is oxidized fully to CO_2 and H_2O. Unconverted fuel can exit from the reducer either in the form of unconverted coal char or partially converted volatiles. The unconverted coal char will be carried over to the oxidizer, resulting in a contaminated H_2 stream. The partially converted volatiles, will reduce the energy conversion efficiency of the process, as well as contaminate the CO_2 stream from the reducer. Although a higher iron oxide conversion is desirable, optimization of the reducer should emphasize coal conversion efficiency rather than iron oxide conversion.

The char gasification enhancer can enhance the reaction rate between the oxygen carrier and the coal; however, the extent of reaction can be limited by the thermodynamic equilibria. The thermodynamic analysis for syngas conversions in Chapter 4 shows that a supply feed of Fe_2O_3 above the stoichiometric amount is required to oxidize coal fully to CO_2 and H_2O. Furthermore, a countercurrent moving bed reducer will be more effective than a fluidized bed reducer. To examine the reducer performance when solid fuels such as coal are used, an ASPEN PLUS simulation is performed. For illustration purposes, pure carbon (graphite) is used to represent coal. Figure 5.19 shows the relationship between the carbon distribution in the end products and the Fe_2O_3/carbon molar ratio using two different reducer designs.

Figure 5.19(a) shows that more than 6 mol of Fe_2O_3 needs to be fed to oxidize fully 1 mol of carbon to CO_2 in a fluidized bed reducer. Based on the same iron oxide conversion definition used in Chapter 4, the fluidized-bed reducer yields a conversion for Fe_2O_3 of 11.11%. In comparison, a countercurrent moving-bed reducer requires only 0.7 mol of Fe_2O_3 to convert 1 mol of carbon to CO_2, which corresponds to a conversion for Fe_2O_3 of 95.2% and a weight ratio of Fe_2O_3 to carbon of 9.33:1. The dramatic increase in reduction of the iron oxide particles illustrates a desirable gas–solid flow pattern for using countercurrent moving beds over fluidized beds.

Fates of Polutants in Reducer. The previous figure illustrates the effect of reducer design with coal approximated by pure carbon. The actual composition of coal, however, is far more complex than pure carbon. This section analyzes the fate of pollutants such as sulfur and mercury in coal. Here, Illinois

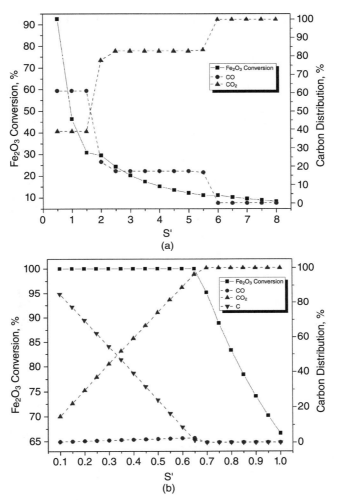

Figure 5.19. Iron oxide conversion and carbon distribution with respect to different Fe_2O_3/carbon ratios at 900°C and 30 atm for (a) a reducer with perfect mixing and (b) a countercurrent moving-bed reactor.

#6 coal with a composition shown previously in Table 4.7 is used as the fuel. The mercury content in coal is assumed to be 100 ppb (by weight). All mercury in the coal is assumed to be in elemental form. At 900°C and 30 atm, the maximum conversion of Fe_2O_3 that ensures a full conversion of coal to CO_2 and H_2O is calculated to be 73.8%. Table 5.3 shows the mass balance of the reducer using the ASPEN PLUS model.

As shown in Table 5.3, all mercury and chloride compounds will exit the reducer as a part of the gaseous stream. Although a small amount of sulfur

TABLE 5.3. Reducer Mass Balance Based on the ASPEN PLUS Model at 900°C

Species	Inlet Flow Rate (kmol/s)	Outlet Flow Rate (kmol/s)
CO	0	6.06E–03
CO_2	0	5.302
H_2	2.232	2.46E–03
H_2O	0.617	2.840
Hg	4.99E–07	4.99E–07
S	0.0783	1.30E–11
H_2S	0	1.03E–05
COS	0	7.83E–07
SO_2	0	5.08E–03
SO_3	0	1.28E-08
N_2	0.0446	0.0446
O_2	0.215	1.76E–11
Cl_2	4.09E–03	2.21E–11
HCl	0	8.18E–03
H_3N	0	2.11E–08
CH_4	0	3.13E–13
NO	0	3.85E–10
NO_2	0	1.25E–16
N_2O	0	1.05E–14
Fe_2O_3	5.6	0
Fe_3O_4	0	0
$Fe_{0.947}O$	0	4.385
$Fe_{0.877}S$	0	0.0732
Fe	0	6.983
C	5.308	0
Ash[a]	9.7	9.7
Total Flow (kg/s)	994.2763	994.2763

[a]The molar weight of ash is set to 1.

will escape from the reducer along with the exhaust gas in the form of SO_2, 93.5% of the sulfur in the coal will react with iron oxide in the presence of reducing agents, forming $Fe_{0.877}S$. The solid sulfur compound will be carried over to the oxidizer along with the reduced iron oxide particles.

Effect of Temperature. The sensitivity analysis is carried out using the ASPEN PLUS model discussed in Section 5.4.2. The result of the analysis is shown in Figure 5.20. As shown, a higher reaction temperature favors the endothermic coal–Fe_2O_3 reaction from both kinetic and thermodynamic viewpoints. As shown in Figure 5.20, with an Fe_2O_3/coal ratio identical to the case shown in Table 5.3, carbon cannot be converted fully at temperatures below 850°C. Therefore, the reducer needs to be operated at above 850°C to maximize fuel conversion.

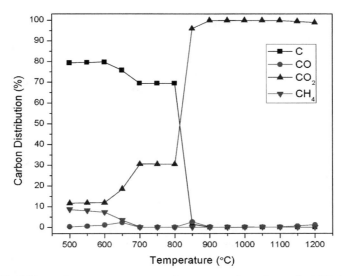

Figure 5.20. Effect of temperature on carbon conversions in coal at 30 atm with an Fe_2O_3-to-coal ratio of 8.94:1 by weight.

A practical concern on the reducer operating temperatures is its effect on ash handling. An operating temperature significantly exceeding 900°C is expected to make coal ash sticky and, hence, affect the solids flow in the reactor. As will be discussed in Section 5.5, a temperature higher than 1,000°C may decrease the overall process efficiency. Thus, it is desirable to operate a reducer at temperatures of ~900°C.

Effect of Pressure. The coal–Fe_2O_3 reaction in the CDCL reducer generates gaseous products. Thus, a lower operating pressure favors the coal conversion as shown in Figure 5.21 for the reaction at 850°C. Considering both the kinetics and the equilibrium conversion, the suitable operating pressure range for the CDCL reducer operation is determined to be 1–30 atm.

Effect of Steam and CO₂. Although steam and CO_2 can be used as char reaction enhancers, an excessive amount of steam and/or CO_2 can negatively affect the Fe_2O_3 conversion as a result of their capability of oxidizing Fe/FeO. Figure 5.22 illustrates the effect of steam and CO_2 on the Fe_2O_3 conversion. The operating conditions of the reducer are identical to those shown in Table 5.3. It is shown that an injection of a small amount of CO_2 or steam into the reducer will not lead to a drastic decrease in the Fe_2O_3 conversion.

Experimental Testing of Reducer Operations
Bench-scale tests were carried out in a moving bed reactor at the Ohio State University (OSU) based on the configuration described in Chapter 4. Given in the following sections are representative test results obtained from the

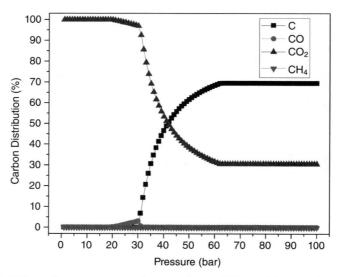

Figure 5.21. Effect of pressure on coal conversion at 850°C, with an Fe_2O_3 to coal ratio of 8.94:1 by weight.

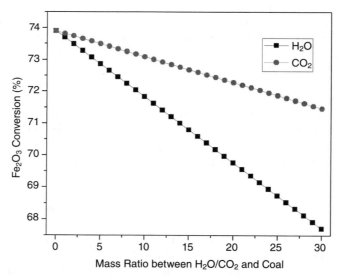

Figure 5.22. Effect of steam and CO_2 on the Fe_2O_3 conversion at 900°C and 30 atm, with an Fe_2O_3 to coal ratio of 8.94:1 by weight.

operational configuration of a moving bed given in Figure 5.17, where the reducer can be divided into two stages with Stage I conducting coal-volatile conversion and Stage II performing char gasification and iron oxide particle reduction. The results corroborate the fixed bed testing.

Reducer Stage I Testing. Iron oxide is active in cracking coal volatiles into methane,[38] which is the most stable hydrocarbon formed from the cracking of coal volatiles up to 1,030°C.[39] The reactions between Pittsburgh #8 coal volatiles/tar and the composite Fe_2O_3 particles in a fixed bed reveal that 87% of the volatiles either were cracked to methane or were oxidized to CO_2/H_2O by the composite particles within a gas residence time of 4.6 seconds at 850°C. Thus, if the composite particles are capable of oxidizing methane, then they will be able to oxidize coal volatiles.

The methane conversion profile in a given countercurrent moving-bed reactor is shown in Figure 4.27 of Chapter 4. This methane conversion profile indicates the Stage I reducer operation behavior. As shown in the figure, more than 99.8% of methane is converted to CO_2 and H_2O. Thus, the results reflect that iron-oxide composite particles are capable of close-to-fully oxidizing coal volatiles into CO_2 and H_2O.

Reducer Stage II Testing. The oxidation of coal chars obtained from both bituminous and lignite coal using the composite particles yields a more than 90% char conversion. The oxidization of low-volatile anthracite coal using the composite particles yields 95% conversion, as shown in Figure 5.23. The figure also shows a high particle conversion and a high CO_2 outlet concentration. During the testing, the solids residence time in the bed is 60–100 minutes, and

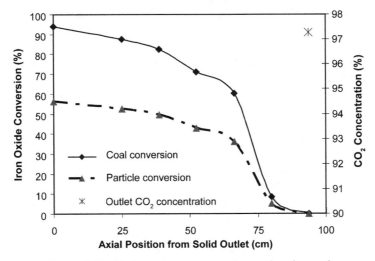

Figure 5.23. Reducer test results using anthracite coal.

TABLE 5.4. Summary of the Reducer Demonstration Results using Coal, Coal Char, and Volatiles

Type of Fuel	Coal Volatiles (CH_4)	Lignite Char	Bituminous Char	Anthracite Coal
Configuration Tested	Both	I	II	I
Fuel Conversion (%)	99.8	94.9	90.5	95.5
CO_2 Concentration in Exhaust (% Dry Basis)	98.8	99.23	99.8	97.3
Conversion Enhancer Used	H_2	CO_2	O_2 and CO_2	H_2

char gasification enhancers such as H_2 and/or CO_2 are present. The test results are given in Table 5.4.

As shown in Table 5.4, a countercurrent moving bed reducer with an iron oxide-based oxygen carrier can convert 90%–95% of coal char into concentrated CO_2 and H_2O. The X-ray diffraction (XRD) analysis of the composite particles obtained after the reducer testing reveals the formation of $Fe_{0.877}S$, which is consistent with the thermodynamic analysis. Although the coal and particle conversions are lower than those predicted from the thermodynamic analysis, the optimization of design and operating conditions such as char reaction enhancer flow rate, operating temperature, and gas and solid contact time can improve the reducer performance.

5.4.3 Performance of the Oxidizer and the Combustor

Oxidizer

The oxidizer in the CDCL Process is similar to the oxidizer in the SCL Process, which was discussed in Chapter 4. Steam is sent to the oxidizer to convert Fe and FeO to Fe_3O_4 while producing hydrogen. Although the unconverted steam readily can be condensed, a lower steam conversion will lead to a larger energy consumption for steam generation. Thus, it is essential to maximize the steam-to-hydrogen conversion. Similar to the reducer, a countercurrent moving-bed reactor is an effective design for the oxidizer. Figure 5.24 compares the theoretical steam to hydrogen conversion in a moving bed and a fluidized bed operated at 700°C and 30 atm using pure iron particles. At a given steam-to-iron molar flow rate, the moving bed results in a significantly higher conversion.

In addition to the gas–solid contacting mode, the composition of the Fe/FeO particle is also an important factor in the steam to hydrogen conversion in the oxidizer. Figure 5.25 shows the effect of Fe content on the steam conversion in a countercurrent moving bed oxidizer. As shown in the figure, the increased presence of Fe in the particles can drastically improve the steam to hydrogen conversion. When pure FeO is used, only 26.4% of the steam can be converted to hydrogen; however, when the particles contain more than

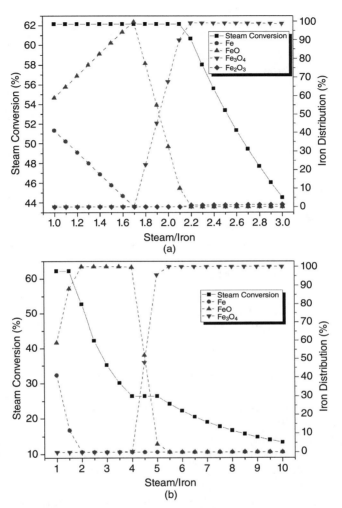

Figure 5.24. Steam to hydrogen conversion for (a) countercurrent moving bed oxidizer and (b) fluidized bed oxidizer. Reactor operating conditions are: 700°C, 30 atm.

33% metallic iron (by mole), the steam conversion is increased to 62.2%. Any further increase in the iron content will not lead to a higher steam conversion because the gas composition already have reached the equilibrium point with Fe. Therefore, the reduced Fe_2O_3 from the reducer should contain at least 33.5% Fe to achieve an optimum steam to hydrogen conversion, which corresponds to a conversion for Fe_2O_3 of 49.0% or higher. Thus, the performance of the reducer and that of the oxidizer closely are related and contribute synergistically to the overall looping process efficiency. Contrast this situation with the early steam-iron processes in which the iron oxide reduction in the

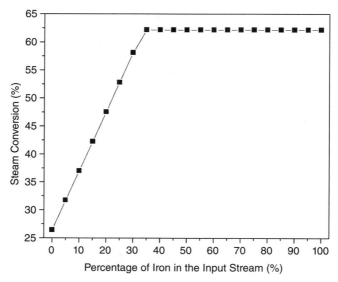

Figure 5.25. Relationship between the Fe/FeO composition and the steam-to-hydrogen conversion in a countercurrent moving bed oxidizer operated at 30 atm and 700°C. x axis denotes the molar percentage of metallic iron in the Fe/FeO mixture.

fluidized bed or two-stage fluidized-bed reducer was low, thereby leading to a low steam conversion in the oxidizer. This earlier situation does not represent an ideal synergistic case of the looping process operation.

The reaction temperature also has a significant effect on the steam-to-hydrogen conversion in the oxidizer. A lower reaction temperature thermodynamically would favor the exothermic reaction between steam and Fe/FeO. For instance, the reaction temperature of 500°C would lead to a steam-to-hydrogen conversion of 82.2% compared with a 62.2% conversion at 700°C. Furthermore, a lower operating temperature will reduce the capital cost of the oxidizer, as lower cost materials can be used. However, the low reactivity of the iron ore renders it challenging to perform the steam-iron reaction below 600°C. By comparison, the composite Fe_2O_3 particles obtained through particle optimization described in Chapter 2 can undergo the steam-iron reaction at 500°C or even lower with a satisfactory reaction rate.

The fate of sulfur in the oxidizer operation is another important issue that needs to be accounted for. At 700°C and 30 atm, the oxidizer can convert all Fe and FeO to Fe_3O_4, with a steam conversion rate of 62%. Furthermore, about 9.9% of $Fe_{0.877}S$ carried over to the oxidizer will react with steam to produce Fe_3O_4 and H_2S based on the thermodynamic analysis. This reaction leads to a H_2S concentration of 430 ppmv in the hydrogen product (dry basis). The remaining $Fe_{0.877}S$ will be carried over to the combustor along with Fe_3O_4 to react with air.

Combustor

The early steam-iron processes did not include a combustor, but it is crucial for the SCL and the CDCL processes to include a combustor unit. The combustor plays an important role in the fuel conversion and the energy management of the process. The early steam-iron processes directly circulated the solid products from the oxidizer to the reducer. Because steam oxidation only can regenerate iron to Fe_3O_4, complete oxidation of fuels to CO_2 and H_2O cannot be achieved in these processes. For example, when Fe_3O_4 is used in the reducer to convert coal at 900°C, the exhaust gas will have a CO concentration of 11.2% at a minimum. To compare, nearly 100% fuel conversion can be achieved when Fe_2O_3 is used. The CDCL Process also can be operated by sending part of the reduced particles from the reducer outlet directly to the combustor for heat generation, whereas the remaining particles are processed through the oxidizer for hydrogen generation. In this manner, the CDCL system can generate any combination of hydrogen and electricity products from coal. More information regarding the combustion of reduced iron oxide particles is given in Chapters 3 and 4.

NO_x and SO_x are the main contaminants from the combustor. All $Fe_{0.877}S$ introduced to the combustor will be oxidized fully to Fe_2O_3, SO_2, and SO_3. At high combustor operating temperatures, NO_x also will be formed. It is estimated by ASPEN PLUS simulation that when the combustor is operated at 30–32 atm and 1,150°C, the NO_x level is ~300 ppm, and the concentrations of SO_2 and SO_3 are 1.4% and 500 ppm (by volume), respectively. Furthermore, note that because of the slow reaction rates for thermal NO_x formation, the actual NO_x level in the combustor may be significantly lower than the equilibrium value predicted by ASPEN PLUS. These compounds can be separated from the flue gas using commercial flue gas cleaning devices. A further account of the pollutants is given in the following section.

5.4.4 Fate of Pollutants and Ash

This section discusses issues such as ash separation, pollutant control, and the tolerance of particles to pollutants that are relevant to the CDCL Process operation.

Ash Separation

Determining an appropriate size of iron oxide particles requires extensive consideration of—in addition to particle reactivity—the fluidization characteristics during pneumatic conveyance in the entrained bed combustor and the behavior during gas–solid countercurrent flow in the moving bed reducer and oxidizer. Considering these fluidization effects, 0.7–8 mm is the appropriate size for iron oxide particles. Particles in this size range are substantially larger than those of pulverized coal (50–250 μm). Note that typical fly ash particles will be about an order of magnitude smaller than the iron particles (under the operating conditions of the reducer, oxidizer and combustor, and most of the ash particles will be in fly ash form, 0.5–100 μm). Thus, most of the ash particles can be separated from the iron oxide particles based on the size difference using such devices as sieves and/or cyclones. To test the fly ash separation, fly ash obtained from a pulverized coal combustion power plant was mixed with cylindrical Fe_2O_3 composite pellets of 5-mm diameter and 3-mm height. Redox reactions of the mixture then are performed in the countercurrent moving-bed reactor, shown in Figure 4.27, prior to mechanical ash separation from the pellets using a 2.8-mm sieve. It was determined that 75.8% of the ash was separated from the pellets with ease after 15 seconds of mechanical sieving. Given such an ash separation efficiency, the ash content in the CDCL system is estimated to be 0.54 wt% with respect to the overall solids content in the reactor. Such a low ash content is expected to have little effect on CDCL operation. The test results also imply that mechanical ash separation methods are feasible.

Tolerance of Particles to Contaminants
Understanding the blinding effect of the oxygen carrier with coal contaminants is important in the assessment of the viability of oxygen carrier particles for the CDCL Process. The blinding effect refers to the deactivation of the particles from poisoning or fouling. The blinding effect, along with particle attrition, directly affects the quantity of the fresh particle makeup and, hence the process economics. As discussed in Chapter 2, the particle attrition rate is ~0.57%/cycle. To assess the blinding effects induced by the coal contaminants, the same batch of iron oxide particles was used to react with different types of coal, coal char and ash. The particles were regenerated with air after each reduction experiment, and the reactivity of the particles that endured three redox cycles were characterized in a TGA and a differential bed. No notable decrease in the particle reactivity was evidenced from these tests. The blinding effect of the particles needs to be tested further with more cycles to ascertain whether iron oxide particles are robust enough for the CDCL Process. More study is required to determine the optimum design of particles to improve contaminant tolerance.

Fate of Contaminants
Sections 5.4.2 and 5.4.3 cover the fates of the contaminants with regard to CDCL reactor operation. This section specifically looks through a process point of view to define a pollutant control strategy for the CDCL Process. The SO_x, NO_x, and mercury compounds generated from the CDCL Process can be captured with ease using existing pollutant control methods. Based on thermodynamic analysis, 93.5% of the sulfur in coal can be captured by composite particles in the form of $Fe_{0.877}S$. The remaining sulfur will be released from the reducer along with CO_2 in the form of SO_2. The SO_2 can be sequestered along with CO_2.[40] $Fe_{0.877}S$ in the reducer will be carried over to the oxidizer to react with steam. In the oxidizer, 9.9% of $Fe_{0.877}S$ will be oxidized with steam, forming Fe_3O_4 and H_2S. The H_2S concentration in the oxidizer product gas

stream is 430 ppmv, which can be removed using traditional scrubbing techniques such as methylldiethanolamine (MDEA) or Selexol. The remaining $Fe_{0.877}S$ will be carried over to the combustor to form Fe_2O_3, SO_2, and SO_3. The SO_x concentration in the exhaust gas is estimated to be 1.5%, which can be stripped using an existing flue gas desulfurization (FGD) unit in the pulverized coal (PC) power plant. Because of the presence of nitrogen in air, the combustor also generates NO_x. The NO_x concentration from the combustor exhaust gas stream is estimated to be 300 ppm, which is much lower than the output from a PC boiler. The existing selective catalytic reduction (SCR) method can be used to capture NO_x. An ASPEN PLUS process simulation showed that 100% of the mercury in coal will be emitted in elemental form from the reducer. The concentration of mercury will be ~43 ppb (by wt.) when Illinois #6 coal is used. The CO_2 stream containing mercury may be directly sequestered. If not, an activated carbon bed can be used to remove mercury from the CO_2 stream before it is sequestered.

At the present stage of the CDCL Process development, the fate of the contaminants is mainly predicted based on thermodynamic analysis. Experimental data will be needed to substantiate all results.

5.4.5 *Energy Management, Heat Integration, and General Comments*

Stoichiometrically, 1 mol of carbon in coal can be converted to 2 mol of hydrogen according to

$$C + 2H_2O(g) \rightarrow CO_2 + 2H_2 \qquad \Delta H = 178 \text{ kJ/mol at } 298.15 \text{ K} \qquad (5.4.4)$$

In the conversion process, a portion of the energy from coal needs to be used to support steam generation, air separation, hydrogen product separation and purification, and the endothermic steam–carbon reaction. A practical coal-to-hydrogen process will deliver a thermal efficiency far lower than 100%. In addition to the energy utilization listed, the efficiency will be reduced as a result of exergy degradation from the less-than-perfect operating efficiency for such units as heat exchange devices and the limitation on heat integration. Thus, process intensification strategies are needed to minimize parasitic energy consumption and energy loss, which enhance the overall energy conversion efficiency. This section discusses such strategies used in the CDCL Process, and they are extendible to other coal conversion processes. These strategies include: (1) minimization of air separation, (2) minimization of the energy consumption for CO_2 separation, (3) reduction of the steam usage, and (4) optimization of the heat-integration scheme.

Air Separation

The traditional gasifier consumes a significant amount of oxygen. As discussed in Section 4.2.1, cryogenic distillation for air separation requires an extremely low temperature, and extensive gas compression and recompression for the separation of O_2 from N_2, which is energy-intensive. Thus, oxidation using oxygen carriers would be a more economical approach than the use of pure oxygen because using oxygen carriers eliminates the step of O_2 separation. Furthermore, using one chemical loop and, hence one type of looping particles as in the Coal-Direct Chemical Looping (CDCL) Process will be more economically feasible than using two chemical loops and, hence two types of looping particles as in the ALSTOM Hybrid Combustion-Gasification Process. The heat required in the CDCL Process and, thus the oxygen needed can be compensated for completely via the CDCL Configuration II, as discussed in Section 5.3.2.

CO₂ Separation
CO_2 separation is another energy-consuming step in traditional coal conversion processes. Various CO_2 separation methods for combustion flue gas and syngas applications are available, as discussed in Chapter 1 and Chapter 3. Basically, there are two approaches: the high-temperature sorbent method and the low-temperature solvent method. The high-temperature sorbent-based processes such as the CO_2 Acceptor Process and the HyPr-Ring Process use $CaO/Ca(OH)_2$ to capture CO_2 through the carbonation reaction. Regeneration of the spent sorbent involves intricate heat-integration steps. In the CDCL Process, both chemical looping combustion and chemical looping gasification are adopted to provide process versatility in product generation and CO_2 separation. This gasification scheme can circumvent the energy-intensive CO_2 separation approaches encountered in the traditional processes. Moreover, the CO_2 from the CDCL Process is already at a high pressure, and thus the energy consumption in CO_2 compression also can be reduced.

Minimization of Steam Usage
Hydrogen is produced from coal by a reaction with steam, either directly or indirectly, in a traditional process. The latent heat and the sensible heat for steam generation represent a significant portion of the total energy consumption of the process. Because of the equilibrium and kinetics limitations, excessive steam beyond the stoichiometric requirement usually is needed to reach a desired H_2 yield. Although the heat in the excess steam partially can be recovered, a significant increase in the process entropy occurs. For iron-based chemical looping processes such as the CDCL Process, the extent of reduction of the iron oxide particles in the reducer is the key factor affecting the steam requirement. Through the coal and iron oxide particle conversion enhancement strategies discussed in Sections 5.4.1 and 5.4.2, the steam usage in the CDCL Process is minimized.

Optimization of the Heat-Integration Scheme
Heat integration is of direct relevance to the efficiency of an energy conversion process. The optimization of heat-integration schemes generally is a complex issue. When an exothermic reaction in a process system takes place at a temperature higher than another reaction in the process system that is endother-

mic, the heat generated from the exothermic reaction then can be used readily to support the endothermic reaction. In the heat-integration sense, the heat generated by the exothermic reaction is stored in the reaction products generated from the endothermic reaction. Thus, in this integration, the exergy in heat is recuperated. As another illustration, the heat-integration scheme for Configuration II of the CDCL Process is given for the overall reaction of the process as

$$xC + yH_2O + (x - y/2)O_2 \rightarrow xCO_2 + yH_2 + \Delta H \qquad (5.4.5)$$

For simplicity, the composition of coal is considered as pure carbon. Principally, the heat of a reaction can be adjusted by varying the ratio of x to y. Thus, at the condition when the enthalpy change ΔH is zero, all chemical energy in the carbon is converted to H_2, considering that the enthalpy of devaluation for CO_2, H_2O, and O_2 is negligible as compared with carbon and H_2. For a real process, however, heat loss is inevitable. Furthermore, parasitic energy within the process is used to generate excess steam for hydrogen production, to produce electricity for gas compression and particle circulation, and to generate heat for pollutant separation. Therefore, an optimal heat integration strategy that maximizes the H_2 yield will minimize the absolute value of ΔH.

In the CDCL Process, the combustion of a fully or partially reduced iron oxide is highly exothermic and can take place at temperatures that are much higher than those for the endothermic coal oxidation reaction. Therefore, the heat released in the combustion of the reduced iron oxide is used to compensate for the endothermic coal oxidation reaction. Moreover, the heat generation from the combustion of particles directly from the reducer and from the oxidizer can be easily altered by adjusting the percentage of the particles from the reducer and the oxidizer to the combustor, making the process flexible for hydrogen and electricity cogeneration. The overall material and energy integration scheme for the CDCL Process is shown in Figure 5.26.

5.5 Process Simulation on the Coal-Direct Chemical Looping Process

This section presents a case study of the energy conversion efficiency of the CDCL Process system for H_2 production using ASPEN PLUS process simulation. The advantages of the energy management strategies in the looping processes are illustrated.

5.5.1 ASPEN Model Setup

Model Assumptions
The ASPEN simulation model is used to illustrate Configuration II of the CDCL Process. Hydrogen is the desired product in this case. Thus, the power generated from the system only is used to offset parasitic energy consumption.

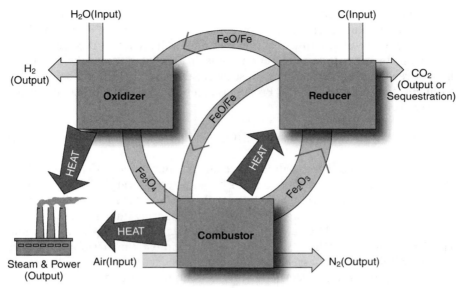

Figure 5.26. Material flow and energy flow in a CDCL Process.

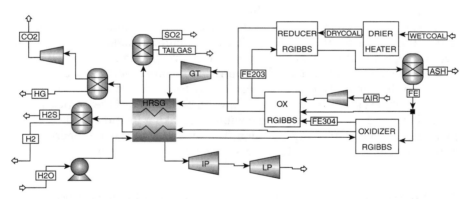

Figure 5.27. Process flow diagram of the ASPEN PLUS model for the CDCL Process optimized for hydrogen production.

The process flow diagram is shown in Figure 5.27. The assumptions used in the simulation are identical to the ones given in Section 4.5.1, except that:

1. All reactions reach thermodynamic equilibrium.
2. The solids are made up of 48.9% Fe_2O_3 and 51.1% supporting materials, mainly SiC (by weight).
3. Steam is used as the char reaction enhancer.

4. The heat loss in the CDCL system is 0.5% of the HHV of coal.

5. The coal is dried to 3% moisture prior to entering the reducer.

In setting up the ASPEN simulation, it is important to verify whether the correct physical properties are represented for the system. ASPEN PLUS retrieves parameters from different databanks, and the COMBUST, INORGANIC, SOLIDS and PURE databanks were selected for the CDCL Prcess simulation. Revised physical property data, as discussed in Chapter 4, are used for FeO and Fe_3O_4. PR-BM is the property method used for the global system, whereas STEAM-TA is the method used for steam generation and conversion units such as heat recovery steam generation (HRSG) and steam turbines.

5.5.2 Simulation Results

Mass and energy flows obtained from the ASPEN simulation for the CDCL Process are discussed in the following sections.

Reducer
Coal first is pulverized and dried to a moisture content of 3%. The coal powder then is introduced to the middle section of the reducer at a rate of 132.9 t/h. The hot Fe_2O_3 and SiC particles (1,250°C) from the combustor enter the reducer from the top at a rate of 2,312.2 t/h (with 48.9% Fe_2O_3 and 51.1% SiC). The sensible heat of the particles partially is used to compensate for the heat needed in the reducer. In addition, the sensible heat in the combustor flue gas (1,250°C, 373.4 t/h) also is transferred to the reducer through the heat exchangers. The reducer operates at 870°C and 30 atm. In the reducer, Fe_2O_3 is reduced to a mixture of Fe (26.1 wt.% based on the overall solids), $Fe_{0.947}O$ (wüstite, 15.6 wt.%), and $Fe_{0.877}S$ (Pyrrhotite, 0.35 wt.%). The overall exit solids flow rate from the reducer is 2,063.8 t/hr. An exhaust gas stream of 370.8 t/h is produced with 68.7% CO_2 and 30.6% steam from the top of the reducer. The exhaust gas also contains 0.58% N_2 and 0.14% SO_2. The hot gas is sent to an HRSG for heat recovery. It then is cleaned, condensed, and compressed to 150 atm for sequestration.

Oxidizer
The oxidizer is operated at 720°C. Approximately 70.0% of the reduced Fe/FeO particles are used for hydrogen generation in the oxidizer. The remainder is sent directly to the combustor. Steam at 240°C and 32 atm is introduced at the bottom of the oxidizer at a flow rate of 286.0 t/h. Excess heat is removed by generating high-pressure steam. In this reaction, Fe and FeO are oxidized to Fe_3O_4, whereas H_2O is reduced to H_2. A gas stream of 19.9 t/h H_2 (62.1 mol%) is produced from the oxidizer. The product gas stream first is sent to the HRSG and then an acid gas removal (AGR) system. The purified hydrogen

(>99.995%) is compressed to 60 atm. The solid stream is sent to the combustor.

Combustor

Fe_3O_4 exiting the oxidizer, together with 30.0% of the solids discharged from the reducer, is conveyed pneumatically with air to the reducer. During conveyance, all reduced particles are reoxidized to Fe_2O_3, and a significant amount of heat is released. As a result, both the solids and the spent air are heated to ~1,250°C. The sensible heat carried by the particle and a portion of the hot spent air is used to meet the reducer heat requirement. The remaining hot spent air is used for feedstock preheating and steam generation. The spent air, after heat exchange, will go through SCR and FGD scrubbers before being vented to the atmosphere. Ash in the solid stream is separated from the hot solids by a cyclone before entering the reducer.

Tables 5.5 through 5.8 give the simulation results from the ASPEN model. As shown, CDCL can produce 19.9 t H_2 from 132.9 t coal while generating sufficient electricity for parasitic energy consumption. This energy conversion corresponds to a theoretical efficiency of 78.3% (HHV) for hydrogen production. There are alternative heat management schemes for the CDCL Process. For example, an expander can be used after the combustor for power generation, resulting in a higher efficiency. Furthermore, the heat required by the reducer fully can be provided by the high-temperature solids from the combustor. The efficiency for this case is estimated at 79% (HHV) with 72% of the

TABLE 5.5. Mass Input–Output of the CDCL Process

	Input				Output				
Stream	Coal	Water	Air	Recycled Solids	Off Gas	CO_2	H_2	Ash	Other
Temperature (°C)		25		1,250	120	40	40	800	25
Pressure (atm)		1		30	16	135	60	30	1
Mass Flow (t/hr)	132.9	178	477.7	2,312.2	373.4	311.8	19.9	12.9	70.6

TABLE 5.6. Energy Input-Output of the CDCL Process

	Coal Feedstock	Air Compressor (Electricity Consumption)	Steam Turbine (Electricity Generation)	H_2 Product	Waste Heat
Energy (MW)	1000	67.3	−67.5	−782.9	−216.9

TABLE 5.7. Heat and Energy Requirements in the CDCL Process

Unit	Reducer	Oxidizer	Combustor	HRSG
Energy (MW)	50.4	−46.2	0	344.1

TABLE 5.8. Power Balance in the CDCL Process

	Input Compressor			Output Steam Turbine			
Unit Operations	Air	H₂	CO₂	HP	IP	LP	Net
Power (MW)	45.8	12.6	8.9	−17.2	−30.6	−19.7	−0.2

efficiency coming from hydrogen and the remaining 7% from electrical power generation.

Overall, the energy conversion efficiency for the CDCL Process is nearly 20% higher than the traditional coal gasification WGS process for H_2 production with CO_2 capture discussed in Chapter 4. The improved efficiency in the CDCL Process results from the improved energy management of the process.

5.6 Concluding Remarks

Chemical looping techniques can process carbonaceous solid feedstock directly or indirectly through a variety of methods. The indirect chemical looping techniques use syngas obtained from the gasifier or other gaseous fuels as the reducer inlet fuel as discussed in Chapter 4. This chapter covered direct chemical looping processes using solid fuels as the feedstock. Among all energy conversion processes, those using direct chemical looping techniques represent the greatest potential in energy conversion efficiency. Many issues, however, remain to be addressed, particularly with respect to the interactive characteristics between coal and oxygen carriers and the subsequent effects on the looping reactor and process system operation. The direct chemical looping gasification of solid fuels often can be realized by one of the following two approaches: (1) direct gasification with *in situ* CO_2 capture using high-temperature sorbents or (2) direct gasification using an oxygen carrier. Calcium-based chemical looping gasification processes use the first approach, whereas iron based chemical looping gasification processes use the latter approach.

Calcium-based chemical looping gasification processes share a common approach for which $CaO/Ca(OH)_2$ sorbents are used to enhance fuel conversion to H_2 by *in situ* CO_2 removal. They differ, however, in the manner under which the spent $CaCO_3$ sorbent is regenerated. The regeneration of calcium-based sorbents for effective application in a looping system continues to pose challenges for heat integration in sustaining the highly endothermic calcination reaction. Endothermic calcination heat can be provided by the combustion of reactant fuels or product fuels as used by the CO_2 Acceptor Process, the HyPr-Ring Process, and the ZECA Process. It also can be provided by high-temperature inert heat transfer carrier particles and/or oxygen-carrier particles as used by the ALSTOM Hybrid Combustion-Gasification Process and the GE Fuel-Flexible Process. Other important factors for calcium-based gasification processes include the sorbent reactivity, sorbent recyclability, and sorbent reaction kinetics. These factors will affect the sorbent loading in the reactors, the CO_2 capture rate, the fuel/sorbent ratio, the makeup sorbent amount and, hence, the process system economics.

The CDCL Process using solid fuels and metal oxide oxygen carriers in a fluidized-bed reducer, as applied to chemical looping combustion, is discussed in Section 3.3.5. Here, the CDCL Process using solid fuels and iron-based oxygen carriers in a countercurrent gas–solid flow reducer as applied to chemical looping gasification is illustrated. These processes are conducted in a similar manner to the Syngas Chemical Looping Process and show promise. The solids flow in the CDCL Process can be in a moving bed or in a series of fluidized beds. The CDCL Process and the Syngas Chemical Looping Process differ in heat-loading requirements in the reducer and in the composition of solid and gas pollutants in the reducer and the oxidizer. The CDCL Process can be operated in two configurations depending on the reducer heat-integration scheme. Understanding the blinding effect of the oxygen-carrier with coal contaminants is critical for the accurate assessment of the viability of the oxygen-carrier particles to be used in the CDCL Process. The blinding effect and particle attrition rate directly affect the recyclability of the particles and, hence the quantity of the particles makeup.

The ASPEN PLUS simulation projects the fates of the pollutants including SO_x, NO_x, mercury, and H_2S in the looping reactors. The simulation indicates that sulfur in coal can be bound by the iron oxide particles in the form of $Fe_{0.877}S$. The simulation also reveals a high conversion efficiency of coal to hydrogen or to electricity in the CDCL Process.

References

1. Massachusetts Institute of Technology, "The Future of Coal: Options for a Carbon-Constrained World," MIT, Cambridge, MA (2007).
2. National Energy Technology Laboratories, "CO_2 Sequestration—CO_2 Capture" (2007).
3. Lewandowski, D., and D. Gray, "Potential Market Penetration of IGCC in the East Central North American Reliability Council Region of the U.S.," Paper Presented at the Gasification Technologies Conference, San Francisco, CA (2001).

4. Kanniche, M., and C. Bouallou, "CO$_2$ Capture Study in Advanced Integrated Gasification Combined Cycle," *Applied Thermal Engineering*, 27(16), 2693–2702 (2007).

5. Rubin, E. S., C. Chen, and A. B. Rao, "Cost and Performance of Fossil Fuel Power Plants with CO$_2$ Capture and Storage," *Energy Policy*, 35(9), 4444–4454 (2007).

6. Damen, K., M. van Troost, A. Faaij, and W. Turkenburg, "A Comparison of Electricity and Hydrogen Production Systems with CO$_2$ Capture and Storage: Part A. Review and Selection of Promising Conversion and Capture Technologies," *Progress in Energy and Combustion Science*, 32, 215–246 (2006).

7. National Energy Technology Laboratory, "Carbon Sequestration: CO$_2$ Capture" (2006).

8. Woods, M. C., P. J. Capicotto, J. L. Haslbeck, N. J. Kuehn, M. Matuszewski, L. L. Pinkerton, M. D. Rutkowski, R. L. Schoff, and V. Vaysman, "Cost and Performance Baseline for Fossil Energy Plants," DOE/NETL-2007/1281 (2007).

9. Stiegel, G. J., and M. Ramezan, "Hydrogen from Coal Gasification: An Economical Pathway to a Sustainable Energy Future," *International Journal of Coal Geology*, 65(3–4), 173–190 (2006).

10. Feng, W., P. J. Ji, and T. W. Tan, "Efficiency Penalty Analysis for Pure H$_2$ Production Processes with CO$_2$ Capture" *AIChE Journal*, 53(1), 249–261 (2007).

11. Dobbyn, R. C., H. M. Ondik, W. A. Willard, W. S. Brower, I. J. Feinberg, T. A. Hahn, G. E. Hicho, M. E. Read, C. R. Robbins, J. H. Smith, et al., "An Evaluation of the Performance of Materials and Components Used in the CO$_2$ Acceptor Process Gasification Pilot Plant," DOE Report No. DE85013673 (1978).

12. Lin, S. Y., M. Harada, Y. Suzuki, and H. Hatano, "Process Analysis for Hydrogen Production by Reaction Integrated Novel Gasification (Hypr-Ring)," *Energy Conversion and Management*, 46(6), 869–880 (2005).

13. Lin, S. Y., "Hydrogen Production by Reaction Integrated Novel Gasification," Presented at the 3rd International Workshop on *In-Situ* CO$_2$ Removal, ISCR, Ottawa, Canada (2007).

14. Lin, S. Y., M. Harada, Y. Suzuki, and H. Hatano, "Comparison of Pyrolysis Products between Coal, Coal/CaO, and Coal/Ca(OH)$_2$ Materials," *Energy & Fuels*, 17(3), 602–607 (2003).

15. Lin, S. Y., M. Harada, Y. Suzuki, and H. Hatano, "Hydrogen Production from Coal by Separating Carbon Dioxide During Gasification," *Fuel*, 81(16), 2079–2085 (2002).

16. Lin, S. Y., M. Harada, Y. Suzuki, and H. Hatano, "Gasification of Organic Material/ CaO Pellets with High-Pressure Steam," *Energy & Fuels*, 18(4), 1014–1020 (2004).

17. Lin, S. Y., M. Harada, Y. Suzuki, and H. Hatano, "CO$_2$ Separation During Hydrocarbon Gasification," *Energy*, 30(11–12), 2186–2193 (2005).

18. Lin, S. Y., M. Harada, Y. Suzuki, and H. Hatano, "Continuous Experiment Regarding Hydrogen Production by Coal/CaO Reaction with Steam (Ii) Solid Formation," *Fuel*, 85(7–8), 1143–1150 (2006).

19. Wang, Y., S. Y. Lin, and Y. Suzuki, "Effect of CaO Content on Hydration Rates of Ca-Based Sorbents at High Temperature," *Fuel Processing Technology*, 89(2), 220–226 (2008).

20. Wang, Y., S. Y. Lin, and Y. Suzuki, "Study of Limestone Calcination with CO_2 Capture: Decomposition Behavior in a Co2 Atmosphere," *Energy & Fuels*, 21(6), 3317–3321 (2007).

21. Wang, Y., S. Y. Lin, and Y. Suzuki, "Limestone Calcination with CO_2 Capture (Ii): Decomposition in CO_2/Steam and CO_2/N_2 Atmospheres," *Energy & Fuels*, 22(4), 2326–2331 (2008).

22. Ziock, H. J., K. S. Lackner, and D. P. Harrison, "Zero Emission Coal Power, a New Concept," 2001 (2001).

23. Backham, L., E. Croiset, and P. L. Douglas, "Simulation of a Coal Hydrogasification Process with Integrated CO_2 Capture" Paper Presented at the 7th Greenhouse Gas Control Technologies (GHGT) Conference, Vancouver, BC (2004).

24. Slowinski, G., "Some Technical Issues of Zero-Emission Coal Technology," *International Journal of Hydrogen Energy*, 31(8), 1091–1102 (2006).

25. Nawaz, M., and J. Ruby, "Zero Emission Coal Alliance Project Conceptual Design and Economics," Paper Presented at the 26th International Conference on Coal Utilization and Fuel Systems, Clearwater, Florida (2002).

26. Andrus, H. E., G. Burns, J. H. Chiu, G. N. Liljedahl, P. T. Stromberg, and P. R. Thibeault, "Hybrid Combustion-Gasification Chemical Looping Power Technology Development," ALSTOM Technical DOE Report for Project DE-FC26-03NT41866 (2006).

27. Andrus, H. A. E., Jr., J. H. Chiu, G. N. Liljedahl, P. T. Stromberg, P. R. Thibeault, and S. C. Jain, *Alstom's Hybrid Combustion-Gasification Chemical Looping Technology Development*, Proceedings of the 22nd Annual International Pittsburgh Coal Conference, Pittsburgh, Pennsylvania, September 11–15, (2005).

28. Fan, L.-S., and C. Zhu, Principles of Gas-Solids Flows, Cambridge University Press, New York, NY (1998).

29. Andrus, H. E., J. H. Chiu, G. N. Liljedahl, P. T. Stromberg, and P. R. Thibeault, "Hybrid Combustion-Gasification Chemical Looping Coal Power Technology Development Phase I—Final Report," ALSTOM Technical DOE Report for Project DE-FC26-03NT41866 (2004).

30. Rizeq, R. G., R. K. Lyon, V. Zamansky, and K. Das, "Fuel-Flexible Agc Technology for Production of H_2 Power, and Sequestration-Ready CO_2," Proceedings of the International Conference on Coal Utilization & Fuel Systems, Clearwater, Florida March 5–8 (2001).

31. Rizeq, R. G., J. West, A. Frydman, R. Subia, V. Zamansky, H. Loreth, L. Stonawski, T. Wiltowski, E. Hippo, and S. B. Lalvani, "Fuel-Flexible Gasification-Combustion Technology for Production of H_2 and Sequestration-Ready CO_2," Annual Technical Progress Report of Oct. 1, 2002 Sept. 30, 2003, DOE Award No. DE-FC26-00FT40974 (2003).

32. Rizeq, R. G., J. West, A. Frydman, R. Subia, R. Kumar, and V. Zamansky, "Fuel-Flexible Gasification-Combustion Technology for Production of H_2 and Sequestration-Ready CO_2," Annual Technical Progress Report of Oct. 1, 2001–Sept. 30, 2002, DOE Award No. DE-FC26-00FT40974 (2002).

33. Rizeq, R. G., J. West, R. Subia, A. Frydman, P. Kulkarni, J. Schwerman, V. Zamansky, J. Reinker, K. Mondal, L. Stonawski, et al., "Fuel-Flexible Gasification-Combustion Technology for Production of H_2 and Sequestration-Ready CO_2," Final Technical Report of Oct. 1, 2000–Aug. 31, 2004, DOE Award No. DE-FC26-00FT40974 (2005).

34. Stultz, S. C., and J. B. Kitto, Steam, Its Generation and Use, 40th edition, Babcock & Wilcox, Lynchburg, VA (1992).

35. Zhao, H., L. Liu, B. Wang, D. Xu, L. Jiang, and C. Zheng, "Sol–Gel-Derived $NiO/NiAl_2O_4$ Oxygen Carriers for Chemical-Looping Combustion by Coal Char," *Energy & Fuels*, 22(2), 898–905 (2008).

36. Scott, S. A., J. S. Dennis, A. N. Hayhurst, and T. Brown, "In Situ Gasification of a Solid Fuel and CO_2 Separation Using Chemical Looping," *AIChE Journal*, 52(9), 3325–3328 (2006).

37. ASPENTech, "Getting Started Modeling Processes with Solids," Aspen Technology, Cambridge, MA (2006).

38. Simell, P. A., J. K. Leppalahti, and J. B. S. Bredenberg, "Catalytic Purification of Tarry Fuel Gas with Carbonate Rocks and Ferrous Materials," *Fuel*, 71(2), 211–218 (1992).

39. Gueret, C., M. Daroux, and F. Billaud, "Methane Pyrolysis: Thermodynamics," *Chemical Engineering Science*, 52(5), 815–827 (1997).

40. Bachu, S., and W. D. Gunter, "Acid Gas Injection in the Alberta Basin, Canada: A CO_2 Storage Experience," *Geological Society Special Publication*, (223), 225–234 (2004).

CHAPTER 6

NOVEL APPLICATIONS OF CHEMICAL LOOPING TECHNOLOGIES

A.-H. A. Park, P. Gupta, F. Li, D. Sridhar, and L.-S. Fan

6.1 Introduction

As discussed in the previous chapters, the Type I chemical looping process (see Chapter 2) uses a metal/metal oxide particle with a specially designed support and promoters that can undergo multiple (>100) reduction/oxidation cycles while maintaining a high oxygen-carrying capacity. These particles, in oxidized form, are capable of reacting with different kinds of carbonaceous fuels, such as coal, biomass, syngas, hydrocarbons, and wax, after which the particles are reduced to the metallic form. At the reduced state, particles can be oxidized to the original state by air, O_2, CO_2, or steam. Thus, these engineered chemical looping particles allow the efficient conversion of various carbonaceous fuels to heat CO, H_2, syngas, or any combination of these products. These particles also can be used in the production of steam, electricity, chemicals, or gaseous and liquid fuels. Furthermore, the particle reaction rate, because of the presence of promoter and support, can be an order of magnitude faster than the metal/metal oxide in its pure form. In addition, as mentioned in earlier chapters, the redox process in two different reactors also provides a built-in CO_2 separation feature.

The Type II chemical looping process (see Chapter 2) uses metal oxide/metal carbonate particles to capture CO_2, sulfur, and halide impurities simultaneously over multiple cycles while maintaining a high capture capacity (0.5-g CO_2 captured per gram of metal oxide). The metal oxide particles are capable of capturing CO_2, sulfur and halide impurities from flue gas and fuel gas streams generated from a variety of feed stocks, including coal, natural gas, and biomass to form a mixture of solids mostly consisting of the metal carbonate. The metal carbonate can be calcined to produce a sequestration-ready CO_2 stream. This chemical looping process integrated with a gasification system can produce electricity, hydrogen, chemicals, and liquid fuel and with a combustion system can produce electricity with a very low CO_2 footprint.

The high efficiency and flexibility of these chemical looping processes allow a wide range of novel applications for this technology. These applications include onboard hydrogen production, carbon dioxide capture, solid oxide fuel cell, direct solid fuel cell, enhanced steam methane reforming, tar sand digestion, liquid fuel production, and chemical looping with oxygen uncoupling, all of which are discussed in this chapter.

6.2   Hydrogen Storage and Onboard Hydrogen Production

Because hydrogen is an important clean energy source for the future, various hydrogen production technologies, including the chemical looping process, currently are being developed. The goal is to provide technically and economically feasible ways of generating hydrogen at a large scale. One key issue related to the application of hydrogen as a carbon-free, pollutant-free energy carrier is hydrogen storage. In particular, hydrogen has the potential to be the transportation fuel of the future, because it does not lead to CO_2 emissions that cause global warming. Unless hydrogen is used in a stationary energy conversion system, it has to be transferred to another site, generated in a distributed generation system or generated onboard a vehicle. However, economically storing or generating hydrogen onboard a vehicle at a high density (both volumetric and gravimetric) is a challenge.[1-12] Thus, the development of high-capacity hydrogen storage materials has been one of the focal areas for energy research in recent years. The key factors to be considered for the development of hydrogen storage and/or onboard hydrogen generation systems are as follows: hydrogen capacity, cost, durability/operability, hydrogen charging/discharging rates, fuel quality, the environment, safety, and health.[12] In what follows, a new onboard hydrogen generation system derived from chemical looping technology is introduced, as well as other, more extensively studied, hydrogen storage methods.

6.2.1   Compressed Hydrogen Gas and Liquefied Hydrogen

Two of the simplest ways to store hydrogen are as a compressed gas and in liquefied form. Current commercially available high-pressure tanks are capable of storing hydrogen gas at 340 atm (5,000 psi) or 680 atm (10,000 psi).[11] At 340 atm, the specific volume of hydrogen is 42 L/kg. Thus, for a tank containing 5 kg of hydrogen, this corresponds to a volume of 210 L, without accounting

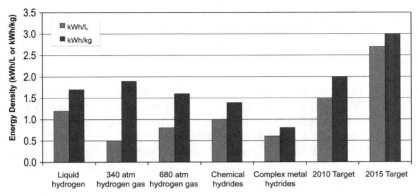

Figure 6.1. Gravimetric and volumetric energy densities of various hydrogen storage.[13]

for the volume of the tank walls. Although the hydrogen is stored in pure form, when the weight of the tank is factored in (e.g., 86.3 kg for 4.7 kg of fuel at 204 atm), the gravimetric fuel capacity of high-pressure tanks is approximately 5 wt.%, which is significantly smaller than that of conventional liquid fuels.[3] As shown in Figure 6.1, the gravimetric and volumetric energy densities of compressed hydrogen gas at 340 atm and 680 atm still are substantially lower than the goals set by the DOE FreedomCAR Partnership.[13] Considering the nonideal behavior of hydrogen gas at high pressure, the increase in pressure beyond 680 atm would not improve significantly the energy content of this storage option. For example, doubling the pressure from 680 atm to 1,360 atm would only increase the volumetric hydrogen gas density by 30%,[12] whereas higher pressure requires a large increase in compression power. At 340 atm, it is estimated that 8.5% of the energy content of the hydrogen would be consumed in compression.[14] Unlike conventional tanks for liquid fuels, which can be constructed in different shapes and dimensions to best use the available space, high-pressure tanks for hydrogen storage only can be cylindrical. This also limits the application of compressed gas as a hydrogen storage option. However, compressed hydrogen gas currently is a mature option for vehicular applications, and further research is ongoing for the purpose of cost reduction and lower tank weight design.[13]

Another method of storing pure hydrogen is in the liquefied form. According to its phase diagram, hydrogen exists as a liquid at below −253°C.[13] Liquid hydrogen has a much higher density than compressed hydrogen with a specific volume of 14 L/kg, which significantly reduces the size of the onboard fuel tank. The major issues associated with liquefied hydrogen are the energy requirements for liquefaction and the dormancy period. The average energy consumption during liquefaction of hydrogen is 30% of the LHV (lower heating value of hydrogen), which is much greater than the energy for hydrogen compression to 340–680 atm.[13] The dormancy period involves the gradual warming-up of hydrogen in the storage tank and the successive boil-off.

Research has suggested that for a tank containing 4.6 kg of fuel, approximately 4% of the hydrogen will be boiled off per day.[14,15] Thus, cryogenic tanks with advanced insulation are being developed to overcome the dormancy issues.[16,17] In particular, a high-pressure vessel, lined with aluminum and equipped with a composite outer wrap, a multilayer vacuum insulation, and an outer vacuum vessel, has been tested extensively and implemented in a demonstration vehicle.[16,17] This vehicle can be fueled using either compressed hydrogen gas or liquid hydrogen depending on the desired driving distance.[13]

A hybrid hydrogen tank is a relatively new idea consisting of a typical compressed hydrogen tank that is filled partially with low-temperature reversible hydrogen absorbing alloys.[18] Typical reversible hydrides such as LaNi and FeTi can store hydrogen with volumetric densities of 6.7–7.7 L/kg at 100°C, which is several times greater than compressed hydrogen[13] in terms of volumetric density. Because of its high volumetric energy density, this combined storage option can allow more compact hydrogen storage without sacrificing the advantages of a compressed hydrogen system.

Compressed, liquefied, and hybrid tanks have the major advantage of instantaneously providing pure hydrogen at high rates, whereas other systems involving chemisorption or physisorption of hydrogen require a significant temperature swing to regenerate hydrogen on demand. Refueling also is relatively simple and fast because no chemical reactions are involved with the refilling of compressed and liquefied tanks, and the hydrogenation of the hydrides in hybrid tanks is considered to be rapid.

6.2.2 Metal Hydrides

Metal-hydride-based hydrogen storage is not a new technology and has been researched extensively during the past 30 years. Metal hydrides store hydrogen chemically in a crystalline structure that can pack more hydrogen into a given volume than liquid hydrogen. The reversible equilibrium reaction that involves hydrogen storage and regeneration can be written as follows:

$$MH_{n,\text{solid}} + \text{Heat} \leftrightarrow M_{\text{solid}} + (n/2)H_2 \qquad (6.2.1)$$

However, because hydrogen is very light and the metals usually are fairly heavy by comparison, the hydrogen storage per unit weight is not as high. Hydrides involving light elements such as Li, Be, B, N, Na, Mg, and Al generally have acceptable hydrogen storage per unit weight;[19–26] but for binary hydrides of these elements, the binding energy is too strong to be practical. A high binding energy means that the hydride will only release hydrogen at high temperatures.[24,26] For example, MgH_2, one of the least stable binary hydrides among the light metals, has a theoretical hydrogen capacity of 7.6 wt.%, but its equilibrium hydrogen pressure only reaches 1 atm at about 300°C.[26]

AlH_3 is a covalently bonded solid with 10.1 wt.% theoretical hydrogen storage that is actually thermodynamically unstable at room temperature.[11]

Because it is unstable at room temperature, it can release hydrogen at temperatures below 100°C. No currently known process could produce AlH_3 efficiently or on a large scale, and this recycling of Al to AlH_3 would have to be done offsite. Many other hydrides exhibit high hydrogen capacities and may have other favorable properties, but they have the same drawback as AlH_3 in that there is no known way to recharge the spent "fuel" efficiently or on a large scale.[19] However, current examples of a few known hydride reactions include hydrolysis of MgH_2 with water to produce $Mg(OH)_2$ and hydrogen, and catalytic hydrolysis of $NaBH_4$ (which can be stored as a 25% solution in water) to produce aqueous $NaBO_2$ and hydrogen as shown in the following chemical reaction[11]:

$$NaBH_4 + 2H_2O \rightarrow NaBO_2 + 4H_2 \qquad (6.2.2)$$

These sorts of hydrides may still find specialty applications where economics are not the primary consideration, but they currently are not suitable for mobile applications. If, however, a simple and efficient process for regeneration of any of these compounds was developed, they could in fact meet the requirements for both volumetric and gravimetric energy density.

A hydride that both can release hydrogen and be regenerated under mild conditions is called a reversible hydride. These hydrides usually have equilibrium hydrogen pressures of 1 atm at mild temperatures (~0–60°C). With the proper catalysts, these hydrides can be generated from their metal alloys at room temperature and high pressure, and will release hydrogen at 1–2 atm at higher temperatures (still below 100°C).[19] Early reversible metal hydrides include transition metal alloys such as $FeTiH_2$ and MNi_5H_6, where M can be various transition metals. Interestingly enough, the latter compound already is used in Ni–MH rechargeable batteries found in many digital cameras and other electronics. Unfortunately, these alloys typically contain only 1–2 wt.% hydrogen and therefore, they are not suitable for mobile applications.[20]

For a long time, sodium alanate ($NaAlH_4$) was considered to be in the same category as $NaBH_4$ and other irreversible hydrides despite having an equilibrium hydrogen pressure of 1 atm at ~30°C.[20] This changed in 1997 when Bogdanovic and Schwickardi reported that $NaAlH_4$ could be regenerated under relatively mild conditions when doped with certain transition metals.[20] So far, the best results have been obtained using Ti as the dopant.[22] Typical conditions for Ti-catalyzed rehydrogenation are 100°C and 100 atm. A balance must be achieved between the kinetics, which become more favorable as the temperature increases, and the thermodynamics, which become less favorable as the temperature increases. $NaAlH_4$ has a theoretical hydrogen storage capacity of 5.5 wt.%, and a reversible hydrogen storage of up to 5 wt.%. When the weight of the tank itself and the other components of the storage system are accounted for, the total system level capacity is expected to be only 2.5 wt.%. Current research tanks have a capacity of only 1 wt.%.[11]

Other techniques that are worth noting, but are not yet well developed, include lithium imide/lithium amide systems and destabilized mixtures of multiple hydrides. Lithium imide can be hydrogenated to lithium amide with 6.5 wt.% reversible storage at around 250°C.[20] Other than the relatively high temperature required, this system also has the drawback of producing small amounts of ammonia, which poison polymer electrolyte membrane (PEM) fuel cells even at trace levels. Destabilized mixtures of hydrides are formed by mixing a high-capacity, stable hydride with a compound that forms a stable intermediate with the hydride's dehydrogenated form. One such mixture is a blend of MgH_2 and $LiBH_4$, which has been reported to have 9 wt.% reversible storage capacities, but this is only achieved at a high temperature and with slow kinetics.[20] This particular blend achieves a high capacity, because both components have fairly high hydrogen capacities. Future research may produce more destabilized mixtures with more favorable kinetics and thermodynamics.

6.2.3   Bridged Metal-Organic Frameworks

The recent development of fundamentally new classes of nanoporous materials based on metal organic structures has been a significant development in the field of gas storage.[27] Metal-organic frameworks (MOFs) are solids that consist of inorganic groups, usually $[ZnO_4]^{6+}$, bonded to linear aromatic carboxylates that form a robust and highly porous cubic framework.[28] Of particular interest for hydrogen adsorption are the isoreticular MOFs, or IRMOFs, whose surface area and gas adsorption capacities are larger than other microporous materials such as zeolites and carbon nanotubes.[29] MOFs are promising candidates for gas storage applications because they can be synthesized at high purity, with high crystallinity, potentially large quantities, and at low cost. Perhaps most importantly, MOFs can have an almost endless variety of structures and functional groups, leading to the possibility of rational design of sorbent materials tailored to specific applications.[28]

In recent years, IRMOFs by themselves have not produced spectacular results in terms of hydrogen adsorption at room temperature. For example, IRMOF-1 (or MOF-5) adsorbed hydrogen up to 4.5 wt.% at 78 K (−195°C), but only adsorbed 1.0 wt.% at room temperature and 20 bar (2 MPa).[28] When MOFs are mixed with an active-carbon bridge, however, uptake at room temperature increased dramatically thanks to a phenomenon known as hydrogen spillover. Hydrogen spillover can be described as the dissociative chemisorption of hydrogen on one site, such as a metal particle, and the subsequent transportation of atomic hydrogen onto another substrate, usually carbon or alumina.[30] In regard to the IRMOF structure, hydrogen rapidly dissociates and bonds to the metal/active carbon catalyst (Pt/AC), but slowly diffuses from the catalyst toward the MOF structure. It has been shown that surface diffusion of hydrogen atoms is the rate-determining step in hydrogen spillover.[30] However, in the presence of carbon bridging, a secondary-spillover medium, the overall chemisorption process occurs faster. For example, Li and Yang have shown that the uptake of hydrogen onto carbon-bridged IRMOF-8 at ambient conditions is eight times greater than that of pure IRMOF-8 under the same conditions, and that the processes are completely reversible/rechargeable at room temperature.[31]

Despite continued investigations and research on the kinetics of hydrogen spillover, the kinetics study of surface diffusion of hydrogen atoms under the ambient condition has not been reported. Moreover, the mechanistic details of hydrogen spillover still are not well understood.[30] Thus, future investigation is needed to advance this hydrogen storage concept.

6.2.4 Carbon Nanotubes and Graphite Nanofibers

Carbon nanotubes (CNTs) generated considerable excitement when initial reports were published showing high storage capacities. Dillon et al. reported initial research on the storage potential of CNTs.[32] They studied the adsorption of hydrogen on soot that contained 0.1–0.2 wt.% single-walled nanotubes (SWNTs) and extrapolated their data to determine a hydrogen storage capacity of 5–10 wt.%. Lee et al. predicted a hydrogen storage capacity of up to 14 wt.% in SWNTs based on theoretical calculations.[33] Multiwalled nanotube (MWNT) capacity was found to vary from 2.7 wt.% to 7.7 wt.%.[33]

More recent studies have not been able to achieve the results predicted by Lee et al.[33] Research conducted by Iqbal and Wang[34] found hydrogen adsorption of 2.5–3.2 wt.% for SWNTs charged electrochemically. Reversible hydrogen storage of 1.5 wt.% was achieved with desorption taking place at 70°C. Dillon et al. conducted further research on hydrogen adsorption on MWNTs and found that although no hydrogen was adsorbed by clean MWNTs or iron nanoparticles under near-ambient conditions, MWNTs that contained significant quantities of iron nanoparticles were able to store 0.035 wt.% hydrogen.[32] They concluded that the adsorption characteristics must result from an interaction between the MWNTs and the iron nanoparticles. Other studies have confirmed that the presence of metal nanoparticles is essential to hydrogen adsorption in both SWNTs and MWNTs. Most recent studies have found adsorption in the range of 1–4 wt.% for SWNTs and lower for MWNTs.[34]

Graphite nanofibers (GNFs) are a nanostructured form of carbon consisting of layers of graphite formed into fibers. The individual layers of graphite may be parallel (tubular), perpendicular (platelet), or at an angle (herringbone) to the fiber axis. The length and diameter of individual GNFs can vary depending on many factors, but typical values include lengths of 50 μm and diameters of 250 nm.[30] GNFs can be formed by several processes, but the most common processes are thermal decomposition of hydrocarbons, typically acetylene, ethylene, or benzene,[4,35,36,37] and catalytic graphitization of electrospun polymer fibers, typically poly(vinylidene fluoride).[38]

The first research on the use of GNFs for hydrogen storage was published in 1998. Initial research done by Chambers et al.[4] achieved hydrogen adsorption varying from 11 wt.% for tubular GNFs to 67 wt.% for herringbone GNFs. Subsequent hydrogen desorption was reported to be as high as 58 wt.%.

Adsorption took place at room temperature and at pressures ranging from 44 to 112 atm. Desorption took place at room temperature and atmospheric pressure. Ahn et al.[36] conducted research on GNFs and other forms of activated carbon and found much lower storage capacities at room temperature and at 77 K (−196°C). GNFs were found to adsorb approximately the same amount of hydrogen as other forms of activated carbon, with the best capacity being ~1 wt.% at 77 K for GNFs. Fan et al.[37] reported a hydrogen storage capacity of 10–13 wt.% for GNFs. More recent research done by Gupta et al.[35] found a hydrogen storage capacity of 17 wt.% for GNFs grown by thermal decomposition of acetylene on Pd sheets. Hong et al.[38] formed GNFs from electrospun poly(vinylidene fluoride) nanofibers and reported a storage capacity of 0.11–0.18 wt.%.

GNFs present a very attractive option for hydrogen storage due to their high storage capacity and their ability to both adsorb and desorb hydrogen at room temperature. However, the actual hydrogen storage capacity of these materials varies greatly, and the factors affecting this variance are not yet understood. All studies so far have focused on producing milligram or gram quantities of GNFs, but a commercial system for use in a typical automobile may require anywhere from 3 to 50 kg of GNF per vehicle. No process currently exists for the mass production of GNFs, and it is not known whether such a process could be economical. In addition, the ability of GNFs to withstand multiple hydrogenation/dehydrogenation cycles is not well understood. One study found that the capacity decreased by 30% after multiple cycles,[37] whereas another study found the capacity actually increased over the first 10 cycles and remained constant thereafter.[35] Although desorption of hydrogen from GNFs is very fast,[4] adsorption takes hours to complete.[4,37]

6.2.5 Onboard Hydrogen Production via Iron Based Materials

An onboard hydrogen production process such as one derived from the chemical looping strategy could be an option for the H$_2$ storage techniques dicussed in the previous sections. When an iron based looping medium is used, this process allows the storage and production of hydrogen onboard a vehicle. This hydrogen storage system is based on the reaction of steam with Fe in order to produce hydrogen as described in the reaction:

$$3Fe + 4H_2O \leftrightarrow Fe_3O_4 + 4H_2 \qquad (6.2.3)$$

Thermodynamics dictates that for stoichiometric conversion of this reaction, steam must be made to react with Fe in a countercurrent fashion. This is easily achieved in a series of fixed bed reactors. TGA experiments show that a temperature above 400°C is sufficient to drive the reaction to completion. Although such an operating temperature is considered to be high relative to aforementioned metal hydride based hydrogen strorage materials, and it requires a large amount of heat input for the hydrogen stored to be released, the exothermic nature of the steam iron reaction can potentially minimize the heat input for onboard hydrogen production using iron and steam. ASPEN simulations of Reaction (6.2.3) show that if steam at 300°C is made to react with Fe at 25°C, the resultant heat of reaction will lead to an adiabatic temperature rise to 600°C. Hence, if a source of steam is available, Reaction (6.2.3) can potentially be carried out without the need for extra heating devices.

For the system to work, there needs to be a place to generate steam from water. A number of sources of heat may be utilized to achieve this.

Source 1: Hydrogen is produced at high temperatures, but for use in a PEM fuel cell the temperature of the hydrogen needs to be brought down to about 100°C. This heat may be used to generate steam.

Source 2: Additionally, more steam may be generated by using the heat from the oxidization of Fe_3O_4 to Fe_2O_3 in air.

$$2Fe_3O_4 + \frac{1}{2}O_2 \rightarrow 3Fe_2O_3 \qquad (6.2.4)$$

Source 3: In case a hydrogen internal combustion engine (ICE) is used, heat can be absorbed from the engine using a suitable coolant (e.g., mineral oil) with a boiling point above 100°C to generate steam in a separate heat exchanger. If permitted by the specifications of the ICE, water may be directly injected to cool the engine resulting in steam production.

Source 4: The high-grade heat content of the exhaust from a H_2 ICE can be utilized to generate steam on board the vehicle.

Source 5: Use some of the Fe stored on board to react with air. This will allow for the combustion of Fe to Fe_2O_3, releasing a large amount of energy. For this purpose, a dedicated fixed bed can be put into the vehicle for heat generation. Water can be passed through this bed to capture the heat and generate steam.

$$4Fe + 3O_2 \rightarrow 2Fe_2O_3 \qquad (6.2.5)$$

Source 6: The fuel cell device utilizing the hydrogen generates a large amount of heat. This heat may be integrated back to generate steam. A PEM fuel cell typically works in the 70–150°C range. The excess heat may be used to generate low grade steam.

Source 7: The hydrogen produced is put into a storage vessel. This vessel mainly provides hydrogen to the fuel cell, but part of the hydrogen stored may be routed back and combusted to generate heat for producing steam. For initial startup of the vehicle, a dedicated small hydrogen tank may be installed in the vehicle to switch on when the hydrogen storage tank does not have enough hydrogen pressure.

Source 8: The PEM fuel cell converts only about 90% of the hydrogen input. The remaining hydrogen may be combusted to provide heat for steam.

Source 9: Part of the electricity produced by the fuel cell and stored in the battery may be utilized to electrically heat the water using resistance heating to generate steam.

The water required for H_2 production may be recovered from the fuel cell exhaust in an air-cooled condenser that removes heat by convection into the ambient air when the vehicle is in motion. Greater than 90% of the water necessary is expected to be recovered in this manner. Additionally, in case it is desired, the exhaust may be routed through the air-conditioning system of the vehicle for further condensation, but this may impose an energy penalty on the system.

Once the Fe particles have been oxidized in the vehicle (to Fe_3O_4 or Fe_2O_3 as the case may be), they can be reduced back to Fe by utilizing a reaction with fuels:

$$[(2x+y/2-z)/3]Fe_2O_3 + C_xH_yO_z \rightarrow xCO_2 + (y/2)H_2O + [(4x+y-2z)/3]\,Fe \tag{6.2.6}$$

The fuels may be a gaseous fuel like natural gas, producer gas, or syngas, a liquid fuel like gasoline, or a solid fuel like coal, or wood. In the case where Fe_3O_4 is generated from the vehicle, Reaction (6.2.4) can be carried out first to convert it to Fe_2O_3 so that the CO_2 generation from Reaction (6.2.6) can be enhanced. This reaction will also help increase the temperature of the particles so that Reaction (6.2.6) can readily proceed.

Reaction (6.2.6) will likely need to be conducted outside the vehicle at a central fuel station. Such a centralized station may be built outside a city where the spent particles can be regenerated and distributed to retail outlets. Figure 6.2 shows the overall scheme for using iron to produce hydrogen on a vehicle. The design of the particle regeneration rectors given in the figure involves the flow of reactive fuel countercurrent to the flow of the spent particles as described in Section 4.4.1. The regenerated iron particles exit from the reactor and can be used for producing hydrogen on board the vehicle. The hydrogen generation can be carried out in a packed bed reactor or a monolithic bed reactor as shown in Figure 6.3. The steam can pass through the reactors at high temperatures and react with iron to form hydrogen. For a packed bed reactor shown in Figure 6.3(a), the particle formulation can be cast into small pellets of 1–5 mm in size. These pellets can be randomly packed into a metallic vessel to establish a packed bed. It is desired that the metallic vessel is of a rectangular cross section for close packing in a module, though circular cross section can also be used. The designs for the monolithic bed reactors are shown in Figures 6.3(b) and 6.3(c). The monoliths in the design are of ceramic structures with channels/holes present over the length of the structure. This

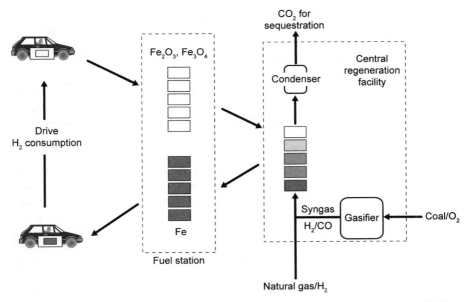

Figure 6.2. Overall scheme of using Fe modules to produce hydrogen on a vehicle.

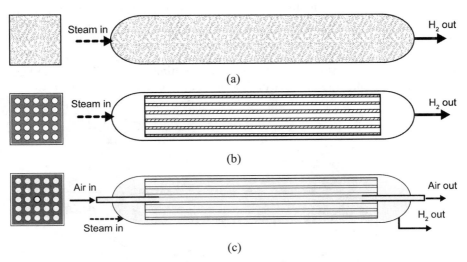

Figure 6.3. Designs for reactors with Fe containing media: (a) packed bed of small pellets; (b) monolithic bed with straight channels for steam; and (c) monolithic bed with channels for steam and air.

structure allows the packing of iron particles in a confined space. With suitable design on end caps, it is possible to route fluids through different channels. It also allows air to pass through some channels, while steam flows through other channels, as shown in Figure 6.3(c). This functionality could provide the heat, released from Reactions (6.2.4) and (6.2.5), for generation of steam used for its reaction with iron. In all designs, refractory lining or vacuum jacket could be used to prevent heat losses from the reactors. It is noted that in module design, iron can also be embedded in the ceramic channel wall materials. In this case, both the reduction and regeneration reactions take place directly on the module.

An important aspect of the iron particle regeneration process is that the gaseous products from Reaction (6.2.6) will mainly be steam and CO_2. After condensation of steam, the CO_2 rich stream can be easily separated and sequestered using the same carbon sequestration techniques intended for stationary power plants. Hence, the process can potentially save the cost for CO_2 capture. Further, since no CO_2 is emitted from the vehicle, the overall scheme provides a viable method for CO_2 emission control when a suitable CO_2 sequestration technique is available.

A number of such reactors may be required for generating an adequate quantity of H_2 for use in a transportation vehicle. Also, if the reactors are connected in series, the H_2 purities obtained can be very high. Such a series reactor assembly forms the key component of a module that can be used for onboard hydrogen production applications. Figures 6.4, 6.5 and 6.6 present some configurations for the modules made up of the reactors shown in Figure 6.3. The reactors are connected using metallic tubing welded to the end caps. The reactor assembly is enclosed in a suitable metallic box/container with insulation to prevent loss of heat to the surroundings. Depending on the heat integration scheme, the number of fluid inlet and outlet points may be varied.

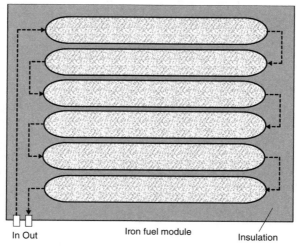

In Out   Iron fuel module   Insulation

Figure 6.4. H_2 production module using a series of fixed-bed reactors.

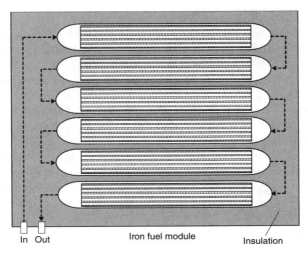

Figure 6.5. H₂ production module using a series of monolithic-bed reactors.

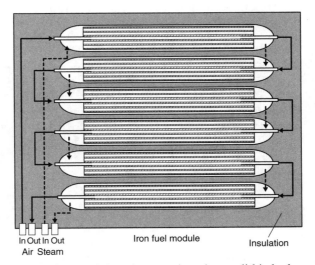

Figure 6.6. H₂ production module using a series of monolithic-bed reactors with air injection to provide heat for steam formation.

In the simplest scheme, as shown in Figures 6.4 and 6.5, there are only one inlet (steam) point and one outlet (H₂) point. The fluid flow lines can be connected to or disconnected from the module using quick-connect couplings. The module can easily be loaded to or unloaded from a transportation vehicle.

Such modules may be directly purchased from a retail store. These may also be sold through "Iron Stations" or the "Fuel Station," the future equivalent of gas stations, as shown in Figure 6.2. It is note that even though Reaction (6.2.3) takes place at a temperature above 400°C, within the reactors, only the region proximate to where the reaction front is developed will be hot.

System Integration Onboard a Transportation Vehicle
A number of scenarios can be considered in regard to the integration of the
Fe modules with other components in a transportation vehicle. Two most
relevant scenarios are presented in the following. Other scenarios may be
conceived by combining the concepts discussed in these scenarios.

Scenario 1: Use of H_2 ICE Heat to Produce Steam. The heat generated, but
not utilized for shaft work in a H_2 ICE, is enough to generate the steam
required to produce hydrogen. This heat is taken away from the engine using
the coolant and the exhaust, and may be recovered to produce steam. Figure
6.7 shows such integration. A minimal amount of water makeup may be
required if the exhaust water can be recovered using a condenser. Such a
condenser may be cooled using the water reservoir or using convective heat
transfer to air. Once started, the system is self sustaining on heat until the time
all of the Fe in the module is consumed.

*Scenario 2: Use of Oxidation Reactions of Fe and Fe_3O_4 to Provide for Steam
Formation Heat.* This scenario does not require integration with the fuel cell

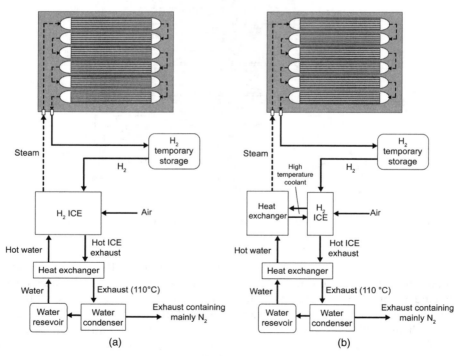

Figure 6.7. System integration for use of Fe module with a H_2 ICE using (a) water to
cool the ICE and (b) a high-temperature coolant to cool the ICE.

or the H_2 ICE. Hence, this scenario can be used to provide hydrogen with stand-alone Fe module usage. Air is supplied along with steam into the Fe modules, as shown in Figure 6.8. This air then oxidizes the Fe/Fe_3O_4 to Fe_2O_3, releasing heat that takes the unreacted Fe particles to reaction temperatures of higher than 400°C. The configuration shown in Figure 6.8(c) involves first sending air through one packed reactor. Water is then added and being

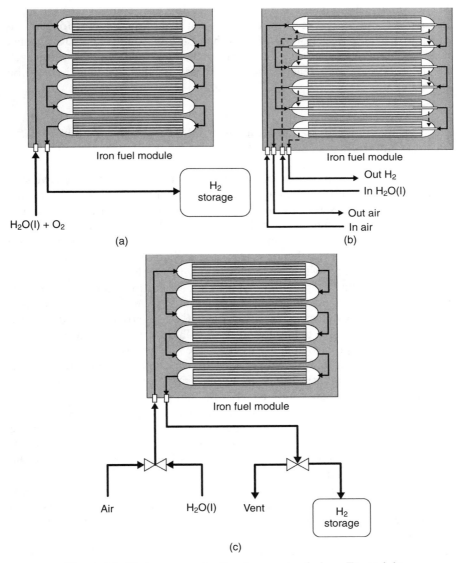

Figure 6.8. Hydrogen production from a stand-alone Fe module.

converted to steam. Steam reacts with Fe in the remaining reactors to produce hydrogen. For this operation, one reactor bed in the Fe to Fe_2O_3 oxidation mode with air is required to produce hydrogen from five Fe reactors.

Meeting DOE Targets
The energy density of the Fe composite particles, when fully reduced, is calculated to be up to 1.52 kWh/L or 1.17 kWh/kg of the particles. It is comparable to the DOE 2010 target of 1.5 kWh/L for the volumetric energy density. Since the raw material for producing the composite particle is inexpensive and its synthesis procedure is not elaborate, the cost for this on-board hydrogen production option has the potential to meet the DOE target of $2/kWh set for 2015. The potential challenges to this option, however, include relatively high reaction temperatures and delicate heat integration requirement.

6.3   Carbonation–Calcination Reaction (CCR) Process for Carbon Dioxide Capture

As mentioned in Chapter 1, carbon dioxide capture is the most expensive step of the overall threefold carbon management step, which consists of separation, transportation, and sequestration. The ongoing research and development (R&D) on carbon management includes mapping the strategy for CO_2 separation. Such strategies include improving the energy-intensive low-temperature amine scrubbing process, demonstrating the chilled ammonia process, developing further oxycombustion technology (in which high-purity oxygen is used for combustion), and employing reactive CO_2 separation using dry, solid sorbents such as limestone, potassium carbonates, lithium silicates, and sodium carbonates, which yield a sequestration-ready CO_2 gas stream upon decomposition.[39] Other processes being investigated include low-temperature pressure swing adsorption processes using hydrotalcite.[40] Thus, cost-effective carbon-capture technologies play an important role in CO_2 mitigation for current plant operation. In a typical flue gas stream (dry basis) generated from coal combustion power plants, the concentration of CO_2 is low, representing approximately 15% of the flue gas stream. Low CO_2 partial pressures, combined with the extremely high flue gas generation rate, make CO_2 capture from PC power plants an energy-intensive step. An ideal CO_2 capture technology would incorporate effective process integration schemes to minimize the parasitic energy required for CO_2 separation.

One process alternative uses a calcium-based solid sorbent at a high temperature to capture CO_2 and SO_2 simultaneously. The CCR Process, which is an outgrowth of two other processes developed at The Ohio State University: the Ohio State Carbonation Ash Reactivation (OSCAR) Process[41] and the Calcium-Based Reaction Separation for CO_2 (CaRS–CO_2) Process,[42] uses calcium oxide to react with the CO_2 and SO_2 present in the flue gas stream at

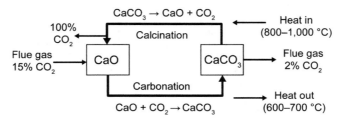

Figure 6.9. The carbonation–calcination–reaction-based CO_2 capture process.

a high temperature (450–750°C). Similar to the Calcium Looping Process for hydrogen production described in Chapter 4, the calcium oxide can be derived from multiple sources including hydrated lime, natural limestone, or reengineered calcium carbonate, such as Precipitated Calcium Carbonate (PCC) (see Chapter 2).[43,44]

The concept of capturing and separating CO_2 from flue gas using limestone-based sorbents is depicted in Figure 6.9. The flue gas comes into contact with calcium oxide (CaO), which reacts with the CO_2 and SO_2 in the flue gas to form calcium carbonate ($CaCO_3$) and calcium sulfate ($CaSO_4$). During the carbonation and sulfation reactions, which typically take place between 600°C and 700°C, heat is released. The reacted sorbent is regenerated in a separate step by decomposing $CaCO_3$ at higher temperatures (greater than 850°C) to yield CaO and CO_2. The thermal stability of calcium sulfate ensures that it does not decompose and remains as calcium sulfate. By calcining the sorbent in a proper environment (see Chapter 2), the degeneration of the reactivity of regenerated calcium sorbent can be minimized while yielding a pure or concentrated gas stream of CO_2 that can be compressed and transported for sequestration. The process thus can take a flue gas that contains typically 10–15% CO_2 and convert it into a nearly pure (>90%) CO_2 stream.

Reactions (6.3.1) and (6.3.2) summarize the basic chemistry involved in the process:

Carbonation: $CaO(s) + CO_2(g) \rightarrow CaCO_3(s)$ $\Delta H = -178$ kJ/mol   (6.3.1)

Calcination: $CaCO_3(s) \rightarrow CaO(s) + CO_2(g)$ $\Delta H = +178$ kJ/mol   (6.3.2)

As shown here, the carbonation process is exothermic, whereas calcination requires heat input. The heat integration is thus essential, in that the heat released at temperatures of 600–700°C during the carbonation process needs to be recovered and used, whereas additional heat energy is necessary for the calcination reaction.

Sorbent reactivity has a significant effect on the CO_2 removal efficiency. Therefore, the type of sorbent has a considerable effect on the overall CO_2 removal, as shown in Figure 6.10. The data were obtained from an entrained bed reactor using a flue gas stream generated by coal combustion at 9 kg/h,

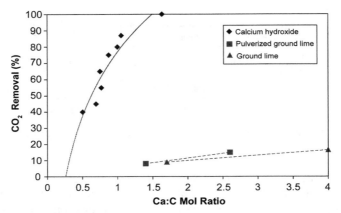

Figure 6.10. Effect of Ca:C molar ratio on the CO_2 removal efficiency for three Ca-based sorbents.

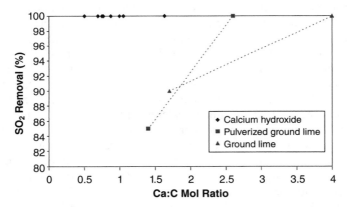

Figure 6.11. Effect of Ca:C molar ratio on the SO_2 removal efficiency for three Ca-based sorbents.

which will be illustrated in further detail later. Clearly, the superior performance of calcium hydroxide is evident as compared with other inexpensive calcium-based sorbents in removing CO_2.

As mentioned, sulfur dioxide also is removed simultaneously from the flue gas stream. Figure 6.11 shows the effect of the Ca:C molar ratio on SO_2 removal. Even at low Ca:C molar ratios, virtually all the SO_2 is captured effectively and removed by the calcium hydroxide sorbent. This is a result of the significantly lower concentration of SO_2 as compared with CO_2 in the flue gas stream. A 1:1 Ca:C molar ratio corresponds to an 80:1 Ca:S molar ratio.

Parasitic energy consumption is related strongly to the sorbent conversion efficiency in the looping operation. Figure 6.12 shows such a relationship for PCC, which is a reengineered calcium carbonate sorbent (see Chapter 2)

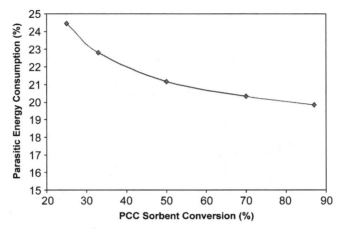

Figure 6.12. Relationship between the parasitic energy consumption and PCC sorbent molar conversion efficiency.

with a reactivity slightly higher than the hydrate-based sorbent. It is shown that the parasitic energy consumption for the CCR Process can be significantly lower than that for such low-temperature CO_2 capture processes as the MEA process, as discussed earlier, when the sorbent conversion efficiency reaches 30%–70%.

Figure 6.13 delineates the heat integration strategies for retrofitting the CCR Process to an existing boiler for 90% CO_2 capture without significant modifications to the coal-based power plant. The flue gas that leaves the economizer of the boiler is routed to the CCR Process system for CO_2 capture. This flue gas from the economizer (stream 1) is combined with the flue gas used to generate the heat required to operate the indirectly fired calciner. Heat is extracted from the total flue gas mixture (stream 2), which contains all the CO_2 emitted by the entire plant, before being sent into the carbonator system for carbonation and sulfation. CO_2 is removed in the carbonator system, and the CO_2 free flue gas (stream 3), which is at ~650°C, is cooled before it is sent into the air preheater followed by the electrostatic precipitator (ESP). In the carbonator, SO_2 and traces of heavy metals including selenium and arsenic also are removed, rendering the CCR Process a multipollutant control process. The spent sorbent from the carbonation system is sent to the calciner to regenerate the calcium oxide (CaO) sorbent for subsequent cycles while yielding a pure CO_2 stream. The sulfated sorbent and fly ash are removed from the system by means of a purge stream, whereas the makeup fresh sorbent is introduced into the system. The amount of solids removed through the purge stream is entirely dependent on the concentration of fly ash and calcium sulfate in the solids stream, which are both coal dependent. The heat of carbonation can be as high as one third of the total thermal capacity of a power

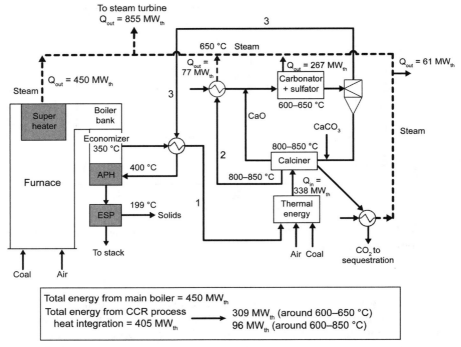

Figure 6.13. Conceptual schematic of Carbonation–Calcination Reaction (CCR) Process integration for a 300-MW$_e$ (900-MW$_{th}$) coal-fired power plant depicting heat integration strategies.[42,45]

plant. In the CCR Process, steam is generated using high-quality heat available from three different sources: (a) the carbonation system (450–750°C), (b) the hot flue gas generated for calciner operation (greater than 850°C), and (c) a pure CO_2 stream from the calciner (greater than 850°C). This steam can be used in a secondary steam turbine system for additional electricity generation or in the existing plant steam cycle by offsetting the boiler load and in driving various feed water pumps in the plant. Figure 6.13 indicates that the heat balance from each unit in the process gives rise to the thermal energy produced for steam generation of 855 MW$_{th}$, reflecting a loss of 45 MW$_{th}$ or 5% in the CCR Process. This small energy loss is for an ideal system without CO_2 compression. With the inclusion of the energy requirement for CO_2 compression to 150 atm (2,200 psi) and other process heat losses, the parasitic energy consumption increases from 5% to 20%–24%, as given in Figure 6.12.

The CCR Process is being demonstrated in a 120-KW$_{th}$ subpilot plant, as shown in Figure 6.14, located at OSU. The OSU system consists of a 9-kg/h coal-fired stoker furnace, a rotary calciner, sorbent and ash injection systems, baghouse, fans, particulate control devices, data acquisition and control systems,

Figure 6.14. Subpilotscale-CCR Process facility located at The Ohio State University.

and associated instrumentation. Because most residual ash drops out from the stoker furnace, makeup fly ash can be injected using a screw feeder to simulate gas from a typical pulverized coal-fired boiler. The gas that exits the furnace is first cooled to the required carbonation temperature of 600–700°C prior to the sorbent injection point. Sorbent then is injected to remove CO_2 and SO_2 reactively from the flue gas. The reacted/spent sorbent then is separated from the flue gas and sent to the calciner for regeneration. The CO_2-depleted flue gas is cooled to approximately 50°C by aspirating ambient air into the gas stream. The cooled flue gas is then sent into a baghouse where the remaining sorbent and fly ash are separated. The calciner regenerates the spent sorbent, which is then conveyed back into the process for a subsequent CO_2 capture cycle. A screw feeder system is used for sorbent delivery.

One of the main difficulties associated with the process is solids handling. Lime, limestone, and calcium hydroxide are all cohesive particles. Sorbent attrition and agglomeration, and electrostatic charge effects can cause operational difficulty. This is specifically true in the case of calcium hydroxide because of its natural particle size ($D_{50} < 10\,\mu m$) and tendency to form a cementitious solid when exposed to low moisture concentrations. However, OSU has developed techniques to ensure proper feeding and flow of sorbent through the reactor system. Figure 6.15 shows the inside of the U-bend section of the test loop. The figure shows that the section is clear from any sorbent clogging after a week of operation.

Another well-documented problem with using naturally occurring limestone is that the sorbent's ability to capture CO_2 decreases over multiple cycles.[46–49] However, based on the single-cycle study with a hydration step added in between each cycle, the results, shown in Figure 6.16, indicate that no decay in sorbent reactivity occurs over five cycles.[50] If limestone calcination occurs between 900°C and 1,200°C, the lime product will significantly sinter

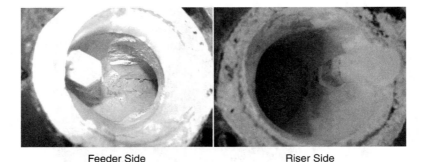

Feeder Side   Riser Side

Figure 6.15. Internal view of the U-bend section of the subpilot-scale facility (given in Figure 6.14) after one week of operation.

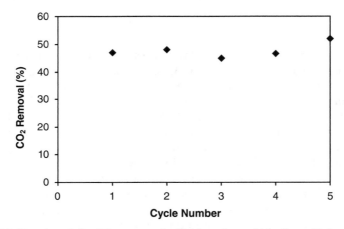

Figure 6.16. Results of the CO_2 removal efficiency from Ohio State University cyclic testing of calcium hydroxide at a Ca:C molar ratio of 0.75.

such that carbon dioxide can no longer diffuse effectively into the pore volume of the calcium oxide particle; however, the hydration reaction still can be effective. Once regenerated into calcium hydroxide, the sorbent effectively can capture CO_2.

A similar process is being demonstrated at CANMET Energy Technology Center in Canada.[51–54] The Ca:C molar ratios used for the CANMET Process range from 4 to 25.[51] These molar ratios are significantly higher than those for the CCR Process, which can remove virtually all the CO_2 in the flue gas stream with a Ca:C molar ratio of 1.5 and a residence time on the order of a second.[50] It is thus expected that, compared with the CCR Process, the CANMET Process will require higher solids circulation and higher parasitic energy consumption for a given rate of the CO_2 removal.

6.4 Chemical Looping Gasification Integrated with Fuel Cells

The high efficiencies and flexibility to produce desired products, coupled with the integrated environmental benefits in terms of a readily sequestrable CO_2 stream, make chemical looping gasification of coal an attractive technology for energy conversion and management.[55-58] As discussed in Chapter 5, the Coal-Direct Chemical Looping (CDCL) Process is very attractive because it is capable of converting as much as 80% of the thermal energy of coal into hydrogen. Here, energy conversion schemes that effectively extract chemical energy from the fuels are proposed by integrating the fuel cell with chemical looping.

6.4.1 Chemical Looping Gasification Integrated with Solid-Oxide Fuel Cells

Figure 6.17 illustrates an electricity generation scheme in which a chemical looping gasification system is integrated strategically with commercial solid-oxide fuel cells (SOFC) to minimize the exergy loss.

Under this configuration, the hydrogen-rich gas produced from the chemical looping oxidizer is introduced directly to the SOFC anode side for power generation. The exhaust of the SOFC anode, which is a hydrogen lean gas with

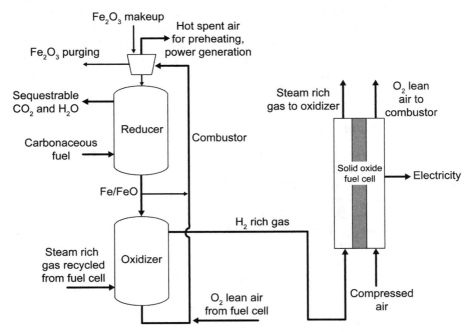

Figure 6.17. Chemical looping gasification integrated with a solid-oxide fuel cell.[59]

TABLE 6.1. Coal-to-Electricity Process Configurations and Process Efficiencies

Process Configuration	Conventional IGCC	CDCL–Combined Cycle	CDCL–SOFC
Efficiency (%HHV)	30–35	47–53	64–71
CO$_2$ Capture Rate (%)	90	100	100

a significant amount of steam, is recycled back to the chemical looping oxidizer for hydrogen generation. As shown in the figure, a closed loop between the chemical looping oxidizer and the SOFC anode is formed through the circulation of the gaseous mixture of steam and H$_2$. The steam and H$_2$ mixture essentially acts as a "working fluid" for power generation. Although some recompression, purging, and makeup may be required for the working fluid, most steam is being circulated in the closed loop and will not be condensed. Therefore, steam condensation and reheating, a step that leads to significant exergy loss in the conventional power generation processes, can be minimized. The integration of the chemical looping oxidizer and the SOFC anode also eliminates the need for a gas turbine, which is required in a typical SOFC combined-cycle system. To enhance process efficiency, the chemical looping combustor can be configured such that the oxygen-lean exhaust air from the SOFC cathode is used to combust the Fe$_3$O$_4$ to Fe$_2$O$_3$. By doing so, the output of the high-grade heat is increased and the compression work for the air is reduced. Table 6.1 compares the efficiencies of the conventional IGCC process, the CDCL combined cycle, and the CDCL integrated with SOFC. As shown in the table, the CDCL–SOFC scheme has the potential of doubling the efficiency of state-of-the-art power generation processes. The significantly improved energy conversion efficiency results from the advanced energy integration scheme between chemical looping and the SOFC.

6.4.2 Direct Solid Fuel Cells

A more advanced electricity generation scheme includes the integration of a chemical looping reducer and a direct solid fuel cell. By modifying the electrochemical oxidation of supported Fe to supported Fe$_2$O$_3$, thus generating electricity, a system integrating a chemical looping reducer and a direct solid fuel cell can be developed as illustrated in Figure 6.18.

In this scheme, reduced metal particles are fed directly into a solid-oxide fuel cell that can process solids directly. Particles are reduced in the fuel reactor and then introduced to the fuel cell to react with oxygen or air at 500–1,000°C to produce electricity. The oxidized particles are recycled back to the fuel reactor to be reduced again. It is desirable for particles to be conductive when metal is at both the oxidized and the reduced states.

The Ohio State University (OSU), led by L.-S. Fan, and the University of Akron (UA), led by S. S. C. Chuang, are developing chemical looping solid-

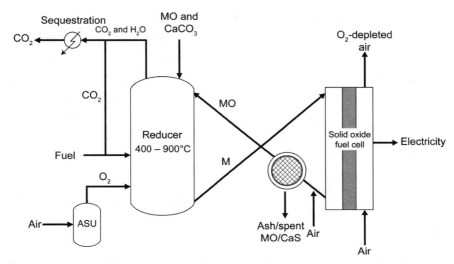

Figure 6.18. Direct solid-oxide fuel cell applications for chemical looping.

oxide fuel-cell systems. The preliminary study of the solid-oxide fuel cell at UA indicates that the reduced Fe on Fe–Ti–O can serve as a fuel, producing $45\,mA/cm^2$ at $0.4\,V$ and $800°C$. The resultant oxidized Fe–Ti–O composite remained in the powder form and did not adhere to the anode surface of the fuel cell. The results suggest that an integrated fuel cell with chemical looping could be a viable approach for generation of electricity and a nearly pure CO_2 stream from coal.

The step for the generation of the electricity from Fe on Fe–Ti–O involves the following reactions[60,61]:

$$\text{Cathodic} \qquad O_2 + 4e^- \rightarrow 2O^{2-} \qquad\qquad (6.4.1)$$

$$\text{Anodic} \qquad 2Fe + 3O^{2-} \rightarrow Fe_2O_3 + 6e^- \qquad (6.4.2)$$

The overall reaction corresponds to an ideal cell potential (i.e., open-circuit voltage) of $0.996\,V$ at $800°C$. The overall reaction is

$$2Fe + 3/2\,O_2 \rightarrow Fe_2O_3 \qquad \Delta G = 384 \text{ kJ/mol} \qquad (6.4.3)$$

The observation of electricity generation from the Fe–Ti–O composite indicates that the O^{2-} can reach the reduced Fe to carry out the anodic reaction. Because this is one of the first attempts at using solid metal as the fuel for the fuel cell, the results appear promising. This direct solid fuel cell concept can be extended to other types of oxygen carriers such as Ni and Cu. Table 6.2 compares the ideal cell potentials of using these metals as fuels. Fe/Fe_2O_3 is the oxygen carrier that gives the highest ideal cell potential. CuO is not included, because it is not stable at $800°C$; therefore, it will not be effective

TABLE 6.2. The Ideal Cell Potentials (Open-Circuit Voltage)

	n	700°C		800°C		900°C	
		ΔG (kJ)	E (V)	ΔG (kJ)	E (V)	ΔG (kJ)	E (V)
$2Ni + O_2 \rightarrow 2NiO$	4	310	0.803	290	0.751	270	0.699
$4Cu + O_2 \rightarrow 2Cu_2O$	4	220	0.570	200	0.518	180	0.466
$4/3Fe + O_2 \rightarrow 2/3Fe_2O_3$	4	407	1.053	384	0.996	363	0.941
$2Fe + O_2 \rightarrow 2FeO$	4	405	1.049	390	1.010	370	0.958
$3/2Fe + O_2 \rightarrow 1/2Fe_3O_4$	4	405	1.049	390	1.010	370	0.958

for coal chemical looping. Examining the anodic reaction suggests the need to build a pathway to facilitate the electrochemical oxidation of supported Fe.

Compared with the chemical looping–SOFC system discussed in Section 6.4.1, the direct solid fuel cell scheme is more challenging because of the relatively immature direct solid fuel cell technology. For the successful development of this direct solid fuel cell, the following issues need to be addressed.

1. The effectiveness of the direct solid fuel cell for generating electricity by using the supported Fe as a fuel.
2. The effectiveness of the supported Fe_2O_3 produced from the fuel cell for reaction with coal in the chemical looping process.

The overall potential of integrating a fuel cell with chemical looping for electric power generation should continue to be evaluated in more detail.

6.5 Enhanced Steam Methane Reforming

Chemical looping combustion integrated with the steam methane reforming process (CLC–SMR) was studied by Rydén and Lyngfelt at Chalmers University of Technology in Sweden.[62] The utilization of chemical looping combustion (CLC) in steam methane reforming (SMR) represents another potential application for the chemical looping strategy.[62] A schematic flow diagram of the CLC–SMR Process is shown in Figure 6.19.

The CLC–SMR Process uses a combustor that is identical to that used in the CLC Process discussed in Chapter 3. The CLC–SMR Process also uses a reducer that is unique in that it integrates the function of the reducer used in the CLC Process to that of a reformer. The integrated reducer/reformer reactor is composed of a low-velocity bubbling fluidized bed with reformer tubes that are placed inside the reactor. The tubes are filled with reforming catalysts and behave as fixed-bed reactors. In the operation of the reformer, steam and methane are introduced into the reformer tubes at a ratio of ~3:1. By

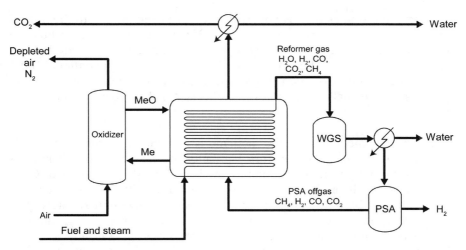

Figure 6.19. Schematic flow diagram of the chemical looping combustion integrated with the steam methane-reforming process (CLC–SMR).[62]

maintaining the reformer tubes at desirable reaction temperatures, typically 700–950°C, the methane is reformed into syngas, which is subsequently converted to hydrogen. This hydrogen stream is then purified through the WGS reaction and pressure swing adsorption (PSA) in the units downstream in the process.

The SMR reaction is highly endothermic as is the reduction reaction of the oxygen carrier particles in the reducer. Both reactions take place in the reducer/reformer reactor. The oxygen carrier reduction reaction occurs exterior to the reformer tube, whereas SMR occurs inside the reformer tube. The fuel for the oxygen carrier reduction is the PSA tail gas and recycled methane. The reduction reaction product gases, CO_2 and steam, then exit the reducer. The reduced oxygen carrier particles are transported from the reducer to the combustor, where they are oxidized with air. The high-temperature particles resulting from the exothermic oxidation reaction provide the necessary heat for reduction and reforming reactions in the reducer/reformer reactor. In the CLC–SMR Process, the CLC system is operated at atmospheric pressure, whereas the SMR reformer tubes are operated at typical reformer pressures (15–40 atm).

As shown, the methane-reforming reaction scheme in the CLC–SMR Process is practically identical to that in the traditional SMR Process with the exception of the heat integration scheme for the reformer. In the traditional SMR Process, the heat required for steam methane reforming is provided by combustion of the fuel exterior to the reformer tubes. In contrast, the CLC–SMR Process uses high-temperature oxygen carrier particles as a heat transfer medium.

To date, the studies on this novel scheme have been limited to theoretical analysis such as heat and mass balances. Experimental testing results are not yet available. Theoretical analysis indicates that the CLC–SMR Process has the potential to achieve a higher H_2 yield than the conventional SMR Process. Therefore, it offers a viable approach for an innovative reforming operation. Some challenges, however, still exist for this process as illustrated below.

1. Approximately 70% of the methane is reformed and converted to H_2 and CO_2 in the WGS reactors. Thus, most of the carbon in this process is separated by the traditional methods such as PSA rather than CLC. Although all of the carbon in the feedstock exits from the reducer/reformer, the CLC system is used to separate only approximately 30% of the total carbon in this process. Thus, the energy savings through carbon capture using the CLC system may not be significant.

2. Both the SMR reaction and the oxygen carrier particle reduction reaction, when nickel- or iron-based oxygen carriers are used, are highly endothermic. Thus, the heat required in the integrated reducer/reformer reactor will even be higher than that required in the traditional SMR reformer. The large heat requirement renders the heat integration in the CLC–SMR Process a challenging task.

3. Because high-temperature oxygen carrier particles are the sole heat carrier in the integrated reducer/reformer reactor, there will be a large solids circulation rate and likely a large temperature difference between the combustor and the reducer/reformer. An estimated solids circulation rate of 6,700 t/h for a 300-MW$_{th}$ CLC–SMR Process indicates a significant solid flow rate issue.

4. Heat transfer using fluidized oxygen carrier particles could pose an increased erosion problem on the reformer tubes.

6.6 Tar Sand Digestion via Steam Generation

In addition to producing hydrogen, reduced metal particles can be burned with air in an oxidation reactor, and the heat generated could be extracted using water to generate high-temperature steam. Figure 6.20 shows heat and steam generation by the chemical looping process. The steam could be used either for electricity generation or for various industrial applications. For example, steam can be used to extract heavy oil from tar sands. Providing steam via chemical looping technology is attractive for tar sand digestion, because less carbon is emitted throughout the overall process. This will better allow for this abundant energy source to be tapped with significantly reduced greenhouse gas emissions.

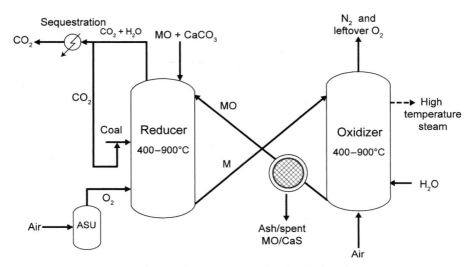

Figure 6.20. Heat and steam generation by the looping process.

6.7 Liquid Fuel Production from Chemical Looping Gasification

Even beyond hydrogen, heat, and electricity generation the chemical looping gasification strategy can be used to achieve liquid fuel synthesis with high yield and low exergy loss. In Section 4.6, a scheme is illustrated in which the Syngas Chemical Looping Process is integrated with the indirect coal-to-liquid (CTL) Process to enhance the liquid fuel yield. In this section, an advanced liquid fuel generation scheme is introduced in which the coal gasifier and air separation unit are avoided in the liquid fuel generation process. The chemical looping gasification system produces CO_2, H_2, and heat/power from the carbonaceous fuel. The heat and power produced compensate for the parasitic energy requirements. Under this scheme as shown in Figure 4.33, part of the CO_2 generated from the reducer is sequestered, while the remaining CO_2 reacts with H_2 generated from the oxidizer in a reactor that simultaneously performs the reverse WGS and Fischer–Tropsch (F–T) reactions:

$$CO_2 + H_2 \leftrightarrow CO + H_2O \tag{6.7.1}$$

$$CO + 2H_2 \leftrightarrow -(CH_2)- + H_2O \tag{6.7.2}$$

The Fischer–Tropsch reaction of this type, also called CO_2 hydrogenation, has been studied since the 1990s to identify novel applications of CO_2 for greenhouse gas control.[63–65] It was determined based on these studies that such an F–T reaction is feasible when an iron-based F–T catalyst is used.

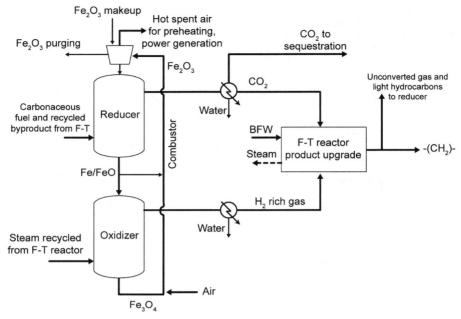

Figure 6.21. Chemical looping gasification for liquid fuel synthesis.[66]

Compared with the conventional CTL processes discussed in Chapter 4, the chemical-looping-based reaction scheme given in Figure 6.21 has several distinct advantages.

1. The unique chemical looping gasification strategy effectively gasifies the fuel with minimal exergy loss (see Section 1.6).

2. Both the endothermicity of the fuel gasification reactions and the exothermicity of the F–T reaction are reduced; moreover, the F–T by-products such as steam and unconverted fuel are readily used in the chemical looping gasification system. Therefore, the exergy loss for the overall process is reduced further via an improved energy management scheme.

3. The fuel gasification and pollutant control schemes are simpler. For example, the elaborate partial syngas shift, COS hydrolysis, and acid gas separation steps in the conventional indirect CTL process are avoided.

4. The CO_2 by-product is concentrated and readily sequestrable.

A case study is performed to evaluate the exergy loss of the conventional CTL process and the chemical-looping-gasification-based CTL process using

TABLE 6.3. Performance Comparisons of Conventional CTL and Chemical-Looping-Gasification-Based CTL

	Conventional CTL	Chemical Looping-CTL
Exergy Loss (MW)[a]	835	**503**
Marginal Change in Exergy Loss (%)	n/a	**−39.8**

[a]For an 80,000-bbl/d (36,288-kg/d) plant.

assumptions similar to those given in Section 1.6. The results are given in Table 6.3. As shown in the table, through the utilization of the chemical looping gasification strategy, the exergy loss of the CTL process can be reduced by nearly 40%. Thus, the novel chemical-looping-based liquid fuel synthesis scheme has the potential to improve significantly the efficiencies of conventional CTL processes.

6.8 Chemical Looping with Oxygen Uncoupling (CLOU)

To combust carbonaceous fuels with metal oxide oxygen carriers, the chemical looping with oxygen uncoupling (CLOU) Process, initiated by Lewis and Gilliland[67] and extensively tested at the Chalmers University of Technology,[68–71] uses a reaction scheme different from that of the typical chemical looping process discussed thus far. The oxygen carrier in the reducer of a typical chemical looping process releases its oxygen directly through a reduction reaction with the fuel. In contrast, the oxygen carrier in the reducer of a CLOU Process releases the oxygen through decomposition to form molecular gaseous oxygen, which then oxidizes the fuels. When the CLOU reaction scheme is implemented, it yields a more effective process in oxidizing, particularly, the solid fuels in the reducer than does the typical chemical looping processes. The specific reactions that take place in CLOU include:

$$Me_xO_y \rightarrow Me_xO_z + (y-z)/2O_2 \tag{6.8.1}$$

$$C_nH_{2m} + (n+m/2)O_2 \rightarrow nCO_2 + mH_2O \tag{6.8.2}$$

$$Me_xO_z + (y-z)/2O_2 \rightarrow Me_xO_y \tag{6.8.3}$$

The first two reactions [Reactions (6.8.1) and (6.8.2)] occur in the reducer, and the third reaction [Reaction (6.8.3)] occurs in the oxidizer. For the first two reactions to occur effectively, the equilibrium partial pressure of the oxygen for Reaction (6.8.1) must be high at the reaction temperature. This unique requirement narrows the potential options for oxygen carriers. Three metal oxides satisfy this requirement; they are CuO, Mn_2O_3, and Co_3O_4. The

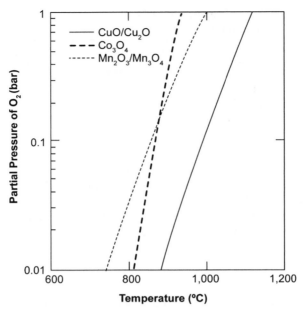

Figure 6.22. Partial pressure of gas-phase O_2 over metal-oxide systems as a function of temperature. CuO/Cu_2O (—); Co_3O_4/CoO (– – –) Mn_2O_3/Mn_3O_4 (- - -).[68]

equilibrium oxygen partial pressures for these three metal oxides as a function of temperature are given in Figure 6.22. It is shown that an increase in temperature leads to an increase in the oxygen partial pressure. When the heat of reaction for both Reactions (6.8.1) and (6.8.2) is considered, the reactions of the metal oxides CuO/Cu_2O and Mn_2O_3/Mn_3O_4 with the fuel are exothermic and are thus more suitable for CLOU than the reaction of metal oxide Co_3O_4/CoO with the fuel that is endothermic.[68] Exothermic reactions are beneficial as the oxygen decomposition is enhanced with an increase in temperature.

Mattisson et al.[68] tested the CLOU Process for methane conversion using a copper-based oxygen carrier of 125–180 µm, synthesized by freeze granulation at 60% CuO and 40% ZrO_2. The test was performed in a 30-mm-ID quartz fluidized-bed reactor operated under batch mode with reduction and oxidation carried out by turns in the same reactor. Nitrogen gas was used between the oxidation and reduction steps to prevent any mixing of gases. The total mass of the oxygen carrier used was 10 g, and the fuel flow rate was 600 mL/min. The experiment was performed at 950°C. The results indicate that the oxygen carrier decomposition changes with the reactor temperature during the reduction step. During the oxidation step, the CO/CO_2 peaks were identified at the initial stage of air oxidation. This result reveals the deposition of carbon, which only occurs in the presence of metallic copper. It was there-

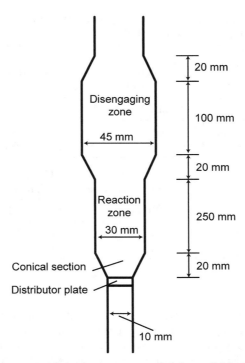

Figure 6.23. Laboratory fluidized-bed reactor used for solid-fuel CLOU testing by Chalmers University.[68]

fore concluded that the CuO-based oxygen carrier was reduced to metallic copper during the reduction step.

After the gaseous fuel testing, studies using various solid fuels were conducted using a laboratory fluidized bed reactor, as shown in Figure 6.23. In the figure, a conical section above the distributor plate was used to provide proper mixing of the solid fuel and the oxygen carrier. A disengaging zone above the reaction zone prevented the entrainment of small particles from the bed. In all the solid fuel tests, 15 g of oxygen carrier was used and 0.1 g of solid fuel was introduced during each reduction cycle. The size range of the fuel, that is, 180–250 μm, is slightly greater than that of the oxygen carrier. The gas flow rate during oxidation and reduction was maintained at 900 mL/min for the air and nitrogen flows.

Mattisson et al.[69] used a copper-based oxygen carrier (40% CuO and 60% ZrO_2) of 125–180 μm, synthesized by freeze granulation, to determine the rate of conversion of petroleum coke. Experiments were designed to study the influence of the reducer temperature on the conversion rate of petroleum coke. The results indicated a strong dependence of the conversion rate of petroleum coke on the reducer temperature. The time for 95% conversion

TABLE 6.4. Solid Fuels Testing Results Using the CLOU Concept: Copper-Based Metal Oxide vs. Iron-Based Metal Oxide[71]

	Mexican Petroleum Coke		South African Coal		Indonesian Coal	
Type of Fuel	CLOU	CLC	CLOU	CLC	CLOU	CLC
Time for 95% Conversion (s)	41	648	40	612	30	282
Rate of 95% Conversion (%/s)	2.3	0.15	2.4	0.16	3.2	0.34

	Columbian Coal		German Lignite		Swedish Wood Char	
Type of Fuel	CLOU	CLC	CLOU	CLC	CLOU	CLC
Time for 95% Conversion (s)	51	606	25	84	28	378
Rate of 95% Conversion (%/s)	1.9	0.16	3.8	1.13	3.4	0.25

decreased from 130 s to 20 s with an increase in temperature from 885°C to 985°C.[69] A similar conversion of petroleum coke was observed after 15 minutes when iron-based oxygen carriers were used in a typical CLC scheme, high-lighting the significant conversion rate observed while using the CLOU scheme.[70] Leion et al.[71] tested the conversion rates for six different solid fuels using the copper-based oxygen carrier. The results were compared with those when the CLC scheme is used, and the outcome is summarized in Table 6.4. All the experiments were carried out at 950°C. The oxidation of the oxygen carriers was carried out using 10% O_2 balanced with nitrogen. The table highlights the increased reaction rates observed when adapting the CLOU scheme for solid fuels. Leion et al.[71] also stated that the methane and CO emissions from the CLOU process are lower than those observed in the typical CLC process.

The activity involving CLOU testing using metal oxides other than copper oxide is limited. The application of the copper-based oxygen carrier is, however, constrained because of the low melting point of copper, as discussed in Chapter 2. Furthermore, as the extent of decomposition of copper oxide is low, a very high circulation rate of copper oxide is thus required for process applications. So far, the only testing performed has been in a laboratory-scale fluidized-bed reactor.

6.9   Concluding Remarks

This chapter presents examples of novel applications of chemical looping processes for H_2 storage and onboard H_2 production, CO_2 capture in combustion flue gas, power generation using fuel cells, SMR, tar sand digestion, and liquid fuel production. Each application employs a unique feature of the chemical looping strategy that characterizes process intensification, leading to a highly efficient process system. In these particular applications, metal oxide can be used to assist in tar sand digestion and steam methane reforming. The reduced metal oxide from the reducer can be used for the following applications: onboard H_2 production by reacting it with steam, power production by combusting reduced metal oxide in the fuel cell anode, and generation of steam or heat by reacting it with air. In processes such as CLOU, the metal oxide, instead of being reduced by the fuel, can be decomposed to yield molecular gaseous oxygen in the reducer at a high temperature. It is followed, in the reducer, by combustion of molecular gaseous oxygen with fuel to form CO_2 and steam. When this reaction scheme is implemented, the CLOU process is more effective in combusting the gaseous or solid fuels in the reducer than does the typical chemical looping process. The use of the calcium sorbent for CO_2 capture through the carbonation–calcination–hydration looping reactions provides a high-temperature, sorbent-based CO_2 separation method that is a viable alternative to the traditionally low-temperature, solvent-based separation methods. All these chemical looping examples represent energy conversion systems on the cutting edge of efficiency and technology where CO_2 separation, pollutant control, and product generation can be achieved readily. Employing the fundamental chemical looping concept, other chemical looping schemes also can be conceived, leading to highly efficient process applications.

References

1. Carpetis, C., "Estimation of Storage Costs for Large Hydrogen Storage Facilities," *International Journal of Hydrogen Energy*, 7(2), 191–203 (1982).
2. DeLuchi, M. A., "Hydrogen vehicles: An Evaluation of Fuel Storage, Performance, Safety, Environmental Impacts and Cost," *International Journal of Hydrogen Energy*, 14(2), 81–130 (1989).
3. Kukkone, C. A., and M. Shelef, "Hydrogen as an Alternative Automotive Fuel," *Alternate Fuel*, R. M. Bata, National Research Council and National Academy of Engineering. Washington, DC, National Academies Press (1992).
4. Chambers, A., C. Park, R. T. K. Baker, and N. M. Rodriguez, "Hydrogen Storage in Graphite Nanofibers," *Journal of Physical Chemistry B*, 102(22), 4253–4256 (1998).
5. Michel, F., H. Fieseler, G. Meyer, and F. Theißen, "On-Board Equipment for Liquid Hydrogen Vehicles," *International Journal of Hydrogen Energy*, 23(3), 191–199 (1998).
6. Newson, E., TH. Haletter, P. Hottinger, F. Von Roth, G. W. H. Scherer, and TH. H. Schucan, "Seasonal Storage of Hydrogen in Stationary Systems with Liquid Organic Hydrides," *International Journal of Hydrogen Energy*, 23(10), 905–909 (1998).
7. Verbetsky, V. N., S. P. Malyshenko, S. V. Mitrokhin, V. V. Solovei, and Y. F. Shmal'ko, "Metal Hydrides: Properties and Practical Applications–Review of the Works in CIS-Countries," *International Journal of Hydrogen Energy*, 23(12), 1165–1177 (1998).

8. Crabtree, G. W., M. S. Dresselhaus, and M. V. Buchanan, "The Hydrogen Economy," *Physics Today*, 5 (12), 39–44 (2004).

9. U.S. Department of Energy, "The 2007 Annual Progress Report," Hydrogen Program, http://www.hydrogen.energy.gov/annual_progress07.html (2008).

10. Satyapal, S., J. Petrovic, C. Read, G. Thomas, and G. Ordaz, "The U. S. Department Energy's National Hydrogen Storage Project: Progress towards Meeting Hydrogen-Powered Vehicle Requirements," *Catalysis Today*, 120(3–4), 246–256 (2007).

11. Pant, K. K., and R. B. Gupta, "Fundamental and Use of Hydrogen as a Fuel," in Hydrogen Fuel: Production, Transport, and Storage, edited by R. B. Gupta, CRC Press, Boca Raton, FL (2008).

12. Satyapal, S., and G. Thomas, "Targets for Onboard Hydrogen Storage Systems: An Aid for the Development of Viable Onboard Hydrogen Storage Technologies," in Hydrogen Fuel: Production, Transport, and Storage, edited by R. B. Gupta, CRC Press, Boca Raton, FL (2008).

13. Gao. M., and R. Krishnamurthy, "Hydrogen Transmission in Pipelines and Storage in Pressurized and Cryogenic Tanks," in Hydrogen Fuel: Production, Transport, and Storage, edited by R. B. Gupta., CRC Press, Boca Raton, FL (2008).

14. Burke, A. F., and M. Gardiner, "Hydrogen Storage Options: Technologies and Comparisons for Light-Duty Vehicle Applications," UCD-ITS-RR-05-01, Institute of Transport Studies, University of California, Davis (2005).

15. Arnold, G., and J. Wolf, "Liquid Hydrogen for Automotive Application Next Generation Fuel for FC and ICE Vehicles," *Journal of the Cryogenic Society of Japan*, 40(6), 221–230 (2005).

16. Aceves, S. M., J. Martines-Frias, and O. Garcia-Villazana, "Analytical and Experimental Evaluation of Insulated Pressure Vessels for Cryogenic Hydrogen Storage," *International Journal of Hydrogen Energy*, 25(11), 1075–1085 (2000).

17. Aceves, S. M., G. D. Berry, J. Martinez-Frias, and F. Espinosa-Loza, "Vehicular Storage of Hydrogen in Insulated Pressure Vessels," *International Journal of Hydrogen Energy*, 31(15), 2274–2283 (2006).

18. Akiba, E., "Research and Development of Hydrogen Storage Technologies," *Journal of the Japan Institute of Energy*, 85(7), 510–516 (2006).

19. Heung, L. K., and G. G. Wicks, "Silica Embedded Metal Hydrides," *Journal of Alloys and Compounds*, 293–295, 446–451 (1999).

20. Bogdanovic, B., and M. Schwickardi, "Ti-doped NaAlH₄ as a Hydrogen-Storage Material–Preparation by Ti-Catalyzed Hydrogenation of Aluminum Powder in Conjunction with Sodium Hydride," *Applied Physics A*, (72), 221–223 (2001).

21. Heung, L. K., "Using Metal Hydride to Store Hydrogen," U.S. Department of Commerce, National Technical Information Service, Springfield, VA (2003).

22. Browman, R. C., Jr., S.-J. Hwang, C. C. Ahn, and J. J. Vajo, *NMR and X-ray Diffraction Studies of Phases in the Destabilized LiH-Si System*, Materials Research Society Symposium Proceedings, 837, N3.6.1–N3.6.6 (2005).

23. Li, S., and P. Jena, *Electronic Structure and Hydrogen Desorption in NaAlH₄*, Materials Research Society Symposium Proceedings, 837, N2.5.1–N2.5.8 (2005).

24. Bruster, E., T. A. Dobbins, R. Tittsworth, and D. Anton, *Decomposition Behavior of Ti-Doped NaAlH₄ Using X-Ray Absorption Spectroscopy at the Titanium K-Edge*, Materials Research Society Symposium Proceedings, 837, N3.4.1–N.3.4.6 (2005).

25. Araujo, C. M., R. Ahuja, and J. M. O. Guillen, "Role of Titanium in Hydrogen Desorption in Crystalline Sodium Alanate," *Applied Physics Letters*, 86(25), 251913 1–3 (2005).

26. Schuth, F., B. Bogdanovic, and M. Felderhof, "Light Metal Hydrides and Complex Hydrides for Hydrogen Storage," *Chemical Communications*, 24, 2249–2258 (2004).

27. Garberoglio, G., A. I. Skoulidas, and J. K. Johnson, "Adsorption of Gases in Metal Organic Materials: Comparison of Simulations and Experiments," *Journal of Physical Chemistry B*, 109(27), 13094–13103 (2005).

28. Rosi, N. L., J. Eckert, M. Eddaoudi, D. T. Vodak, J. Kim, M. O' Keeffe, and O. M. Yaghi., "Hydrogen Storage in Microporous Metal-Organic Frameworks," *Science*, 300, 1127 (2003).

29. Jung, D. H., D. Kim, T. B. Lee, T Lee, S. B. Choi, J. H. Yoon, J. Kim, K. Choi, and S.-H. Choi, "Grand Canonical Monte Carlo Simulation Study on the Catenation Effect on Hydrogen Adsorption onto the Interpenetrating Metal–Organic Frameworks," *Journal of Physical Chemistry B*, 110(46), 22987–22990 (2006).

30. Li, Y., F. H. Yang, and R. T. Yang, "Kinetics and Mechanistic Model for Hydrogen Spillover on Bridged Metal–Organic Frameworks," *Journal of Physical Chemistry C*, 111(8), 3405–3411 (2007).

31. Li, Y., and R. T. Yang, "Hydrogen Storage in Metal–Organic Frameworks by Bridged Hydrogen Spillover," *Journal of the American Chemical Society*, 128(25), 8136–8137 (2006).

32. Dillon, A. C., J. L. Blackburn, P. A. Parilla, Y., Zhao, Y.-H. Kim, S. B. Zhang, A. H. Mahan, J. L. Alleman, K. M. Jones, K. E. H. Gillbert, et al. *Discovering the Mechanism of H₂ Adsoprtion on Aromatic Carbon Nanostructures to Develop Adsorbents for Vehicular Applications*, Materials Research Society Symposium Proceedings, 837, N4.2.1–N4.2.7 (2005).

33. Lee, Y. W., R. Deshpande, A. C. Dillon, M. J. Heben, H. Dai, and B. M. Clemens, *The Role of Metal Catalyst in Near Ambient Hydrogen Adsorption on Multi-walled Carbon Nanotubes*, Materials Research Society Symposium Proceedings, 837, N3.18.1–N3.18.6 (2005).

34. Wang, Y., and Z. Iqbal, *Electrochemical Hydrogen Adsorption/Storage in Pure and Functionalized Single Wall Carbon Nanotubes*, Materials Research Society Symposium Proceedings, 837, N4.4.1–N4.4.12 (2005).

35. Gupta, B. K., R. S. Tiwari, and O. N. Srivastava, "Studies on Synthesis and Hydrogenation Behavior of Graphite Nanofibers Prepared through Palladium Catalyst Assisted Thermal Cracking of Acetylene," *Journal of Alloys and Compounds*, 381, 301–308 (2004).

36. Ahn, C. C., Y. Ye, B. V. Ratnakumar, C. Witham, R. C. Bowman, and B. Fultz, "Hydrogen Desorption and Adsorption Measurements on Graphite Nanofibers," *Applied Physics Letters*, 73, 3378–3380 (1998).

37. Fan, Y.-Y., B. Liao, M. Liu, Y.-L. Wei, M.-Q. Lu, and H.-M. Cheng, "Hydrogen Uptake in Vapor-Grown Carbon Nanofibers," *Carbon*, 37, 1649–1652 (1999).

38. Hong, S. E., D.-K. Kim, S. M. Jo, D. Y. Kim, B. D. Chin, and D. W. Lee, "Graphite Nanofibers Prepared from Catalytic Graphitization of Electrospun Poly(Vinylidene Fluoride) Nanofibers and Their Hydrogen Storage Capacity," *Catalysis Today*, 120, 413–419 (2007).

39. White C. M., B. R. Strazisar, E. J. Granite, J. S. Hoffman, and H. W. Pennline, "Separation and Capture of CO_2 from Large Stationary Sources and Sequestration in Geological Formations–Coalbeds and Deep Saline Aquifers," *Journal of the Air & Waste Management Association*, 53(6), 645–715 (2003).

40. Lee, K. B., A. Verdooren, H. S. Caram, and S. Sircar, "Chemisorption of Carbon Dioxide on Potassium-Carbonate-Promoted Hydrotalcite," *Journal of Colloid and Interface Science*, 308, 30–39 (2007).

41. Fan, L.-S., and R. A. Jadhav, "Clean Coal Technologies: OSCAR and CARBONOX Commercial Demonstrations," *AIChE Journal*, 48(10), 2115–2123 (2002).

42. Fan, L.-S., H. Gupta, and M. Iyer, "Separation of Carbon Dioxide (CO_2) from Gas Mixtures by Calcium Based Reaction Separation (Cars-CO_2) Process," U.S. Patent Application Publication Number 2008/0,233,029.

43. Gupta, H., S. A. Benson, L.-S. Fan, J. D. Laumb, E. S. Olson, C. R. Crocker, R. K. Sharma, R. Z. Knutson, A. S. M. Rokanuzzaman, and J. E. Tibbets, "Pilot-Scale Studies of NO_x Reduction by Activated High-Sodium Lignite Chars: A Demonstration of the CARBONOX Process," *Industrial & Engineering Chemistry Research*, 43(18), 5820–5827 (2004).

44. Taerakul, P., P. Sun, D. W. Golightly, H. W. Walker, L. K. Weavers, B. Zand, T. Butalia, T. Thomas, H. Gupta, and L.-S. Fan, "Characterization and Re-Use Potential of By-Products Generated from the Ohio State Carbonation and Ash Reactivation (OSCAR) Process," *Fuel*, 86(4), 541–553 (2007).

45. Fan, L.-S., and H. Gupta, "Sorbent for the Separation of Carbon Dioxide (CO_2) from Gas Mixtures," U.S. Patent 7,067,456 (2006).

46. Grasa, G. S., and J. C. Abanades, "CO_2 Capture Capacity of CaO in Long Series of Carbonation /Calcination Cycles," *Industrial & Engineering Chemistry Research*, 45, 8846–8851 (2006).

47. Chrissafis, K., "Multicyclic Study on the Carbonation of CaO Using Different Limestones," *Journal of Thermal Analysis and Calorimetry*, 89, 525–529 (2007).

48. Fennell, P. S., R. Paciani, J. S. Dennis, J. F. Davidson, and A. N. Hayhurst, "The Effects of Repeated Cycles of Calcination and Carbonation on a Variety of Different Limestones, as Measured in a Hot Fluidized Bed of Sand," *Energy & Fuels*, 21(4), 2072–2081 (2007).

49. Sun, P., C. J. Lim, and J. R. Grace, "Cyclic CO_2 Capture by Limestone-Derived Sorbent During Prolonged Calcination/Carbonation Cycling," *AIChE Journal*, 54, 1668–1677 (2008).

50. Fan L.-S., S. Ramkumar, W. Wang, and R. Statnick, "Separation of Carbon Dioxide from Gas Mixtures by Calcium Based Reaction Separation Process," Provisional Patent Application No. 61/116,172 (2008).

51. Lu, D. Y., R. W. Hughes, and E. J. Anthony, "*In-Situ* CO_2 Capture Using Ca-Based Sorbent Looping in Dual Fluidized Beds," Paper presented at the 9th International Conference on Circulating Fluidized Beds, Hamburg, Germany (2008).

52. Manovic V., and E. J. Anthony, "Parametric Study on the CO_2 Capture Capacity of CaO-Based Sorbents in Looping Cycles," *Energy and Fuels*, 22(3), 1851–1857 (2008).

53. Manovic V., and E. J. Anthony, "Thermal Activation of CaO-Based Sorbent and Self-Reactivation During CO_2 Capture Looping Cycles," *Environmental Science & Technology*, 42(11), 4170–4174 (2008).

54. Manovic V., and E. J. Anthony, "Sequential SO_2/CO_2 Capture Enhanced by Steam Reactivation of a CaO-Based Sorbent," *Fuel*, 87(8/9), 1564–1573 (2008).

55. Jin, H., and M. Ishida, "A New Type of Coal Gas Fueled Chemical-Looping Combustion," *Fuel*, 83, 2411–2417 (2004).

56. Gupta, P., L. G. Velazquez-Vargas, T. Thomas, and L.-S. Fan, *Process of Chemical Looping Combustion of Coal*, 30th Proceedings of the 30th International Technical Conference on Coal Utilization & Fuel Systems, Clearwater, Florida, April 17–21, (2005).

57. Garcia-Labiano, F., J. Adanez, L. F. de Diego, P. Gayan, and A. Abad, "Effect of Pressure on the Behavior of Copper-, Iron-, and Nickel-Based Oxygen Carriers for Chemical-Looping Combustion," *Energy & Fuels*, 20(1), 26–33 (2006).

58. Kronberger, B., A. Lyngfelt, G. Loeffler, and H. Hofbauer, "Design and Fluid Dynamic Analysis of a Bench-Scale Combustion System with CO_2 Separation–Chemical-Looping Combustion," *Industrial & Engineering Chemistry Research*, 44(3), 546–556 (2005).

59. Fan, L.-S., F. Li, L. Zeng, and D. Sridhar, "Integration of Reforming/Water Splitting and Electrochemical Systems for Power Generation with Integrated Carbon Capture", Provisional Patent Application Number 61/240,508 (2009).

60. Chuang, S. S. C., "Catalysis of Solid Oxide Fuel Cell," *Catalysis*, 18, 186–198 (2005).

61. Chuang, S. S. C., "Coal-Based Fuel Cell," WO 2,006,028,502 (2006).

62. Ryden, M., and A. Lyngfelt, "Using Steam Reforming to Produce Hydrogen with Carbon Dioxide Capture by Chemical-Looping Combustion," *International Journal of Hydrogen energy*, 31(10), 1271–1283 (2006).

63. Riedel T., M. Claeys, H. Schulz., G. Schaub, S. S. Nam, K. W. Jun, M. J. Choi, G. Kishan, and K. W. Lee, "Comparative Study of Fischer-Tropsch Synthesis with H_2/CO and H_2/CO_2 Syngas Using Fe- and Co-Based Catalysts," *Applied Catalysis A-General*, 186(1–2), 201–213 (1999).

64. Riedel, T., G. Schaub, K. W. Jun, and K. W. Lee, "Kinetics of CO_2 Hydrogenation on a K-Promoted Fe Catalyst," *Industrial & Engineering Chemistry Research*, 40(5), 1355–1363 (2001).

65. Hildebrandt, D., D. Glasser, B. Hausberger, B. Patel, and B. Glasser, "Producing Transportation Fuels with Less Work," *Science*, 323, 1680–1681 (2009).

66. Fan, L.-S., F. Li, and L. Zeng, "Synthetic Fuels and Chemicals Production with In-Situ CO_2 Capture," Provisional Patent Application Number 61/240446 (2009).

67. Lewis, W. K., and E. R. Gilliland, "Production of Pure Carbon Dioxide," U.S. Patent 2,665,972 (1954).

68. Mattisson, T., A. Lyngfelt, and H. Leion, "Chemical-Looping with Oxygen Uncoupling for Combustion of Solid Fuels," *International Journal of Greenhouse Gas Control*, 3(1), 11–19 (2009).

69. Mattisson, T., H. Leion, and A. Lyngfelt, "Chemical-Looping with Oxygen Uncoupling Using CuO/ZrO_2 with Petroleum Coke," *Fuel*, 88(4), 683–690 (2009).

70. Leion, H., T. Mattisson, and A. Lyngfelt, "The Use of Petroleum Coke as Fuel in Chemical–Looping Combustion," *Fuel*, 86(12–13), 1947–1958 (2007).

71. Leion, H., T. Mattisson, and A. Lyngfelt, "Using Chemical-Looping with Oxygen Uncoupling (CLOU) for Combustion of Six Different Solid Fuels," *Energy Procedia*, 1, 447–453 (2009).

SUBJECT INDEX

AUTHOR INDEX

411